T0263975

# Modelling the Flying Bird

# Modelling the Flying Bird

C.J. Pennycuick
*Senior Research Fellow, University of Bristol*

AMSTERDAM • BOSTON • HEIDELBERG • LONDON
NEW YORK • OXFORD • PARIS • SAN DIEGO
SAN FRANCISCO • SINGAPORE • SYDNEY • TOKYO
Academic Press is an imprint of Elsevier

Academic Press is an imprint of Elsevier
30 Corporate Drive, Suite 400, Burlington, MA 01803, USA
525 B Street, Suite 1900, San Diego, California 92101-4495, USA
84 Theobald's Road, London WC1X 8RR, UK

**Library of Congress Cataloging-in-Publication Data**
A catalog record for this title is available from the Library of Congress

**British Library Cataloguing-in-Publication Data**
A catalogue record for this book is available from the British Library

ISBN: 978-0-12-374299-5

For information on all Academic Press publications
visit our Web site at www.books.elsevier.com

Printed and bound by CPI Group (UK) Ltd, Croydon, CR0 4YY

Transferred to Digital Print 2011

**Cover Image:**
M.C. Escher's "Day and Night" © 2008. The M.C. Escher Company-Holland.
All rights reserved. www.mcescher.com.
We are grateful to The M.C. Escher Company for the use of the image.

# CONTENTS

# PREFACE

Being an interdisciplinary activity, computer modelling of bird flight tends to fall into the chasm between ornithology and engineering. Ornithologists mistrust calculation, while engineers think birdwatching is frivolous. It may seem obvious that aeronautical theory can be adapted to cover bird flight, but when I first attempted to do that, it was seen in ornithological circles as an eccentric activity, with little or no practical use. My earlier book *Bird Flight Performance* was politely received but biologists were unconvinced that they needed it. The present book, which is backed by a far more capable computer programme, is a fresh attempt to show why a physical theory is necessary as a framework for any quantitative discussion of animal flight.

The barrier to communication between ornithologists and aeronautical engineers is due to their different attitudes to numbers. Biologists readily accept that the rate at which a bird needs energy to support its weight in air might be correlated with the wing span, but balk at the idea that this measurement (the distance between the wing tips) actually *determines* the power requirement, and can be used to calculate it for any bird, without the need to measure power or run regressions. There is actually no way to use statistical methods to predict the power requirements of even one species, because several variables are involved. These include the wing span, the forward speed, the strength of gravity, and the density of the air, and each of them affects the power in different ways. All of this, and much more, is represented in classical aeronautical theory, of which the relevant parts have been exhaustively tested over the last hundred years, and I have built the *Flight* programme on this foundation.

Ornithologists sometimes want to use the traditional "wing length" as a substitute for the wing span, but this will not do. The power estimates are not correlations, but absolute numbers, calculated from Newtonian mechanics, and the right input numbers have to be used. The requirement to be aware of the definition of each variable and its physical dimensions is obvious to engineers, but less so to those who have been accustomed to relying exclusively on statistical methods. A statistical package looks for patterns in sets of numbers, and will usually produce a result whatever the numbers mean, or even if they mean nothing at all. The difficulty that many biologists seem to have with aeronautical theory is not in understanding the theory itself, but in adjusting their attitude to numbers away from statistics, and towards the engineering point of view.

Once this difficulty has been overcome, using the *Flight* programme is easy. Users who study the output from its simulations of long-distance migration (Chapter 8) will see a level of detail that statistics-based ecologists cannot even begin to dream about, and some may be rightly sceptical that so much can be calculated from so little in the way of input. The programme has been designed to make it easy to set up and test hypotheses that reflect the underlying assumptions, and it is for experimenters and field observers to determine what level of confidence in its predictions is justified. This testing process is currently being transformed by the ever-increasing capabilities of satellite-trackable transmitters that can be carried by birds, but many kinds of training experiments, in wind tunnels or aviaries, can also be used to test the programme (Chapter 15). It remains important to keep a close connection between the numbers and the real world of the flying bird, and the best way to keep that in focus is to learn to fly oneself.

Colin Pennycuick
Bristol, December 2007

# FOREWORD

In Larry McMurtry's novel *Comanche Moon*, the Kickapoo tracker Famous Shoes, who can track anything over any terrain, is musing over his solitary camp fire somewhere in Texas, circa 1861, listening to the geese migrating overhead in the starlight:

> The mystery of the northward-flying geese had always haunted him; he thought the geese might be flying to the edge of the world, so he made a song about them, for no mystery was stronger to Famous Shoes than the mystery of birds. All the animals that he knew left tracks, but the geese, when they spread their wings to fly northward, left no tracks. Famous Shoes thought that the geese must know where the gods lived, and because of their knowledge had been exempted by the gods from having to make tracks. The gods would not want to be visited by just any-one who found a track, but their messengers, the great birds, were allowed to visit them. It was a wonderful thing, a thing Famous Shoes never tired of thinking about. ... Many white men could not trust things unless they could be explained; and yet the most beautiful things, such as the trackless flight of birds, could never be explained.

People do not fly, obviously, but not all white men in Famous Shoes' time knew this. A few years later two of them, Orville and Wilbur Wright, found out how to fly, and now anyone can learn to do it, with a little effort and perseverance. By living at the right time, my luck has included personally migrating across the Nubian Desert in a Piper Cruiser, and across the Greenland ice cap in a Cessna 182, both busy routes for migrating birds, and that is indeed a wonderful thing. I have migrated into Sweden with the cranes in that same Cruiser, and soared with storks and vultures over the Serengeti in a Schleicher ASK-14. Actually, birds do leave tracks in the air. They do not last long, but a skilled tracker can read them (Chapter 4). Eat your heart out, Famous Shoes. We may never know where the gods live, but some of the things that birds do can be explained and understood, especially if we do them ourselves, and this book is the song that I have made about it.

# ACKNOWLEDGEMENTS

I am confident that the theory behind the *Flight* programme is right, because I have been in the habit of entrusting my life to it as a pilot, and at the time of writing I am still alive. I first learned about the theory of flight from those iron-nerved RAF instructors who taught me (a Zoology graduate) to fly Chipmunks, Provosts and Vampires so many years ago. Their efforts were reinforced, when I later became a gliding instructor myself, by pupils who required me to explain and demonstrate how gliders fly, so forcing me to understand the theory on an intuitive level. When I joined Bristol University as a Zoology lecturer in 1964, my aeronautical education took a more formal turn thanks to Tom Lawson and John Flower of the Aeronautical Engineering Department, who helped me to build a wind tunnel in which pigeons could fly. The pigeons soon demonstrated that aeronautical principles do indeed apply to birds, and I got my first opportunity to convince biologists about this soon afterwards, thanks to the broad-minded Reg Moreau, who was editor of *Ibis* in 1969, and Peter Evans who reviewed my somewhat unconventional manuscript. Naively, I supposed that ornithologists would seize eagerly on the revelations in the paper, and this book is my latest attempt to convince them of the advantages of the physics-based approach.

I owe my introduction to studying the flight of wild birds in the field to Hugh Lamprey, then director of the Serengeti Research Institute, who took a glider to the Serengeti in 1968 and let me fly it, and to Hans Kruuk and Tony Sinclair, who taught me how the Serengeti ecosystem works. A later motor-glider project in the Serengeti supplied the background for the gliding section of the *Flight* programme, and led to a project with Thomas Alerstam in Sweden, in which we followed migrating cranes in my Piper Cruiser. My informal association with Lund University has continued, and reached a high point when Thomas, having risen to be head of department, set up the Lund wind tunnel in the new Ecology Building in 1994. Down in the South Atlantic, John Croxall taught me what I know about albatrosses and the Southern Ocean ecosystem during two memorable trips to South Georgia with the British Antarctic Survey in 1980 and 1994. Meanwhile, a long-term collaboration with Mark Fuller, which started while I was based at the University of Miami and still continues, led to a series of field and laboratory projects in various parts of the USA and the Caribbean, which laid the groundwork for *Flight*'s simulation of long-distance migration.

I am extremely grateful to Geoff Spedding for reading and commenting on drafts of the earlier chapters, especially the parts relating to his own remarkable contributions to the study of bird wakes, to Ulla Lindhe Norberg for a similarly expert review of the chapter on bats and pterosaurs, and to Julian Partridge for reviewing the chapter on information sources and commenting on the rest of the book. Literally hundreds of biologists, aviators, students, professors and others in Britain, Sweden, Africa, America, the Caribbean, the South Atlantic and other places have educated me about different aspects of flight, and thus contributed to the book, wittingly or not. I am deeply grateful to them all, and if I have got it wrong, the fault is mine alone. As always, I have depended on the support and forbearance of my wife Sandy and son Adam to make this project possible. The book is dedicated to the doctors and staff of the Bristol Oncology Centre and Southmead Hospital, Bristol, without whose intervention I would not have lived to write it.

Colin Pennycuick
Bristol, December 2007

# 1

# BACKGROUND TO THE MODEL

The *Flight* computer model, which calculates the rate at which a flying animal requires energy for whatever it is doing, is based on classical aerodynamics. This is itself a branch of Newtonian mechanics, which is basically the same for aircraft and birds. Calculating the mechanical power requires information about wing measurements, which are defined in this chapter. The physiological requirements for fuel and oxygen are calculated as a second step, from the mechanical requirements. This approach requires care with the physical dimensions of variables, introduced in this chapter.

My objective in writing this book is to understand what a bird does when it flies, to explain in physical terms how it does it and to provide tools that can be used to predict quantitatively what *any* bird (not just those that have been studied) can and cannot do. The quest is ambitious but not new. Would-be aeronauts have studied the wings of birds with great care down the centuries, hoping to understand them well enough to copy them, and fly themselves. With hindsight we can see now why Otto Lilienthal's meticulous studies and drawings of the wings of storks (Lilienthal, 1889) produced disappointingly little at the time, by way of insight into how wings work. His difficulty was that he had no theory in the 1880s with which to describe and explain what he saw. Now we have theory aplenty, thanks to the efforts of the world's aeronautical research institutions, and it is time for us birdwatchers to turn the process around, and look at birds through the new eyes that aeronautical engineers have given us.

The book is descriptive in parts, especially in the chapters that introduce the wings of flying vertebrates, but these descriptions will look strange to many biologists, because the conventions of morphology are hopelessly inadequate for describing how wings work. It is not possible to explain what wings do, without introducing concepts that are not a traditional part of a biologist's education. This chapter introduces the aeronautical conventions for describing and measuring wings, adapted for birds, and Chapter 2 is about the characteristics of the environment in which birds fly. Chapters 3 and 4, about the principles of flight, introduce a number of concepts that are familiar to engineers, but not to most biologists, and attempt to give the biological reader an intuitive feel for what these ideas mean. Chapters 5 and 6 describe the wings of birds, bats and pterosaurs, and Chapter 7 is on muscles seen as engines. After that the scope broadens to cover such topics as the simulation of long-distance migration, gliding and soaring, the sensory requirements of flight, the use of wind tunnels and the design of experiments on flight. The evolution of flight comes last, because it is not possible to understand how it happened, without invoking the mechanical principles covered in earlier chapters.

## 1.1 ➤ THE *FLIGHT* MODEL

The skeleton of the book is the *Flight* computer model, a programme that incorporates the concepts introduced in the book, and allows the user to apply them to a chosen bird to answer questions about speed, distance, energy consumption and suchlike performance matters. *Flight* is not a model of a particular bird, nor is it based on regressions describing direct measurements of the quantities that it calculates. It is essentially a set of physical rules which are assumed to be general, in the sense that they can be applied to any bird, real or hypothetical, for which the user can provide the measurements required to define the bird and its environment. *Flight* accepts the user's input describing the bird, and provides a variety of options that determine the assumptions to be made in the calculation. Then it predicts how the bird's performance in flapping or gliding flight, or in long-distance migration, would follow from that particular set of assumptions. It is designed in a way that makes it easy to vary the starting assumptions, which can be seen as hypotheses about how the bird works, and immediately observe the effect of a changed assumption on the predicted performance.

*Flight* is, in effect, a working model of a bird, according to the theory given in the book. It comes with its own online manual and databases of bird measurements, which can be loaded directly into the programme. The book contains many examples that have been calculated with *Flight*, showing how the output follows from the assumptions that underlie the programme, and how it can be used to test hypotheses about how the bird works. *Flight* is available as a free download from http://books.elsevier. com/companions/9780123742995, and also from http://www.bio.bristol. ac.uk/people/pennycuick.htm. These websites are updated from time to time with the latest version of the programme.

## 1.1.1 THE MATHEMATICAL IDIOM

It is easiest to explain what *Flight* does, and the concepts underlying it, in the idiom of aeronautical theory on which it is based, that is, in the language of applied mathematics, but this takes a little getting used to, and it is a known fact that many biologists are somewhat resistant to it. I have tried to make the book accessible to readers who are averse to equations, by structuring each chapter with an equation-free main text that explains what the topic of the chapter is about, and isolating the more technical aspects in boxes. I hope that the main text will convey the gist of the argument to mathematical and non-mathematical readers alike, while those who want to know what *Flight* actually does will find the equations in the boxes. Each box that presents a mathematical argument contains its own local variable list, which applies within that box, but not necessarily elsewhere in the book. The conventions for notation and so on are outlined in Box 1.1 in this chapter. Not all the boxes are mathematical. Some deal with the implications of a particular published experiment, an anatomical digression or some other limited topic.

BOX 1.1 **Mathematical conventions.**

Variable names in this book follow the usual conventions of physics, to the extent that a variable name is a single letter, with subscripts to distinguish between different variables of the same physical type. Variable names are italicised, but subscripts are not. For example, the letter $P$ (for Power) is used to stand for a number of different variables that have the physical dimensions of work/time. Subscripts distinguish different kinds of power from each other. $P_{mech}$, the mechanical power produced by a bird's flight muscles, and $P_{chem}$, the rate at which the bird consumes chemical energy

BOX 1.1 *Continued.*

from fuel, are different variables with the same dimensions. Lower case $p$ is used for "specific power", a related group of variables with different dimensions, power/volume for volume-specific power ($p_v$), and power/mass for mass-specific power ($p_m$).

Acronyms are not used as variable names, because they look like several variables multiplied together. "BMR" is a familiar acronym that is mentioned in the text, but it is not used as a variable name, because it looks like "B times M times R". Basal metabolic rate is a variable with the dimensions of power, and it is denoted by $P_{bmr}$. A notable exception to the one-letter rule is that two-letter variable names are traditionally used in engineering for dimensionless numbers named after famous scientists, notably $Re$ for Reynolds number. Like other variables, $Re$ can be subscripted to distinguish $Re_{wing}$ from $Re_{body}$.

### Capital "*B*" for wing span

The use of particular symbols to represent particular variables is a tradition that builds up over time, but it is not a law. The law, which applies internally in boxes in this book, but not always globally throughout the book, is that the definition of every symbol must be stated in the local context. It is legal (if not always helpful to the reader) to assign any letter you like to a physical variable, regardless of tradition. It sometimes happens that more than one tradition develops in different areas of science, and this can cause serious confusion. A particularly awkward example is lower case $b$, which is traditionally used in aeronautical engineering to denote an aircraft's wing span, the distance from one wing tip to the other. This is the width of the swathe of air that the wing influences as the aircraft or bird flies along, and it is the most important morphological measurement for performance calculations. However, there is another tradition, within aeronautics, in which fluid dynamics theorists consider the air flow around a wing by starting at the centre line, and working outwards to the wing tip. The other wing is not very interesting from this point of view, being merely a mirror-image of the first, and unfortunately it has become traditional in this area of theory to use the same symbol $b$ for the *semi*-span. The *Flight* programme comes from the "$b$ for wing span" tradition, but in recent years, the fluid dynamics tradition has been the source of major advances in wind tunnel studies of bird flight (Chapter 4), in which $b$ denotes the semi-span. Ironically, the two traditions have coexisted peacefully in their homeland, aeronautical science, for three-quarters of a century, but now that both have invaded biology from different directions, there is conflict. The same formula may appear from different sources, apparently differing by a factor of 2 (or 4 if it involves the square of the wing span).

In the hope of reducing the confusion, I have broken with tradition in this book, and used capital $B$ for the wing span, avoiding the use of lower case $b$ for anything. If others would just refrain from using capital $B$ for the semi-span, this might at least eliminate conflicting definitions of the same symbol. The reader may be wondering why $S$ should not be used for wing span. The answer, unfortunately, is that $S$ traditionally denotes area in all areas of aeronautics. $S$ for span would cause even worse confusion.

## 1.1.2 DESCRIBING THE BIRD

It is not practical to describe what every feather and every muscle does when a bird flies. Any model of a bird, whether it is constructed by a programmer or an artist, is limited to those aspects of the original that the chosen medium can realistically represent. The objective of this computer model is to predict as much as possible about the bird's capabilities, from as few assumptions as possible. The description of a particular bird needs to include only those measurements that determine the forces acting on it in level or gliding flight, and neglects other information that would complicate the calculation, without producing a useful improvement in the scope or accuracy of the predictions.

In *Flight*, a bird is described by only three numbers, its mass, wing span and wing area. That may seem a rudimentary description, and so it is. Not even the most clueless birdwatcher would confuse the American Turkey Vulture with the Great Blue Heron, but they are the same bird as far as *Flight* is concerned. I shall show in subsequent chapters that despite the minimal amount of input information that *Flight* needs about the bird, the programme predicts a surprisingly wide variety of measures of flight performance. The reader who wishes to test the accuracy of these predictions against field or laboratory observations need only enter the bird's mass, wing span and wing area into the programme, and run it. Rudimentary as these measurements may be, they are unfortunately not to be found in the traditional "morphometrics" of ornithology, and they cannot be reliably determined from museum specimens. The definitions come from aeronautics, not from ornithology, and are given in Boxes 1.2–1.4 of this chapter, together with the measurement procedures. These procedures are not difficult or arduous, but they may be unfamiliar to some biologists, and they need to be carefully followed.

BOX 1.2 **Body mass and its subdivisions: Definitions.**

The concept of "lean mass" is not used in *Flight*. This is an obsolete term that refers to everything that is not fat, including the flight muscles. It was originally conceived as a constant "baseline" against which other masses, including the fat mass, could be compared, but this became untenable when it was realised that large quantities of protein from the flight muscles are consumed during long migratory flights, and smaller amounts from the airframe. These changes are predicted in *Flight*'s migration calculation.

*Flight* considers that a bird's *empty mass* consists of three components, the *flight muscle mass*, the *fat mass* and the *airframe mass*, which is the mass of everything else in the body, that is not flight muscles or consumable fat.

BOX 1.2 *Continued.*

All three components are reduced by substantial amounts in the course of a long migratory flight, for different reasons, and this is represented in the computation. The *fraction* corresponding to each component is the mass component divided by the all-up mass.

### List of variables defined in this box

| | |
|---|---|
| $F_{fat}$ | Fat fraction |
| $F_{musc}$ | Flight muscle fraction |
| $F_{frame}$ | Airframe fraction |
| $m$ | All-up mass |
| $m_{crop}$ | Mass of crop contents |
| $m_{empty}$ | All-up mass with crop empty |
| $m_{fat}$ | Mass of fat that is consumable as fuel |
| $m_{musc}$ | Mass of flight muscles |
| $m_{frame}$ | Mass of airframe |

### All-up mass ($m$)

The total mass of everything that the bird has to lift (just weigh the bird), including any hardware such as rings and radio transmitters. The all-up mass, together with the strength of gravity (Chapter 2), determines the amount of power required from the flight muscles to support the weight.

### Empty body mass ($m_{empty}$)

The all-up mass, measured with the crop empty. This dates from the early development of *Flight*, when birds carrying heavy loads of food in their crops happened to be a subject of special interest.

### Crop mass ($m_{crop}$)

Wet mass of the crop contents, if any.

$$m = m_{empty} + m_{crop}.$$

$m_{crop}$ is normally assumed to be zero on migratory flights.

### Fat mass ($m_{fat}$)

The mass of stored fat that is available to be used as fuel.

### Fat fraction ($F_{fat}$)

The ratio of the fat mass to the all-up mass (NOT to the lean mass!). The starting fat fraction is directly related to the distance a migrating bird can fly before it runs out of fat, and this (not the fat mass as such) is the number that is needed to represent the stored fuel energy in migration calculations (Chapter 8).

$$F_{fat} = \frac{m_{fat}}{m}$$

### Flight muscle mass ($m_{musc}$)

The combined wet mass of the wing depressor and elevator muscles of both sides. In birds, these are the pectoralis and supracoracoideus muscles.

BOX 1.2 *Continued.*

### Flight muscle fraction ($F_{musc}$)
The ratio of the flight muscle mass to the all-up mass.

$$F_{musc} = \frac{m_{musc}}{m}.$$

Note that as a bird takes on or consumes fat, it also builds up or consumes its flight muscles. The flight muscle *mass* is greater when a bird is fat than when it is thin, but the flight muscle *fraction* varies much less, whether the bird is fat or thin.

### Airframe mass ($m_{frame}$)
The mass that is left after subtracting the fat mass and the flight muscle mass from the empty mass. The "airframe" is perceived as the basic structure of the bird, which has to carry the engine (flight muscles) and the fuel (fat), although actually a small part of the airframe also gets consumed on migratory flights.

$$m_{frame} = m_{empty} - m_{fat} - m_{musc}.$$

### Airframe fraction ($F_{frame}$)
The ratio of the airframe mass to the all-up mass.

$$F_{frame} = \frac{m_{frame}}{m}.$$

The three mass fractions change progressively during a long migratory flight, but they always add up to 1:

$$F_{fat} + F_{musc} + F_{frame} = 1.$$

### Entering masses into *Flight*
First enter the empty mass. This is what you get by weighing the bird with its crop empty. If the effects of carrying a crop load are not important to your calculation, you can consider the crop contents to be part of the airframe. In that case set $m_{crop}$ to zero (the default), and set $m_{empty}$ to the mass that you get by weighing the bird, including any crop contents.

Next, enter the fat mass. The programme will automatically calculate and enter the fat fraction. Alternatively, if you enter the fat fraction first, the programme will calculate and enter the fat mass. Likewise, enter either the flight muscle mass or (preferably) the flight muscle fraction.

To fatten up a computer bird, first enter a higher value for the empty mass, then increase the fat mass by a lesser amount (because additional flight muscle mass is added as well as fat). This is not taken care of automatically by the programme. It is best to use field data for the empty mass and fat fraction of heavy pre-migratory birds. In some circumstances it is possible to estimate the fat fraction from measurements of body mass alone, without resorting to carcase analysis (Chapter 8, Box 8.4).

BOX 1.3 **Wing measurements: Definitions.**

The only two wing measurements that are required by *Flight* are the wing span and the wing area. In addition, there are a number of related variables that are mentioned in the text and calculated by the programme, whose definitions are given below.

**Variables defined in this box**

$B$      Wing span
$c$      Wing chord
$c_m$     Mean chord
$R_a$     Aspect ratio
$S_{wing}$   Wing area

**Wing span ($B$)**

A bird's wing span is the most important morphological variable for flight performance calculations. It is the distance from one wing tip to the other, with the wings at full stretch out to the sides, that is, with the elbow and wrist joints fully extended (Figure 1.1A). Wing span was denoted in my own earlier publications by lower case $b$, following the most usual aeronautical convention, but this has led to some confusion as some authors from the theoretical fluid-dynamics tradition define lower case $b$ as the *semi*-span. This usage occurs in both the aeronautical and the ornithological literature, and is liable to cause major misunderstandings and errors. Hoping to minimise this problem, I denote wing span by capital $B$ in this book, thus breaking with both traditions.

**Wing area ($S_{wing}$)**

The wing area, denoted by $S_{wing}$, is essentially the area that supports the bird's weight when it is gliding. It is defined as the area, projected on a flat surface, of both wings, including the part of the body between the wings (Figure 1.1A). Why include part of the body? Because the bird is supported in normal gliding flight by a zone of reduced pressure which extends from one wing tip to the other. There is no gap in the middle (Figure 1.1B). Measuring the wing area is more complicated than measuring the span, more stressful for the bird and harder to do repeatably. On the other hand, this is a less critical measurement. The wing area is important in gliding performance, because it determines gliding speeds, and also the minimum radius of turn for circling in thermals. However, minor changes in the wing area have only a small effect on performance in flapping flight (Spedding and Pennycuick, 2001).

**Chord ($c$) and mean chord ($c_m$)**

*Chord* is an aeronautical term that dates from the nineteenth century, when people built thin wings, with cross sections that were arcs of circles. Modern aircraft wings are not thin arcs in cross section, but the "chord" is still the distance from the leading edge of the wing to the trailing edge, measured along the direction of the air flow (Figure 1.1A). Ornithological readers will be aware that this term was borrowed at some time in the past for use in bird morphometrics, and assigned a meaning that is unrelated to its aeronautical definition, and of no use for flight performance calculations of any kind. The conventional aeronautical definition of "chord" is the only one used in this book.

BOX 1.3 *Continued.*

The chord of a particular wing, unlike its span, does not have a unique value unless the wing is rectangular, which is unusual. Most wings have a maximum *root chord* where the wing joins on to the body, and taper to a smaller *tip chord*, with the chord diminishing along the span. A few flying animals (butterflies) have negative taper, meaning that the tip chord is greater than the root chord. The *mean chord* ($c_m$), which does have a unique value for the wing, is the ratio of the wing area ($S_{wing}$) to the wing span ($B$):

$$c_m = \frac{S_{wing}}{B}. \tag{1}$$

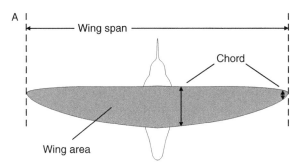

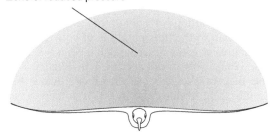

FIGURE 1.1  (A) Definitions of basic wing measurements. The *wing span* is the distance from wing tip to wing tip, and the *wing area* is the projected area of both wings, including the body between the wing roots (grey). These measurements are made with the wings fully extended. It is important that the elbow joint is locked in the fully extended position. The chord, which varies from point to point along the wing, is the distance from the leading edge of the wing to the trailing edge, measured along the direction of the relative air flow. (B) A gliding bird's weight is balanced by the pressure difference between the lower and upper surfaces, multiplied by the wing area. The area of reduced pressure above the wings accounts for most of this pressure difference, and it continues across the body. This is why the area of the body between the wing roots is included in the wing area.

*Flight* calculates the mean chord internally, and uses it for calculating Reynolds numbers (Chapter 4, Box 4.3) and reduced frequencies (Chapter 4, Box 4.4).

BOX 1.3 *Continued.*

### Aspect ratio

The aspect ratio ($R_a$) is the ratio of the wing span to the mean chord, and it expresses the shape of the wing:

$$R_a = \frac{B}{c_m}, \tag{2}$$

or, more conveniently:

$$R_a = \frac{B^2}{S_{wing}}. \tag{3}$$

Wing area is somewhat troublesome to measure (Box 1.4) and not as critical as wing span. If a few wing areas are measured among a sample of birds of the same species, they can be used to get an estimate of the aspect ratio, which may be assumed to be constant for the species. This means that the wings are assumed to be all of the same shape, though not necessarily the same size. Then, if a bird's span has been measured, the aspect ratio can be used to estimate its area by inverting Equation 3:

$$S_{wing} = \frac{B^2}{R_a}. \tag{4}$$

*Flight* will accept either the wing area or the aspect ratio for input. If supplied with one, it will calculate and enter the other automatically, so long as the wing span has already been supplied.

### Tail area

The tail is an accessory lifting surface in birds, and is more analogous in its function to a flap than to the horizontal tail of conventional aircraft. Birds' tails have been represented as an expandable delta wing, behind the main wing (Thomas, 1993). This is not included in *Flight* as most birds only deploy and use their tails at low speeds that are below the range covered by the calculations, and besides, the theory is somewhat conjectural. The tail is usually furled at normal cruising speeds, from the minimum power speed up, and may then be assumed to contribute no lift.

---

The programme will be misled by numbers that mean something different from what it assumes, which is not unusual for numbers identified by the same names in the ornithological literature. It serves no useful purpose for a field or laboratory observer to collect infinitely detailed statistics on variables that do not affect flight performance, and then get wing spans and areas (which do) from bird field guides, museum specimens or published figures from authors who neglected to define exactly what their measurements mean. Body mass is straightforward, but the manner in which the programme subdivides it (Box 1.2) needs to be understood when calculating migration performance. In particular, the concept of "lean mass" is not used in this

BOX 1.4 **Procedures for measuring wings.**

### Measuring wing span

There are two ways to measure the wing span, both of which are quick and easy to do on a live bird, with minimal stress. For a small bird, with both wings in good condition, place the bird on a flat surface, the right way up (not on its back). Stretch both wings out to the sides as far as they will go, with the tips on the surface, and check that the elbow and wrist joints are in their fully extended positions. Place markers, just touching each wing tip. Then fold the wings up, remove the bird, and measure the distance between the markers.

The other way is to measure the semi-span, which is usually easier for large birds. This is the only option if one wing is damaged. Stretch the good wing out as above, and use a tape measure to determine the distance from the backbone to the wing tip. This is the semi-span. Double it to get the span. The measurement is made from the body *centre line* (not the shoulder joint) to the wing tip. The centre line is easy to locate by feeling for the neural spines of the vertebrae, which stand up from the backbone as a sharp ridge. It is important to make sure that the elbow joint is fully extended, by pushing it gently forward until it locks.

### Measuring wing area

The wing area is measured in two stages. First make a tracing of one wing (not forgetting to measure the wing span), and then measure the area from the tracing. A wing tracing that is not accompanied by a wing span measurement is completely useless, and cannot be used for measuring wing area. The best idea is to write all the data about the bird, including the span, directly on the wing tracing. Wings of small birds can be traced in a sketchbook that opens flat, while a roll of parcel paper is good for large birds.

*Tracing the wing*
Put the drawing surface at the edge of a table, and hold the bird with one wing spread on the drawing surface, and its body beside the table edge, but not actually on it (Figure 1.2A). Spread the wing straight out to the bird's side, with the elbow and wrist joints fully extended. Do this with the bird right-way up, not on its back. Find the elbow joint (quite close in to the side of the body), and push it gently forwards until it locks in the fully extended position. Then draw the outline of the wing, following in and out of the indentations between the flight feathers. This results in a "partial wing", which is incomplete (open) at the inner end.

*Finding a partial wing area from the tracing*
First complete the partial wing tracing by drawing a straight line across the open end, parallel to the body centre line. This is the "wing root line". Its exact position is not critical, so choose a position that gives a realistic root chord (defined in Box 1.2). The first job is to measure the area of the partial wing. Of course, there are digital ways of doing this, and it may be worth the trouble of setting one up, if you have hundreds of small wings to measure. If you have to measure occasional warblers, ducks, pelicans etc., low-tech methods are easier, quicker, less error-prone and just as accurate if not more so.

**BOX 1.4** *Continued.*

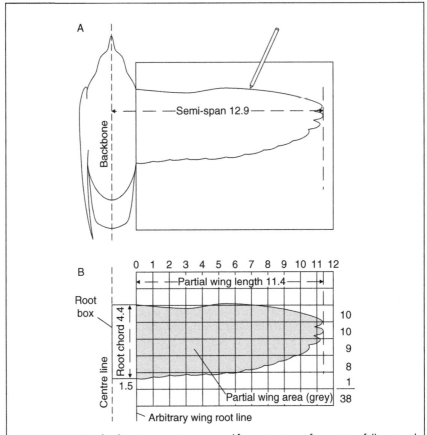

FIGURE 1.2 (A) A bird's wing area is measured from a tracing of one wing, fully spread over a drawing surface. The root end of the wing is left open at a point that is representative of the root chord. (B) The tracing is closed by ruling a straight wing-root line parallel to the body centre line, and the enclosed area (grey) is measured by counting squares on a transparent grid laid over it. The root box extends the wing root to the centre line (backbone), and the combined area is then doubled to get the total wing area (see Box 1.4).

First use a drawing programme to make a rectangular grid of 1 cm × 1 cm squares (or 0.5 cm × 0.5 cm for small birds). Number the lines along all four edges. Print the grid out on acrylic sheet as used for overhead transparencies, and check that the line spacing is indeed what it is supposed to be. Lay the grid over your wing tracing, aligning one edge with the wing root line, as shown in Figure 1.2B. Line up the leading edge of the wing so that it roughly corresponds with one of the horizontal grid lines. Starting from the left edge of the grid in Figure 1.2B, the third row of squares contains 8 full squares (allowing for the fact that the leading edge wanders up and down across the grid line), and some partial squares beyond column 8. If the filled parts of columns 11 and 12 were flipped over, they would fit in

BOX 1.4 *Continued.*

the unfilled parts of columns 9 and 10, making two complete squares beyond column 8. That makes 10 filled squares for the first row of the partial wing (row 3). Row 4 has a bit more than 11 filled squares, and row 5 has a bit less than 11, so count them as 11 each. Row 6 has about 8 filled squares, and all the small parts of the trailing edge in row 7 add up to about 1 filled square. That makes 41 filled squares in all for the partial wing. If the bird is so small that you get less than 100 squares in the partial wing, then it is better to use a grid with smaller squares, so that you have a chance of measuring the area within 1%. That is ample precision for the wing area measurement. In practice, 0.5 cm × 0.5 cm squares are good for small passerines, and 1 cm × 1 cm squares suffice for bigger birds.

*Completing the wing area measurement*
Although the squares in Figure 1.2B are bigger than ideal for the size of the bird, that will not stop us from completing the wing area calculation. We now know that the partial wing (the grey area) is 41 cm$^2$, but this is not the whole area for one side. You have to extend the root end of the wing to the centreline, by adding a "root box". First measure the root chord on the tracing, along the wing root line which you marked in. Then measure the "partial wing length", which is the distance from the wing root line to the tip of the longest primary. You already know the semi-span, having measured it directly on the bird. The width of the root box is the difference between the semi-span and the partial wing length (1.5 cm in Figure 1.2B), and the length of the box is 4.4 cm, the same as the root chord. The area of the root box is therefore 1.5 cm × 4.4 cm = 6.6 cm$^2$. You can now work out the wing area as follows:

| | |
|---|---|
| Partial wing | 41.0 cm$^2$ |
| Root box | 6.6 |
| Area this side | 47.6 |
| | ×2 |
| Both sides | 95.2 cm$^2$ |

The best place to do this little calculation is on the tracing, right beside the partial wing. *Flight* wants the wing span (*B*) in metres (divide cm by 100) and the wing area ($S_{wing}$) in square metres (divide cm$^2$ by 10$^4$). While you are at it, work out the aspect ratio as well ($B^2/S_{wing}$), and write all three results on the tracing:

$$B = 0.258 \text{ m}$$
$$S = 0.00952 \text{ m}^2$$
$$R_a = 6.99$$

An aspect ratio near 7 means that the wing is shaped about like the one in Figure 1.2. If we had got a ridiculous aspect ratio of 70 or 0.7, that would alert us to a mistake in the calculation.

Notice that the measured wing area is not very sensitive to the exact position where you draw the wing root line, to complete the partial wing. If you

BOX 1.4 *Continued.*

move the wing root line outwards a bit, you get a smaller partial wing, but this is compensated by a bigger root box, and *vice versa*. Little or no subjective judgement is required by this method of measuring wing areas, and it is consequently very repeatable between different observers.

**Entering wing measurements into *Flight***
The wing span must be entered first (in metres), and this should be a first-hand measurement—never a guess, or an estimate from some dubious published regression, or a quote from a field guide. The wing area is a less critical measurement. If you have a measured value, then enter it (in square metres). The programme will automatically calculate the aspect ratio, and display it in the box. Check that it is a believable value, and if not, look for wrong units or spurious factors of 10 in the entered wing span and area.

Sometimes you have a good value for the wing span (essential), but no measured wing area. In that case, you can enter the aspect ratio, if you can guess it from other birds that you have measured, whose wings are similar in shape. The programme will then calculate and enter the wing area.

book or in *Flight*, because its use in migration studies is obsolete and misleading. "Mass fractions" are defined in Box 1.2 for components such as the stored fat and flight muscles, and these are the ratio of the mass component to the all-up mass, *not* to the lean mass.

Bats, pterosaurs and even mechanical ornithopters can be described by their mass, wing span and wing area, and *Flight* will predict their performance, interchangeably with birds. For such non-birds (and birds with oddly shaped bodies), it may be necessary to adjust the values of some non-morphological variables, which are set to default values by the programme, but can be changed in the setup screens for different calculations. The reader should not be intimidated by the number of variables that can be adjusted, or by the somewhat arcane nature of some of them. The defaults will do for most practical purposes, but if one such variable (a drag coefficient for instance) is suspected to be the source of an observed discrepancy, it is easy to change the value systematically through several programme runs, keeping all other values the same. The results can be saved as an Excel Workbook, in which the results of each run are saved as a new Worksheet, together with the input from which they were generated. The meanings of those variables that are accessible to the user, and the effects and implications of tweaking their values, are explained in later chapters, and in *Flight*'s online manual.

### 1.1.3 DESCRIBING THE FLIGHT ENVIRONMENT

Besides the three morphological variables that describe the bird, *Flight* also requires values for two further variables (only) that describe the environment in which the bird flies. These are the acceleration due to gravity and the air density, both of which have a major effect on *Flight*'s performance predictions. These variables are discussed in Chapter 2, with methods of entering values into *Flight*. A default value is used for gravity, but this can be changed by the reader who wants to simulate flight elsewhere than here on earth. Air density is often overlooked or ignored by biologists, although not by pilots, who are acutely aware of its effects on flight performance. These effects also apply to birds, and it is essential to supply a realistic value. There is no default value for the air density, and *Flight* will not run until the user selects one of a number of options. For example, the programme will calculate and enter the air density if the user supplies measured values of the ambient pressure and temperature, or it will calculate a hypothetical value that corresponds to a specified height in the International Standard Atmosphere (Chapter 2).

## 1.2 ☙ THE ENGINEERING APPROACH TO NUMBERS

### 1.2.1 CALCULATION AS OPPOSED TO STATISTICS

In physiology, if you want an estimate of the rate of fuel consumption, then you have to measure it directly, or else measure something that you hope is proportional to it, like the rate of oxygen consumption. The result comes out in whatever units happen to be inscribed on the apparatus, such as watts, British Thermal Units per hour, calories per minute or even millilitres of oxygen per hour. If you only have to deal with one type of quantity, an arbitrary choice of units is fine for collecting statistics fodder, and may even serve for very basic calculations, but that is not *Flight*'s approach. The programme does not get its power estimates from regressions based on data of this type; in fact it does not use regressions at all. Instead, it estimates the power from other variables with other dimensions, basically the force that the wings apply to the air, and the speed with which they move. The force in turn comes from the rate at which momentum (mass times speed) is added to the air flowing over the wings. *Flight* then goes on to assume that the power estimated from force times speed must be accounted for by the rate of consumption of fuel energy. The calculation does not depend on any direct measurements of power as such. Unnatural as it may seem to many biologists, no statistics are involved.

The vast literature about measured rates of energy consumption in birds gets barely a mention in this book, because total fuel consumption is the end result of all processes that consume energy. A statistical summary of measurements of this type, on some particular bird, can be used to predict the energy consumption of the same bird, but cannot be transferred to other birds, flying under other conditions. The conditions include the air density, a fact that would be difficult to account for statistically, and seems to be unknown to most physiologists anyway. *Flight* works in the opposite direction from physiological experiments. It starts by simulating the underlying physical processes that result in a requirement for fuel energy, estimates the contribution of each to the fuel requirement, and adds in other assumed requirements (like basal metabolism) to estimate the total fuel consumption. Physiological experiments can be used to test whether the programme's predictions are accurate, but only if the required morphological measurements are carefully made and tabulated, and the local air density during the experiment is measured and recorded.

## 1.2.2 THEORY TELLS YOU WHAT TO MEASURE

If the model predicts that a bird can do something, which you know from observation that it definitely cannot do, that is a discrepancy which needs to be resolved by identifying and amending some wrong value in the input data, or possibly an error in the structure of the model itself. The resolution of discrepancies allows this type of model to be progressively improved and refined, so that greater confidence can be attached to its predictions. The time to think about this is at the planning stage, by examining the output that *Flight* generates, and using it to determine what measurements are needed. Too many experimenters turn to theoretical models as an afterthought, only to find that they have neglected to measure variables on which any kind of performance calculation depends, such as the bird's wing span and the air density, and have made meticulous measurements of quantities that cannot be predicted from any physical model. Before deciding what exactly to measure in a new experiment, please check the programme's input to see what information it needs, and look at its output to see whether you can measure anything that it predicts. The testing of hypotheses, and resolution of discrepancies, is covered at greater length at the end of the book (Chapter 15).

## 1.2.3 STATISTICAL TESTS AND SAMPLE SIZE

Many biologists and journal editors seem to regard copious and intricate statistical tests as an essential part of any kind of biological investigation, but this is not usually appropriate in flight studies. Statistical tests have their uses, for example for checking whether a sample of measurements can be reconciled with a predicted value, but not for generating the predictions in the first place. The predictions come from Newtonian physics, usually involving combinations of variables with different dimensions. The mindset required for collecting measurements that are to be used in this way differs radically from the more usual biological situation, in which arbitrarily defined numbers are collected for use as input data for a statistical package, without regard to their physical meaning, even if they have any.

The idea of "uncertainty" as applied to numbers calculated by the programme is different in concept from the statistical idea of confidence limits, and applies to an output number (such as the power output of the flight muscles) that is calculated by a formula from a set of input numbers (body mass, wing span, air density etc.). The uncertainty of the output is calculated by combining contributions caused by the individual uncertainties of each of the input variables (Spedding and Pennycuick, 2001), a procedure that is standard in physics and engineering, but less familiar in biology. Uncertainty calculations are not a major preoccupation in this book, but the principle is used to draw "uncertainty bands" above and below the power curves that *Flight* generates (power *versus* speed), and on either side of the estimated minimum power speed, as this is an important benchmark number (Chapter 15, Box 15.2).

No statistics are required to determine that a particular bird is able to fly, and the observation that it can do so puts upper or lower limits on some of the numbers that are involved in calculating the power output. One of these, the maximum isometric stress that the flight muscles can generate, has the status of a biological constant in that its value, once determined, is expected to apply to all vertebrates, living or extinct, including pterosaurs. This number is difficult to measure by direct experiment, but its value is estimated in Chapter 7 (Section 7.3.7) from the observation that a particular whooper swan was capable of sustained level flight, and for this a sample size of $N = 1$ is not merely acceptable, but mandatory. The default value in *Flight* for the isometric stress comes from the largest whooper swan for whom the necessary data were available, not from a typical or average swan, for the reasons explained in Chapter 7. It is the individual bird that flies, not the mean of a sample. When comparing the results of wind tunnel experiments with *Flight*'s predictions, it may be possible to repeat the measurements on another bird, but this does not usually bring any benefits in terms of increased precision, because the calculations have to be reconciled separately for each individual, however many there may be. $N = 1$ is fine in most cases.

## 1.3 ▬ DIMENSIONS AND UNITS

It is possible to enter data, run the *Flight* programme, and get results, without having any idea what the programme does or how it works, but there are hazards in this type of approach. At the minimum, it is essential to present input in the units that the programme expects, which are the basic SI units, not multiples or submultiples. Mass is in kilograms not grams, wing span is in metres, not millimetres, and so on. The expected units are stated on the setup screens. Just enter the right number, and press TAB (not Return). It is also a good idea to be clear about the dimensions of every variable in the input and output,

**BOX 1.5 Dimensions and units.**

The "dimensions" of a variable are a statement of its physical nature, for example power can be said to have the dimensions of work/time. The dimensions of all physical quantities required in mechanics can be expressed in terms of only three primary quantities. In physics, the favoured three are traditionally mass (**M**), length (**L**) and time (**T**). In those terms, force has the dimensions of mass (**M**) times acceleration ($\mathbf{LT}^{-2}$), so that the dimensions of work (force times distance) are $\mathbf{M} \times \mathbf{LT}^{-2} \times \mathbf{L}$, making $\mathbf{ML^2T^{-2}}$. The dimensions of power (work/time) are therefore $\mathbf{ML^2T^{-3}}$. The result is the same if you think of power as force ($\mathbf{MLT}^{-2}$) times speed ($\mathbf{LT}^{-1}$). The SI unit of power, with dimensions $\mathbf{ML^2T^{-3}}$, is called the watt (abbreviated W) but that is a convenience to save the trouble of writing out its full name, which is kg m$^2$ s$^{-3}$. All SI units used in mechanics are built up in this way from the kilogram, metre and second. Compound units that have names honouring famous scientists (newton, pascal, joule etc.) are written with a lower case initial letter, although their abbreviations (N, Pa, J) may not be. Table 1.1 is a summary of the dimensions and SI units of the variables used in subsequent chapters of this book.

Equations of the kind found in the boxes in this book are actually sentences, in which the "equals" sign is the verb. Each equation says that the expression that stands on the left of the "equals" sign is the same as the expression on the right-hand side, whether or not they look similar at first sight. If that is so, then the dimensions must be the same on both sides,

BOX 1.5 *Continued.*

TABLE 1.1 Dimensions of variables and SI units.

| Quantity | SI Unit | Dimensions | |
|---|---|---|---|
| Mass | kilogram (kg) | Mass | M |
| Length | metre (m) | Length | L |
| Time | second (s) | Time | T |
| Area | square metre ($m^2$) | Length$^2$ | $L^2$ |
| Volume | cubic metre ($m^3$) | Length$^3$ | $L^3$ |
| 2nd moment of area | metre to the fourth ($m^4$) | Length$^4$ | $L^4$ |
| Frequency | hertz (Hz) | Inverse time | $T^{-1}$ |
| Density | kilogram per cubic metre (kg m$^{-3}$) | Mass/volume | $ML^{-3}$ |
| Moment of inertia | kilogram metre-squared (kg m$^2$) | Mass × length$^2$ | $ML^2$ |
| Velocity | metre per second (m s$^{-1}$) | Length/time | $LT^{-1}$ |
| Acceleration | metre per second-squared (m s$^{-2}$) | Length/time$^2$ | $LT^{-2}$ |
| Force | newton (N) | Mass × acceleration | $MLT^{-2}$ |
| Pressure | pascal (Pa) | Force/area | $ML^{-1}T^{-2}$ |
| Work, energy | joule (J) | Force × length | $ML^2T^{-2}$ |
| Moment, torque | newton metre (N m) | Force × length | $ML^2T^{-2}$ |
| Power | watt (W) | Work/time | $ML^2T^{-3}$ |
| Specific work | joule per kilogram (J kg$^{-1}$) | Work/mass | $L^2T^{-2}$ |
| Specific power | watt per kilogram (W kg$^{-1}$) | Power/mass | $L^2T^{-3}$ |
| Dynamic viscosity | newton sec per square metre (N s m$^{-2}$) | Pressure × time | $ML^{-1}T^{-1}$ |
| Kinematic viscosity | square metre per second (m$^2$ s$^{-1}$) | Area/time | $L^2T^{-1}$ |

and this is often useful as a quick way to check for errors in complicated equations. In Box 7.3 of Chapter 7, this principle is taken further, and used to find an expression that predicts a bird's wingbeat frequency in cruising flight from five variables that affect the wingbeat frequency, but have different dimensions, namely the bird's mass, wing span and wing area, plus the air density and the strength of gravity. Not only does the calculation of wingbeat frequency in *Flight* not involve a regression, actually it does not involve any measurements at all. The basic result comes from considering the dimensions of the variables only. It is not completely unique, but leaves only a small amount of scope for putting more weight on one input variable at the expense of another. Observations of wingbeat frequencies are used only in the final stage, to resolve this residual ambiguity.

and not to waffle (for example) about "flight costs" without specifying whether this means energy, or energy per unit time, or energy per unit distance. There is more information about dimensions and units in Box 1.5, and also in my *Conversion Factors* (Pennycuick, 1988a).

If all of the variables in a physical calculation are expressed in units that belong to an internally consistent family, like the SI, then no conversion factors are needed for the results. Each calculated result comes out in the SI unit with the appropriate dimensions, whatever that may be. Speed is everywhere in flight calculations, and the SI unit is the metre per second. If you insist on measuring speed in kilometres per hour, this introduces an unnecessary conversion factor of 3.6, and if the speed has to be squared or cubed, a massive tangle of conversion factors quickly takes over the calculation. Physiologists will recognise this syndrome.

## 1.4 ⬤ LITERATURE CITATIONS

This book presents the particular model of the mechanics of bird flight that is embodied in the *Flight* programme, together with sufficient background to understand how the programme works, and how to use it for testing the underlying assumptions. The book is not a review, and it does not attempt to provide an exhaustive bibliography, citing everything that has ever been written about flight. Books in which the literature is comprehensively reviewed include Norberg (1990) and Videler (2005). The reference list is short for a book of this length, and contains those publications from which I got information that was used in the book, or built into the *Flight* programme. The list includes some general textbooks on aeronautical engineering (von Mises, 1959; Anderson, 1991) which cover the theoretical and experimental background on which the model is based, and other books and papers from which I got items of theory or observation that I have used in the text. I cite a source if I have used it to build *Flight*, or to explain it in the book, otherwise not. Perhaps eccentrically, I do not even cite my own papers, unless I have used the findings in the book or in the programme.

# 2

# THE FLIGHT ENVIRONMENT

The two most important environmental quantities for flight calculations are the strength of gravity and the density of the air, neither of which is routinely recorded in traditional biological investigations. This chapter outlines the properties of the earth's gravity field and atmosphere, gives practical methods for estimating the air density at the bird's flying height, and introduces the properties, uses and limitations of the fictional International Standard Atmosphere. If the flight environment has changed over geological time, this would affect the interpretation of fossil flying animals.

A bird's flight performance is affected by the chemical composition, humidity and temperature of the air in which it flies, but these effects are physiological, and are not covered by the *Flight* program. *Flight* deals with the physics of flight, and it requires values for just two environmental variables, neither of which is routinely recorded by physiologists or ecologists. These are the strength of gravity and the air density. Flying would be a very different proposition on either of our planetary neighbours Mars and Venus, even if they had oxygen in their atmospheres and benign temperatures, and were able to support life. Both have weaker gravity than ours, which reduces the weight for a given mass, and also the energetic cost of supporting it. Flying would be slow on Venus, and cheap in terms of power, because the atmosphere is two orders of magnitude denser than ours, but fast and expensive on Mars, because the atmosphere there is two orders of magnitude less dense than ours. Here on Earth, gravity is nearly constant, anywhere that a bird is likely to go, but the density of the air varies wildly from place to place, from day to day and (especially) at different heights above sea level.

## 2.1 ➡ THE EARTH'S GRAVITY FIELD

The Newtonian view of gravity is sufficient for flight performance calculations. It says that the earth exerts a force on the apple (its weight) which is proportional to the product of the mass of the earth and that of the apple, and inversely proportional to the square of the distance between their centres of mass. The weight force actually attracts the earth to the apple as well as *vice versa*, but since the earth is much more massive than the apple, we perceive this mutual attraction as a force that causes the apple to accelerate, if dropped, towards the earth. The weight of a meteorite of constant mass (i.e., not burning up) increases as it falls in from space because it is getting nearer to the centre of the earth, and reaches a maximum value at the surface. If it happens to fall down a mine shaft, its weight starts to decrease, because some of the earth's mass is now attracting it from above. Since different objects experience weights proportional to their masses, all experience the same acceleration in free fall, relative to the earth, at the same point in the gravity field. This *acceleration due to gravity* is used as the measure of the strength of gravity. The earth's gravity at the surface is strongest at the poles and weakest at the equator, for two reasons. In the first place, the earth's radius is greater at the equator, and an object at the surface there is further away from the earth's centre than it would be at the poles. Secondly, the rotation of the earth forces an object on the surface to accelerate towards the earth's centre, thereby reducing its weight. This centripetal effect is strongest at the equator, and dwindles to nothing at the poles.

The effects of latitude and height on gravity are combined empirically in Helmert's equation (Box 2.1, Table 2.1). At any latitude, the earth's gravity decreases with height above sea level. Taking the effects of latitude and height together, the acceleration due to gravity varies from 9.83 m s$^{-2}$ at sea level at the poles, down to 9.75 m s$^{-2}$ at a height of 10,000 m above sea level over the equator (Table 2.1). Small "gravity anomalies" are superimposed on this underlying gravity distribution,

BOX 2.1 **The earth's gravity.**

Helmert's equation is a polynomial expression which gives an approximation to the acceleration due to gravity ($g$) in m s$^{-2}$, as a function of latitude
($L$) and height ($h$) in metres above mean sea level:

$$g = 9.80616 - [0.025928\cos(2L)] + [0.000069\cos^2(2L)] - [(3.086 \times 10^{-6})h]$$

Values of $g$ from this formula are tabulated in Table 2.1 for latitudes from
0 to 90 degrees (either north or south), and heights up to 10,000 m above sea
level, which is near the top of the troposphere, and higher than any bird is
known to fly. The formula allows for the earth's angular velocity and ellipsoi-
dal shape, but not for gravity anomalies due to topography, or variations of
density in the mantle and crust.

TABLE 2.1 Earth's surface gravity.

| Latitude (N or S) | Height above mean sea level (m) | | | | | | | | | | |
|---|---|---|---|---|---|---|---|---|---|---|---|
| | 0 | 1000 | 2000 | 3000 | 4000 | 5000 | 6000 | 7000 | 8000 | 9000 | 10000 |
| 0° | 9.780 | 9.777 | 9.774 | 9.771 | 9.768 | 9.765 | 9.762 | 9.759 | 9.756 | 9.753 | 9.749 |
| 10° | 9.782 | 9.779 | 9.776 | 9.773 | 9.770 | 9.766 | 9.763 | 9.760 | 9.757 | 9.754 | 9.751 |
| 20° | 9.786 | 9.783 | 9.780 | 9.777 | 9.774 | 9.771 | 9.768 | 9.765 | 9.762 | 9.759 | 9.755 |
| 30° | 9.793 | 9.790 | 9.787 | 9.784 | 9.781 | 9.778 | 9.775 | 9.772 | 9.769 | 9.765 | 9.762 |
| 40° | 9.802 | 9.799 | 9.795 | 9.792 | 9.789 | 9.786 | 9.783 | 9.780 | 9.777 | 9.774 | 9.771 |
| 50° | 9.811 | 9.808 | 9.804 | 9.801 | 9.798 | 9.795 | 9.792 | 9.789 | 9.786 | 9.783 | 9.780 |
| 60° | 9.819 | 9.816 | 9.813 | 9.810 | 9.807 | 9.804 | 9.801 | 9.798 | 9.794 | 9.791 | 9.788 |
| 70° | 9.826 | 9.823 | 9.820 | 9.817 | 9.814 | 9.811 | 9.808 | 9.804 | 9.801 | 9.798 | 9.795 |
| 80° | 9.831 | 9.827 | 9.824 | 9.821 | 9.818 | 9.815 | 9.812 | 9.809 | 9.806 | 9.803 | 9.800 |
| 90° | 9.832 | 9.829 | 9.826 | 9.823 | 9.820 | 9.817 | 9.814 | 9.811 | 9.807 | 9.804 | 9.801 |

Acceleration due to gravity in m s$^{-2}$ from Helmert's equation.

due to density variations in the earth's mantle and crust. It is usually
assumed that gravity was much the same in past ages as it is now,
although it is not clear that the average density of the mantle has
always been the same as it is now. If the mantle were to expand,
surface gravity would decrease without any change in the earth's mass,
and if it did that in mesozoic times, the existence of very large dinosaurs and pterosaurs would be easier to understand (Box 2.4). The
default value used for gravity in *Flight* is 9.81 m s$^{-2}$, and this will
usually be within half of one per cent of the present value of gravity,
anywhere that a bird is likely to go. The value of gravity in the
programme can be changed by the user, but this is intended for simulating flight on other planets, or on earth at past or future times, rather
than for refining the present value of gravity at particular points.

The effect of acceleration is indistinguishable from that of gravity as far as any physical process is concerned. This includes the power requirements for flight, and applies to all accelerations, not just the one due to the earth's rotation. The basic power curve or glide polar is calculated in later chapters on the assumption that the bird is proceeding at a constant speed in a straight line, relative to the earth's surface, but if it accelerates upwards or downwards, then the acceleration produces effects that are indistinguishable from those due to a change in the strength and/or direction of gravity. For example, in the flapping phase of "bounding" flight, as commonly seen in small passerines, the flight path curves upwards, meaning that the bird accelerates in a direction perpendicular to its flight path. The effect of this acceleration on the bird's requirement for power is the same as that of an increase in gravity, and its power requirements and wingbeat frequency are affected accordingly (Chapter 9, Box 9.1).

## 2.2 ⬤ THE EARTH'S ATMOSPHERE

The earth's atmosphere consists mostly of nitrogen and oxygen, and its continued existence depends on the gravity field being strong enough to retain these gases. The lower layers of the atmosphere are compressed by the weight of the gas molecules above, and the three basic physical properties of the atmosphere, its pressure, temperature and density, all decrease in a regular way with height above the surface. The atmosphere is the source of oxygen for aerobic flight, and the sink for disposing of carbon dioxide, water and heat. The temperature and relative humidity of the air affect the physiology of flight, while its transparency (or lack of it) affects orientation and navigation. However, the only property of the air for which a value is required for the mechanical calculations in *Flight* is its density, which appears in all of the performance equations in the boxes accompanying Chapters 3 and 10. A measured or estimated value must be entered before *Flight* will run. Double-clicking on the box for air density in any of *Flight*'s Setup screens will invoke several easy options for setting the air density, either from direct measurements, or from estimates derived from the International Standard Atmosphere (below), or a combination of the two.

### 2.2.1 THE INTERNATIONAL STANDARD ATMOSPHERE

The International Standard Atmosphere is a convenient fiction representing the physical properties of a typical sample of the atmosphere, somewhere between the tropics and poles, and between fair weather

BOX 2.2 **The International Standard Atmosphere.**

The International Standard Atmosphere is a much-simplified "average" atmosphere, which consists of defining the vertical distribution of a number of physical properties of the air, as functions of height above sea level. The following information is from von Mises (1959). Readers who consult this, and have difficulty with the archaic units that von Mises used, will find conversion factors, and background information about unit systems, in Pennycuick (1988a).

**Variable definitions for this box**
$h$   Height above sea level (ASL)
$p$   Air pressure
$T$   Air temperature (Celsius)
$\lambda$   Air temperature lapse rate
$v$   Air kinematic viscosity
$\rho$   Air density

**Air temperature**
The lowest zone of the International Standard Atmosphere is the troposphere, in which the temperature is 15 °C at sea level, and declines with height at a *lapse rate* ($\lambda$) whose value is constant at 0.0065 °C m$^{-1}$ up to 11,000 m above sea level (ASL). The temperature ($T$) in degrees C at a height $h$ (metres) above mean sea level is:

$$T = 15 - 0.0065h \qquad (1)$$

The top of the troposphere (the tropopause) is at 11,000 m ASL, and the temperature there is $-56.5$ °C. Above the tropopause is an isothermal zone, the stratosphere, where the temperature remains at $-56.5$ °C, although the pressure and density continue to decrease with height. Bird flight is essentially confined to the lower half of the troposphere.

**Air pressure**
The barometric pressure at sea level in the International Standard Atmosphere is 1013.2 millibars. The millibar is the same as the hectopascal (hPa), meaning 100 pascals. The pressure ($p$) at a height $h$ (in metres ASL) is:

$$p = 1013[1 - (2.26 \times 10^{-5})h]^{5.256} \qquad (2)$$

**Air density**
The air density at sea level in the International Standard Atmosphere is 1.226 kg m$^{-3}$. Taking the sea level pressure to be 1013 hPa and the temperature 15 °C (or 288 K), the local air density ($\rho$) at any point in the troposphere can be found from Boyle's Law, using the locally measured values of pressure ($p$) and temperature ($T$):

$$\rho = 1.226(p/1013)[288/(T + 273)] \qquad (3)$$

If the pressure is in hPa (same as millibars) and the temperature is in Celsius, then the density from Equation (3) will be in kg m$^{-3}$. As both the pressure and temperature decrease with height, so also does the air density, steeply at first, and then ever more gradually.

BOX 2.2 *Continued.*

### Kinematic viscosity

The air's kinematic viscosity is the ratio of its viscosity to its density. Kinematic viscosity has the dimensions of area/time, and the SI unit is the square metre per second. It is nearly independent of pressure or density, but increases with decreasing temperature, and thus with height in the troposphere. It is represented by the symbol $v$ (Greek "nu"), and its value in the International Standard Atmosphere may be approximated up to the tropopause by the polynomial:

$$v = 1.466 + 0.09507h + 0.01047h^2 \qquad (4)$$

Note that $h$ in this formula is the height in kilometres (not metres), and that the kinematic viscosity is in units of $10^{-5}\,\mathrm{m^2\,s^{-1}}$. Thus the sea-level value ($h = 0$) is $v = 1.466 \times 10^{-5}\,\mathrm{m^2\,s^{-1}}$. A graph of Equation (4) is shown in Figure 2.1. *Flight* calculates the air's kinematic viscosity internally from the flying height [Equation (4)] and uses it to calculate Reynolds numbers, which are included in the program's output (Chapter 3, Box 3.6).

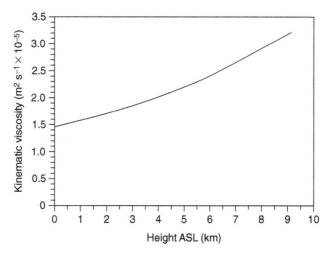

FIGURE 2.1 The kinematic viscosity of air in the International Standard Atmosphere is $1.466 \times 10^{-5}\,\mathrm{m^2\,s^{-2}}$ at sea level, increasing with altitude according to Equation (4) of Box 2.2, which is a second-degree polynomial fitted through figures tabulated by von Mises (1959).

and foul. Its physical properties are functions of height above sea level only (Box 2.2). They do not vary with geographical location, or with the seasons, unlike those of the real atmosphere. The temperature at sea level in the International Standard Atmosphere is 15.0 °C, and the "lapse rate" is constant at 0.0065 °C per metre of height, up to a height of 11,000 m, where the temperature is −56.5 °C. This level is the *tropopause*, the upper boundary of the *troposphere*. Above the tropopause the *stratosphere* begins, where the lapse rate in the International Standard Atmosphere is zero, and the temperature is constant at $-56.5\,°C$. The real atmosphere conforms reasonably well to this tidy picture in the mid-latitudes, but the real tropopause is higher (more like 18 km) over the tropics, and lower (about 9 km) over the poles. The average sea-level temperature is, of course, higher in the tropics than in the polar regions, while the temperature in the stratosphere is around $-80\,°C$ over the tropics, and warmer (around $-50\,°C$) over the poles.

The International Standard Atmosphere was devised for such practical purposes as standardising aircraft altimeters, and it is also useful for hypothetical scenarios, in which (for instance) a bird is assumed to fly at a particular altitude, and estimates of atmospheric variables are needed. It should not be used as a substitute for measuring the air density when observing bird flight in the field or the laboratory, as the local properties of the real atmosphere vary according to the weather, and the climate at any particular location. Monitoring

---

BOX 2.3 **Measuring air density.**

All flight performance calculations involve the air density. The ambient air density at the observer's position can be calculated from the barometric pressure and the air temperature, and these two measurements should always be recorded when observing bird flight, either in the field or in the laboratory.

**Variable definitions for this box**
$h_{ft}$ Height in feet above datum indicated by altimeter
$h$   Height in metres above datum indicated by altimeter
$\Delta h$ Measured height in metres of bird above observer
$p$   Barometric pressure
$T$   Temperature (Celsius)
$\rho$   Air density
$\rho_1$  Air density at the observer's position
$\rho_2$  Air density at the bird's flying height

**Measuring ambient air density directly**
The local air density ($\rho$), in kg m$^{-3}$, can be found from the measured barometric pressure ($p$) in millibars (same as hectopascals) and air temperature ($T$) in degrees Celsius, from the Boyle's Law equation:

$$\rho = 1.226(p/1013)[288/(T+273)] \tag{1}$$

Laboratory-grade instruments are permanently installed in wind tunnels, so that the barometric pressure and air temperature in the test section can be read electronically, and the air density can be calculated from these measurements. For field observations, the barometric pressure can be measured with a small aneroid barometer, or from an electronic barometer wristwatch as used by hikers. A barometer watch should be checked for temperature

BOX 2.3 *Continued.*

compensation, by putting it in the fridge for an hour or two. The pressure reading should not be affected by a 20 °C change in temperature. If it is, get a more expensive barometer watch. Ambient air temperature can be measured with a small liquid or electronic thermometer, taking care to place the sensor in a shady, well-ventilated place that is not too near warm objects like cars. The thermometer needs to be out in the breeze, not strapped to the observer's wrist!

**Measuring air pressure in an aircraft**
When observing birds from an aircraft, the ambient barometric pressure can be obtained from the altimeter, which is actually an aneroid barometer, calibrated to indicate height directly. The pressure-setting adjustment on the altimeter should be set to 1013 mbar (or 29.92 inches of mercury). If the altimeter indicates a height ($h_{ft}$) in feet, what this really means is that the pressure ($p$) in millibars is:

$$p = 1013[1 - (6.88 \times 10^{-6})h_{ft}]^{5.256} \qquad (2a)$$

If the altimeter indicates the height ($h$) in metres, this formula becomes:

$$p = 1013[1 - (2.26 \times 10^{-5})h]^{5.256} \qquad (2b)$$

**Air density for flying birds**
When using tracking radar or an ornithodolite, the bird and the observer are usually at different heights, but the height difference ($\Delta h$) is measured. $\Delta h$ is considered positive if the bird is above the observer, negative if it is below. The air density at the observer's position ($\rho_1$) is found from the ambient temperature and pressure, using Equation. (1) above. A correction is then applied to get the density at the bird's flying height ($\rho_2$), on the basis that the air temperature and pressure decrease with height according to Equations (1) and (2) of Box 2.2, in the manner assumed for the troposphere in the International Standard Atmosphere. In that case:

$$\rho_2/\rho_1 = \{1 - [0.00650\Delta h/(273 + T)]\}^{4.256} \qquad (3)$$

**Data files from radar or optical tracking**
Tracking projects that use tracking radar or optical tracking typically generate a data set in which air speeds (for example) have been observed for birds flying at a range of different (measured) heights. Before such observations can be compared with each other, and/or with a value predicted by *Flight*, they have to be "reduced" to a common altitude, such as sea level. This type of data reduction is discussed in Chapter 15, Box 15.3. It requires an estimate of the air density for each individual observation. The best idea is to record the air pressure and temperature at the tracking site while observations are in progress, and apply the correction of Equation (3), at the time that each observation is recorded. Then each tracking record can include an estimate of the bird's local air density. This can later be used to reduce each observation to sea level, before proceeding with the analysis of the whole data set.

BOX 2.3 *Continued.*

It is not valid to compare speed observations of birds that were flying at different heights, without taking account of the differences in air density (Chapter 15, Section 15.1.5).

### Entering air density and/or altitude into *Flight*

The *Flight* program requires a value for the air density to be entered before it will do any calculations. When you save a bird in the User Birds database, the air density that you used in the last calculation is saved as part of the data for the bird, and will be re-entered when you run the same bird again. Some (but not all) of the birds in the Preset Birds database also have associated air densities, which can of course be changed by the user after entering the bird.

Otherwise, you have to enter a value for the air density. Double-clicking on the box for air density in any Setup screen brings up a small screen, which offers you a choice of five methods for estimating and entering a suitable value. These are based either on direct observation, which should always be used for field or laboratory observations, or on the International Standard Atmosphere, which is useful for hypothetical scenarios. A combination of both methods is provided for tracking-radar and ornithodolite observations. The five options are as follows:

1. *Enter air density directly*
   If you know the air density in kg m$^{-3}$, then just enter it—but usually you don't.
2. *Enter barometric pressure and air temperature*
   If you are observing birds flying at your own level, either in the field or in the laboratory, then record the ambient barometric pressure and air temperature, and enter these values. *Flight* will work out the air density from Equation (1), and enter it into the program.
3. *Enter bird's height above the observer*
   For tracking-radar or optical tracking, in which you are measuring the heights at which individual birds fly above the observer, record the pressure and temperature at the observer's position as above, then enter the bird's height above the observer (negative if it is below you). *Flight* will first work out the air density at the observer's position, and then apply a correction for the height difference by using Equation (3).
4. *Specified height in International Standard Atmosphere*
   This is strictly for hypothetical calculations, as the International Standard Atmosphere is not a reliable guide to conditions in the real atmosphere, which vary widely from day to day and from place to place, according to the weather. Never use this method for field observations, just to save the trouble of measuring the air pressure and temperature (no trouble at all). An easy way to enter an estimate for the air density at a given altitude, on any of the Setup screens, is to enter the altitude (in metres) in the box below the one for air density, and press TAB. *Flight* will calculate the air density that corresponds to that altitude in the International Standard Atmosphere, and enter it. Conversely, if you enter the air density directly, the altitude will be adjusted to suit.

BOX 2.3 *Continued.*

5. *Sea level in the International Standard Atmosphere*
Entering zero altitude sets the air density to 1.226 kg m$^{-3}$. Use this to calculate sea-level performance estimates. If you have a lot of field observations at different heights, then reduce them to sea level before plotting them on a graph, or comparing them to a sea-level performance curve (Chapter 15, Box 15.3).

the air density at the observer's level is simply a matter of recording the air pressure and temperature (Box 2.3). Portable instruments (including wristwatches) that will make these measurements are readily available. The real air density at sea level commonly varies by 5% or more above or below the sea-level value in the International Standard Atmosphere, and the density decreases progressively with height above sea level, all the way to zero in the vacuum of space. Pilots calculate the "density altitude" before attempting to take off at tropical upland airports, meaning the height in the International Standard Atmosphere that corresponds to the prevailing air density at the surface. The density altitude can be hundreds of metres higher than the actual altitude in "hot and high" conditions. Most bird flight occurs in the lower levels of the troposphere, where the air density is 60% or more of the sea-level value.

## 2.2.2 WIND AND WEATHER

Weather in the familiar form of clouds and precipitation is mainly confined to the troposphere. Bradbury (1989) explains how the large-scale processes work, and produce the small-scale weather in which birds and pilots fly. Although solar energy arrives from above, only a small percentage of the incoming radiation is absorbed in the atmosphere, mostly by smoke or aerosol particles. Some of it is reflected back into space by clouds, or by reflective parts of the surface, especially ice and snow fields. The remainder is absorbed and heats up the ground or water, which in turn heats the atmosphere from below. Heat transferred upwards from the surface to the atmosphere is the energy source that drives the weather. It results in horizontal temperature gradients, which, in turn, result in gradients of pressure. One might expect a body of air to accelerate down a pressure gradient, from high pressure to low, and so it does in the tropics, but in the mid-latitudes, somewhat counter-intuitively, the effect of the earth's rotation is to deflect the flow of air so that it flows along the lines of equal pressure (isobars) rather than across them. The result is that in the northern hemisphere the air circulates anti-clockwise around an area of low pressure (depression) and clockwise around an area of high pressure (anticyclone), and *vice versa* in the southern hemisphere. At the surface, friction with the ground deflects the flow, so that the air rises gently in a depression because of the converging flow at the bottom, and conversely the air sinks in an anticyclone because it diverges at the bottom.

### 2.2.3 CONVECTION AND CLOUD FORMATION

If we consider a small "parcel" of air rising through the troposphere, it expands because the pressure decreases with height. If the expansion is *adiabatic*, meaning that no heat passes into or out of the parcel of air, then in the absence of cloud, the temperature decreases with height at the *dry adiabatic lapse rate*, which is about 1 °C per 100 m of height. If the water vapour content of the parcel of air remains constant in terms of mass percentage, then the relative humidity increases, because a lower mass percentage of water vapour is needed to saturate the air as the temperature drops. When the relative humidity reaches 100% (saturated air), cloud forms. The condensation of water vapour into liquid water droplets releases heat, with the result that the *saturated adiabatic lapse rate*, which prevails in cloud, is only about 0.5 °C per 100 m of height. The nominal lapse rate in the International Standard Atmosphere is 0.65 °C per 100 m of height (above), intermediate between the dry and saturated values. Cumulus clouds, growing in clear air, have a well-defined *cloudbase* level, where air coming up from below becomes saturated. The characteristic "cauliflower top" develops as the saturated air in the cloud becomes ever more buoyant, relative to the dry air outside. On larger scales, cloud forms where air is rising gently over a large area, as in a depression, and clears in areas of slowly subsiding air, as in an anticyclone, although there are many factors that modify these basic trends.

## 2.3 ☙ AIR DENSITY IN *FLIGHT*

When you save a bird in the "User Birds" database of *Flight*, the air density is saved together with the bird's morphological measurements, because this is an integral part of the scenario for any performance calculations. When observing bird flight, whether in the laboratory or in the field, it is essential to record the prevailing air density if you want to compare the results with similar observations, on other birds in other places. The air density at the observer's position can be found as explained in Box 2.3, from the readings of a barometer and an air

TABLE 2.2 Mass of the earth's atmosphere.

| | |
|---|---|
| Whole atmosphere | 5.2 |
| Nitrogen | 3.9 |
| Oxygen | 1.2 |
| Carbon dioxide | 0.0026 (rising) |
| Water vapour | 0.12 |

Mass of the earth's atmosphere and its main constituent gases, in units of $10^{18}$ kg, from Budyko et al. (1985).

thermometer, recorded in the field. This value of the density is only valid for the bird if it is flying at the observer's height, as in a wind tunnel. If the bird is flying at a measured height above or below the observer, as in tracking radar observations, then the density at the observer's position can be corrected for the height difference, by assuming that the air density varies with height in the same manner as in the International Standard Atmosphere (Box 2.2). *Flight* will accept a value for the air density in kg m$^{-3}$, or calculate a value from information that you provide. For example, you can supply observed values of barometric pressure and air temperature (and a height difference if applicable), or you can specify a height, and get *Flight* to calculate the air density that would prevail at that height, according to the International Standard Atmosphere (Table 2.2).

## 2.4 ⬝ GRAVITY AND THE ATMOSPHERE IN FORMER TIMES

Today's atmosphere is the end product of thousands of millions of years of evolution, and some of its major features, including its oxygen content, are biological in origin. Reconstructing the chemical composition and physical properties of the atmosphere in past geological eras is difficult, but there are indications that the atmosphere was denser in mesozoic times than it is now (Dudley 1998), possibly a lot denser. Performance calculations for the large pterosaurs that flourished throughout the Jurassic and Cretaceous periods are conjectural for several reasons, among which lack of information about the prevailing air density is not the least important. Prolonged volcanic episodes that would have involved increased outgassing, such as the formation of the Deccan Traps at the end of the Cretaceous, would have injected large amounts of water vapour and carbon dioxide into the atmosphere, and although neither of these gases would have an appreciable effect on the sea-level air density, an indirect effect on the mass of

BOX 2.4 **Constancy of the flight environment.**

The two environmental variables that occur in the equations for speed and power (Chapter 3) are the acceleration due to gravity and the air density. It is often assumed (explicitly or not) in discussions of the flight performance of fossil flying animals that the strength of gravity and the air density were the same when those creatures lived as they are now. This raises some difficulties in the case of flying animals that were bigger than any living birds, namely, the larger Cretaceous pterodactyls (Chapter 6, Section 6.2.5) and the giant Miocene bird *Argentavis*, as scaling considerations indicate that there is an upper limit to the size and mass of an animal that can maintain height by muscle power (Chapter 7, Box 7.4). Also, it seems that present-day swans are very near that limit (Chapter 7, Box 7.5). The favoured explanation for still larger flying fossil animals is that they must have been incapable of level flight, and therefore dependant on soaring. However, if the air density were higher in ancient times than it is now, or the strength of gravity lower, or both, the difficulty would be reduced, and might disappear altogether. Is this possible?

### Variable definitions for this box
$G$     Newton's gravitational constant
$g$     Acceleration due to gravity at the earth's surface
$m$     Mass of a small body resting on the earth's surface
$m_a$     Mass of the earth's atmosphere
$m_e$     Mass of a spherical earth
$p_0$     Atmospheric pressure at sea level
$R$     Gas constant
$r_e$     Radius of a spherical earth
$S_e$     Surface area of a spherical earth
$T$     Absolute temperature
$W$     Weight of a small body resting on the earth's surface
$\rho_e$     Mean density of the earth
$\rho_0$     Air density at sea level

### The mass of the earth is constant
Particles of dust fall into the earth's atmosphere every day from space, sometimes making visible trails as they burn up (meteors). Larger objects (meteorites) reach the surface and produce craters, and a few are large enough to cause widespread devastation on the scale that wiped out the dinosaurs. Kyte and Wasson (1986) estimated that all of these different-sized objects together increase the earth's mass by around 78 million kilograms per year on average, and that this rate has not varied much over past intervals of tens of millions of years. If we go back 100 million years into the Mesozoic, then the earth's mass would have increased by about $8 \times 10^{15}$ kg since then, according to this estimate. However, as the earth's present mass is just under $6 \times 10^{24}$ kg, or $10^9$ times the estimated amount added in the last 100 million years, it is safe to say that the accretion of mass has been negligible for all practical purposes, throughout the time when there have been flying animals. The earth's mass may be considered constant at $5.976 \times 10^{24}$ kg (Beatty et al. 1981).

BOX 2.4 *Continued.*

---

### Radius of the earth and surface gravity

Constant earth mass does not necessarily imply that the earth's radius, surface area or volume have been constant through time. This depends on the density of the solid earth, or of its major components, which might not be constant. Our knowledge of the physical properties of the minerals that make up the earth's interior is based almost exclusively on indirect evidence, because they are inaccessible to direct observation, so that studying the effects of the enormous pressures that must exist deep in the earth, on minerals whose composition is conjectural, is at best an uncertain enterprise. It is conceivable that some mineral components of the mantle might undergo phase changes whereby the application of a very large pressure could cause the atoms within a molecule, or the molecules within a crystal, to be rearranged so as to take up less volume. This would cause a discontinuous increase in density, reversible when the pressure is relieved. Such an effect, if it existed, might be regenerative in either direction. If a layer of rock were lifted by some disturbance such as a rising mantle plume, the weight of the rocks would decrease as they moved further from the earth's centre, and the pressure in the layer would decrease. If this were to trigger a phase change to lower density, positive feedback would set in. Likewise the change to a higher density in descending rocks might also be regenerative, so that a widespread change in density might follow from a small initial disturbance. What would be the effect on surface gravity of a large-scale change in the density of the mantle? For a spherical earth of radius $r_e$, the mass ($m_e$) would be:

$$m_e = (4/3)\pi \rho_e r_e^3 \tag{1}$$

where $\rho_e$ is the earth's mean density. If the mass is constant (above), then the radius can be expressed as a function of the density:

$$r_e = (3m_e/4\pi \rho_e)^{1/3} \tag{2}$$

Higher density leads to a lower radius, in proportion to the cube root of the density. Newton's law of gravitation says that the weight ($W$) of a small body of mass $m$, sitting on the earth's surface is:

$$W = Gm_e m/r_e^2 \tag{3}$$

where $G$ is the gravitational constant ($6.6732 \times 10^{-11}$ N m$^2$ kg$^{-2}$). Since a spherical earth of constant density, or one composed of spherical shells, each of constant density, exerts the same gravitational attraction as a single mass concentrated at the centre, the distance separating the earth from a small body on its surface is the same as the earth's radius [$r_e$ in Equation (3)].

Dividing the small body's weight by its mass ($m$), and substituting for the square of the radius from Equation (2), the acceleration due to gravity at the surface ($g$) is:

$$g = G(4\pi/3)^{2/3} m_e^{1/3} \rho_e^{2/3} \tag{4}$$

BOX 2.4 *Continued.*

If the earth's mean density decreases, then its radius increases, and the strength of surface gravity decreases in proportion to the two-thirds power of the density, for example a 10% reduction in the density would result in a 3.5% increase in the radius and a 6.8% decrease in surface gravity. It will be seen in later chapters that such a reduction in gravity would increase the upper limit for the mass of a flying animal that is able to maintain height by muscle power, and it so happens that a very similar effect would apply to the maximum mass of walking animals, albeit for different reasons (Pennycuick 1987c). If the earth's surface gravity were weaker throughout the Cretaceous than it is now, this might account for the existence throughout the period of pterosaurs that were larger than any modern bird, and also of dinosaurs that were larger than modern elephants, although big dinosaurs first appeared somewhat earlier, in late Jurassic times. Later, in Miocene times, there were dinosaur-sized mammals such as *Baluchitherium*, and also a very large bird (*Argentavis*) which was apparently able to fly (Campbell and Tonni 1983; Chatterjee et al. 2007).

Is it possible that the earth was larger than it is at present, with lower surface gravity, from late in the Jurassic until the Miocene, and then gradually contracted to its present size? This might sound far-fetched, but something similar was proposed by Carey (1976), whose initial observation was that the continents (with their continental shelves) do not fit together very well if moved around on a sphere of the present radius, but fit more snugly on a smaller sphere. Carey proposed that the earth was smaller than at present at the beginning of the Jurassic, when the super-continent Gondwanaland started to break up, and has been increasing in size ever since, and he assembled enough supporting evidence for this expanding-earth theory to fill a thick and densely-argued book. Carey's scenario would need some modification to account for the occurrence of giant animals from the Cretaceous through to the Miocene. The initial expansion would have had to be rapid, overshooting the earth's present size by the end of the Jurassic. The earth would then have had to remain larger than at present until after the Miocene, before contracting to its present size.

In view of the limited amount of direct information that is available about the physics of the earth's deep interior, it would perhaps be rash to dismiss the possibility of further surprises, comparable to the discovery of plate tectonics. It will be remembered that Wegener's theory of continental drift was not taken seriously until long after his death, despite the mass of evidence that he assembled, indicating that drift had occurred. Likewise Carey's expanding-earth theory is not taken seriously today, but his efforts may yet be recognised, if a mechanism comes to light that could account for slow but significant changes in the earth's mean density.

### Mass and density of the atmosphere
The pressure that the atmosphere exerts on the earth's surface is simply the weight of the column of air resting on each square metre of the surface.

BOX 2.4 *Continued.*

If $m_a$ is the mass of gas in the atmosphere, and $S_e$ is the earth's surface area, the surface pressure ($p_0$) is

$$p_0 = m_a g / S_e \qquad (5)$$

This neglects the decrease in $g$ with height above the surface, but the error from this is small, as the atmosphere is not very thick in relation to the radius of the earth. The sea-level air density ($\rho_0$) depends on the absolute temperature ($T$) and the pressure:

$$\begin{aligned} \rho_0 &= p_0 / RgT \\ &= m_a / S_e RT \end{aligned} \qquad (6)$$

where $R$ is the gas constant, whose value is 29.3 m $K^{-1}$. Above sea level, both the pressure and the density decline with height (Box 2.2).

Equations (5) and (6) show that estimates of past atmospheric surface pressure and density require estimates of both the mass of the atmosphere, and the also of the earth's surface area, if the possibility is recognised that this might have changed over time. Although there is an extensive literature on the evolution of the atmosphere, much of it is concerned with changes in the *proportions* of different gases, and it is usually difficult to extract estimates of mass, let alone of the area on which the pressure was acting, and the surface density.

According to Budyko et al. (1985), who did provide mass estimates for individual gases in the past, the mass of the atmosphere as a whole was higher throughout Mesozoic times than it is now. Their estimate of the current mass of the atmosphere is approximately $5.2 \times 10^{18}$ kg, which is only about one millionth of the total mass of the planet. They assume that the mass of nitrogen, the largest component of today's atmosphere, has not varied much, because the known routes by which it is added or removed are very slow. However, there is little evidence one way or the other for changes in nitrogen mass, whereas the mass of oxygen, which can be added and removed rapidly in geological terms, appears to have peaked in the Carboniferous, and again in the Cretaceous. Transient episodes of high atmospheric density are possible, due to volcanism or other causes, and such an episode could have accounted for the brief and otherwise anomalous appearance of the giant pterosaur *Quetzalcoatlus* at the very end of the Cretaceous (Chapter 6, Section 6.2.5).

oxygen in the atmosphere might do this (Box 2.4). A density spike at the end of the Cretaceous might explain the brief appearance of the giant pterosaur *Quetzalcoatlus* in the fossil record at that particular instant in evolutionary time (Chapter 6, Section 6.2.5). If the possibility is allowed that the earth's surface gravity might have been weaker in Cretaceous times than it is now, this would make giant terrestrial dinosaurs easier to account for, in addition to giant pterosaurs (Box 2.4).

# 3

# MECHANICS OF LEVEL FLIGHT

A flying animal's weight is supported by the reaction from air that is continuously accelerated downwards. The non-mathematical main text explains how the *Flight* programme calculates the mechanical power needed to do this, and to propel the animal along, while the equations used by the programme are given in the boxes. The effects of air density and the significance of wing morphology and body shape are considered. The method of deriving the physiological requirements from the mechanical calculations is explained.

Flight is locomotion in a fluid medium that is much less dense than the animal. This is only possible for animals that have special adaptations. Non-flying animals sink if dropped in air, because their bodies are typically at least 800 times denser than air. A gangster thrown off a roof accelerates uncontrollably earthwards, acquiring kinetic energy which is dissipated catastrophically on arrival at ground level, whereas a pigeon suffers no inconvenience from similar treatment. It flies around apparently ignoring gravity, climbs or descends at will, and lands in a controlled manner, where and when it wants to. It can do that because it has wings, whereas the gangster does not. Everybody knows that a bird is supported in the air by its wings, but how exactly the wings perform this magic remained a mystery until the late nineteenth century. It is not enough to imitate the appearance of birds' wings, however carefully, as many a painful experiment down the centuries has demonstrated. A theory is needed.

The classical theory of low-speed aerodynamics describes what a wing is and what it does, in a way that can be used by an engineer to design and build a wing that works, and by a biologist to describe an existing wing that belongs to a flying animal. An authoritative introduction may be found in Anderson (1991), and some older textbooks such as von Mises (1959) still retain their following. Anderson (1997) has also written a fascinating account of the evolution (as biologists would see it) of aeronautical theory and practice, highlighting the many obstacles and difficulties that had to be overcome in the development of practical aircraft, in a way that has direct implications for the evolution of animal flight (Chapter 16). This chapter is about what wings do, as represented in the *Flight* programme, and some further information on how they do it follows in Chapter 4. Chapters 5 and 6 are about the mechanics of the two main categories of wings that have evolved in vertebrates, feathered wings and membrane wings, which are seen as alternative solutions to the same adaptive problems.

## 3.1 ● POWER REQUIRED FOR HORIZONTAL FLIGHT

The basic flapping-flight calculation performed by the *Flight* programme estimates the rate at which a bird's muscles have to do mechanical work (the mechanical power), in order to fly horizontally at a steady speed, and at this level it is not necessary to specify whether the animal is a bird, a bat, a pterosaur or even an artificial ornithopter. The power depends on a small set of attributes that all flying animals have, namely a mass, a wing span, an aspect ratio and a streamlined body to which the flapping wings are attached, and also on the environment in which the animal flies, represented by the air density and the strength of gravity. The programme will predict a number of aspects of a specified animal's flight performance, and it can be used by anyone who can supply the basic measurements needed to describe the bird and its environment, without necessarily understanding how the programme works. The reader who wishes to use the programme to generate hypotheses, to be tested against observations of real flying animals, can do that by tinkering with the many options that the programme provides. For this, it helps to understand the equations that actually do the calculation. These can be found in the boxes that accompany this chapter, while the main text is an informal explanation of how the equations work. Chapter 7 covers the characteristics of the engine (flight muscles) that supplies the work required by the wings, and *Flight* combines this with the power curve calculation to generate a numerical simulation of long-distance migration (Chapter 8).

### 3.1.1 MOMENTUM BALANCE—DOWNWASH SUPPORTS THE WEIGHT

In Newtonian terms, a flying bird is immersed in the earth's gravitational field, which exerts a force, the bird's weight, directed towards the earth's centre. Newton's First Law of Motion says that if the pigeon is maintaining a constant height, and not accelerating upwards or downwards, then the net vertical force acting on it must be zero, i.e. the weight is balanced by an upward force of equal magnitude. This is an aerodynamic force, generated by the air flowing past the wings. It depends on Newton's Second Law, which says that "force" is the same as rate of change of momentum. Air has mass, and the bird generates an upward force on its body by imparting downward momentum to a stream of air passing over its wings. All flying animals and aircraft (except balloons) generate *downwash*, and support their weight in level flight by maintaining *momentum balance*. In steady, unaccelerated flight, this means that the *rate* at which downward momentum is imparted to the air must be equal to the bird's weight.

A bird has to supply work at a steady rate from its flight muscles to impart downward momentum to a stream of air. If the downwash stops, the bird falls. *Power* is the rate of doing work, and it is this unremitting requirement for power to support the weight (the *induced power*) that makes flight fundamentally different from other forms of locomotion. The underwater wing-swimming of auks and diving petrels is not a form flight, despite the superficially similar motion, because the bird's weight is supported hydrostatically by the water, not by downwash. Gliding is a form of flight (also covered by the *Flight* programme) which differs from powered flight in that the work comes from the bird's gravitational potential energy rather than from its muscles, meaning that the flight path has to be inclined downwards relative to the surrounding air. Gliding differs from falling in two respects: the speed can be held constant instead of increasing uncontrollably, and the flight path can be made to descend at a small angle to the horizontal. If speed is to be maintained when the flight path is horizontal, or inclined upwards, then work has to be supplied by an engine, meaning an organ or device that converts fuel energy into work. Ultimately the work done by a bird's muscles comes from oxidising a fuel substrate, but that is a later stage of the calculation. The first stage is to estimate the rate at which mechanical work is needed to support the weight and for propulsion.

## 3.1.2 LIFT, DRAG AND POWER

*Lift* and *drag* are components of the aerodynamic force on the bird, caused by the relative motion between the bird and the air, and they are defined by their directions relative to the incident airflow (not to gravity). Lift is the component of force that is perpendicular to the incident airflow, and drag is the component in line with it. Only the relative motion between the bird and the air (the airspeed) is involved. If the air is motionless relative to the ground, and the bird is flying horizontally, then its view of any suspended particles, such as snow flakes, is that they appear to approach horizontally, whereas if the bird is descending, it sees the incident airflow angled up at it from below. Whatever direction the flow is coming from, lift is the component of force perpendicular to that direction, and drag is the component in line with it. Any upward force that results from downwash may be pure lift (as in level flight), or a combination of lift and drag, depending on the direction from which the incident airflow is coming. In the case of a symmetrical parachute, descending vertically, the aerodynamic force is entirely drag, although it is directed upwards, and supports the weight (Figure 3.1). In a flapping wing, the incident airflow comes from different directions on different parts of the wing, and consequently local contributions to the total lift and drag forces also act in different directions.

A wing is a structure that generates a resultant force that is nearly perpendicular to the local incident airflow, so maximising the lift:drag ratio. The secret of flight, in birds and aircraft alike, is to use lift forces for supporting the weight and for manoeuvring, but to arrange that work is done against much smaller drag forces. Perhaps counterintuitively, the rotor of a hovering helicopter does no work directly against gravity. The work is done by an engine which applies torque to the rotor shaft, or by jets which apply a *horizontal* force at the tip of each blade. The engine spins the blades around horizontally, and does work against their drag, but not against their lift. If the lift:drag ratio of a rotor blade were infinite (aerodynamic force perpendicular to incident airflow), then the helicopter would be able to hover without expending any power. That is not possible with a rotor of finite diameter, because the induced power (needed to generate the downwash) appears as induced drag on the blades. The bird's wing span

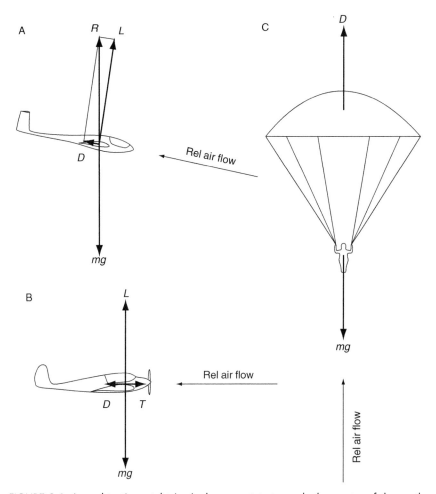

FIGURE 3.1 Any object's weight (*mg*) always points towards the centre of the earth, the direction we call "down". When a glider (A) flies straight at a constant speed, its weight (*mg*) is balanced by the resultant (*R*) of the lift and drag which is nearly (but never quite) perpendicular to the relative airflow. *R* has to point vertically upwards, therefore the lift, (*L*) which is the component of *R* that is perpendicular to the relative airflow, is not vertical in this case, but inclined forwards. The drag (*D*), points backwards and slightly upwards, along the flight path. The addition of a propeller (B) generates a thrust force, which balances the drag. As there is now no net force along the flight path, the weight can be balanced by the lift alone which means that the flight path can be horizontal. The propeller does work against the drag force, which is much smaller than the weight. A parachutist's weight (C) still points downwards, but a pure parachute produces only drag, with no lift, and the flight path is therefore directed straight down. The function of an emergency parachute is to slow down the speed sufficiently, so that the parachutist can survive the impact with the ground. Skydivers' parachutes produce sufficient lift to give the parachutist some ability to steer to a chosen landing spot.

(or a helicopter's rotor diameter), appears in the denominator of the expression for induced power (Box 3.1), meaning that a helicopter with a rotor of infinite diameter, or a bird with infinite wing span, would indeed be able to remain aloft with no induced power. In the real world, the larger the wing span, the less the induced power.

BOX 3.1 **Induced power.**

The induced power is the rate at which the flight muscles have to provide work, in order to impart downward momentum to the air at a sufficient rate to support the weight.

**Variable definitions for this box**

| | |
|---|---|
| $B$ | Wing span |
| $D_i$ | Induced drag |
| $g$ | Acceleration due to gravity |
| $k$ | Induced power factor |
| $m$ | Bird's all-up mass |
| $P_{ind}$ | Induced power in horizontal flight |
| $P_{ind0}$ | Induced power in hovering |
| $r_m$ | Mass rate of flow |
| $S_d$ | Disc area |
| $V_t$ | True airspeed |
| $V_i$ | Induced velocity |
| $\rho$ | Air density |

**Induced power in hovering**

We begin with the special case of a hovering hummingbird, in which the airspeed ($V_t$) is zero. This is quite similar to a hovering helicopter, the difference being that the helicopter's blades sweep out a "wing disc" by rotating steadily in one direction, whereas the hummingbird's wings beat horizontally back and forth, each sweeping out half the disc. Either case can be approximated by an "actuator disc" (Figure 3.3). This is a theoretical thin disc through which the air can freely pass, experiencing an instantaneous increase in pressure as it does so. The air passes through the disc at the induced velocity ($V_i$), and the mass rate of airflow through the disc ($r_m$) is

$$r_m = S_d V_i \rho, \tag{1}$$

where $S_d$ is the disc area and $\rho$ is the air density. To get the rate of change of momentum, the mass rate of flow has to be multiplied by the eventual downward velocity, which is actually not $V_i$ but $2V_i$. This is because the pressure above the disc is below ambient, but the pressure below the disc is higher than ambient by the same amount. The result is that the air accelerates to $V_i$ by the time it reaches the disc, but continues to accelerate after it passes through the disc, eventually reaching $2V_i$ far below the bird. The rate of increase of downward momentum is equal to the weight:

$$2V_i r_m = mg \tag{2}$$

BOX 3.1 *Continued.*

Substituting for $r_m$ as above,

$$2V_i^2 S_d \rho = mg \tag{3}$$

This equation can be rearranged to give the induced velocity in terms of four variables whose values are known (mass, gravity, disc area and air density):

$$V_i = \left[ \frac{mg}{(2S_d\rho)} \right]^{1/2}. \tag{4}$$

The weight is the force with which the bird is pushing downwards, at a speed $V_i$, on the air. Hence the induced power in hovering ($P_{ind0}$) is simply

$$P_{ind0} = mgV_i. \tag{5}$$

Substituting for $V_i$ from Equation (4),

$$P_{ind0} = \left[ \frac{(mg)^3}{(2S_d\rho)} \right]^{1/2}. \tag{6}$$

The "disc area" ($S_d$) is not, as might be thought, the area that is actually swept out by the wings, but the area of the complete circle, of which the wing span is the diameter (more on this below):

$$S_d = \frac{\pi B^2}{4}, \tag{7}$$

where $B$ is the wing span. Substituting this in Equation (6):

$$P_{ind0} = \left[ \frac{2(mg)^3}{\pi B^2 \rho} \right]^{1/2}. \tag{8}$$

This is the induced power in hovering for an ideal actuator disc. It produces a result that is closely related to the chemical power (or rate of oxygen consumption) in terms of four easily measured physical variables, the mass, the acceleration due to gravity, the wing span and the air density. It does not require direct measurements of power as such, but subject to an adjustment (the "induced power factor") to account for the imperfections of real flapping wings (below), it works for a hovering helicopter and, so far as we know, for a hovering hummingbird.

### Induced power in forward flight
If we relax the condition that the forward speed is zero, and allow the bird to fly horizontally at a speed $V_t$, then the mass flow through the wing disc is no

BOX 3.1 *Continued.*

longer due entirely to the induced velocity ($V_i$), as assumed in Equation (1), but to the resultant of $V_t$ and $V_i$:

$$r_m = S_d \sqrt{(V_t^2 + V_i^2)} \rho. \tag{9}$$

As the forward speed builds up, so the required induced velocity decreases. If we skip to a speed where $V_i$ makes a negligible contribution to the mass flow, as compared to the contribution from $V_t$, then Equation (9) for the mass flow ($r_m$) becomes

$$r_m = S_d V_t \rho. \tag{10}$$

The bird still has to accelerate the air to a downward velocity $2V_i$, so Equation (2) for momentum balance becomes

$$2 V_i r_m = mg, \tag{11}$$

and substituting for $r_m$,

$$2 V_t V_i S_d \rho = mg, \tag{12}$$

when the induced velocity in fast forward flight is

$$V_i = \frac{mg}{(2 V_t S_d \rho)}, \tag{13}$$

and the induced power is

$$P_{ind} = mgV_i$$
$$= \frac{(mg)^2}{(2 V_t S_d \rho)}. \tag{14}$$

Substituting $\pi B^2 / 4$ for $S_d$, as above, gives

$$P_{ind} = mgV_i$$
$$= 2 \frac{(mg)^2}{(V_t \pi B^2 \rho)}. \tag{15}$$

This is a provisional expression for the induced power, which needs to be modified to allow for the fact that the flow through the wing disc differs from that through an ideal actuator disc, by introducing the "induced power factor" (below).

The induced power in fast forward flight is inversely proportional to the airspeed, so that it plots as a hyperbola against airspeed. Equation (15) would make the induced power infinity at zero speed, but actually it is $P_{ind0}$ from Equation (8). Equation (15) applies at high speeds, where the mass flow through the disc can be approximated by Equation (10). This approximation is not actually used in *Flight*, as the mass flow is calculated from the resultant of the forward speed and the induced velocity (Equation 9) whether or not the latter is large enough to merit being taken into account.

BOX 3.1 *Continued.*

## The induced power factor

Is the actuator disc realistic? The short answer to that question is: Not very, but it also has its merits. Of course the air does not experience an instantaneous increase of pressure as it passes through the disc, nor does the air accelerated by the wings form a distinct tube, separated by an infinitely thin boundary from the stationary air round about. At best, the boundary will be a thin vortex sheet, and more probably a system of vortices which is not very thin. Some power will go into maintaining those vortices, but only experiment and observation can determine how much. This is horrendously complicated and difficult to study, either theoretically or experimentally, but the complications invariably represent additional power. The actuator disc represents an "ideal" arrangement, a baseline that can never be attained in practice, in which all of the power supplied by the bird goes into supporting the weight, and none into extraneous energy-consuming processes. The real induced power will always be more than that calculated for the actuator disc, by a factor which we can call $k$, the "induced power factor".

$k$ is a number that is more than 1, but not necessarily very much more. Experimental values of $k$ on aircraft wings and helicopter rotors are typically 1.1–1.2. Spedding (1987a) measured $k = 1.04$ in a gliding kestrel. It is not practicable in the present state of knowledge to calculate or measure $k$ for a bird in flapping flight, but this may become feasible in the future. If it does, the best way to use the new knowledge to estimate a bird's power requirements will still be to start with the actuator disc as a baseline, and then multiply this minimal induced power estimate by a measured or calculated value of $k$, rather than guessing a value as one has to do at present. *Flight* sets the default value of $k$ (which can be changed by the user) to 1.2, and inserts it in the numerator of Equation (15), which becomes:

$$P_{\text{ind}} = \frac{2k(mg)^2}{(V_t \pi B^2 \rho)}. \tag{16}$$

*Flight* does not extend the power calculation down to zero speed, but if it did, a similar adjustment would be needed to the induced power in hovering, so that Equation (8) becomes:

$$P_{\text{ind0}} = \left[ \frac{2k^2(mg)^3}{(\pi B^2 \rho)} \right]^{1/2}. \tag{17}$$

Of course, the value of $k$ may vary at different speeds.

## Induced drag

Another way of looking at Equation (16) is to consider that any component of mechanical power can be represented as the product of a drag force and the forward speed. In the case of induced power, we now know the power and the speed, so we can invent a virtual "induced drag" force ($D_i$), given by:

$$D_i = \frac{P_{\text{ind}}}{V_t} = \frac{2k(mg)^2}{(V_t^2 \pi B^2 \rho)}. \tag{18}$$

BOX 3.1 *Continued.*

In the case of a fixed (non-flapping) wing, as in a glider or propeller-driven aircraft, this is the natural way to look at it, as the induced drag is an identifiable component of force, which can be measured in experiments. Equation (16) for the induced power was derived by considering a rotor or a pair of flapping wings, sweeping out a wing disc, whose diameter is the wing span. A fixed wing does not flap or rotate, and does not sweep out any area. If the "swept" area were used to calculate the mass flow, rather than the full circle whose diameter is the wing span as in Equation (10), the induced velocity from Equation (13) would have to be infinite (because $S_d$ would be zero), and the induced power would also be infinite.

The induced drag of a fixed wing is traditionally calculated by a somewhat different route from that taken here, so we can take the standard formula and turn it round, to see what disc area is implied. This turns out to be very simple, as the standard formula for induced drag of a fixed wing is exactly the same as Equation (18). A fixed-wing aircraft, flying steadily along horizontally, supports its weight by imparting a downward induced velocity to a tube of air, whose diameter is the wing span. For a bird, the induced power is given by Equation (16), regardless of whether the wings sweep out the whole of the wing disc, or a part of it, or none of it (gliding). The difference in gliding is not in the amount of the induced power, but in the source of the energy from which it is provided–from the bird's gravitational potential energy instead of from its muscles.

## 3.2 ⬤ THE POWER CURVE CALCULATION IN *FLIGHT*

The power curve is a graph of the power required to maintain horizontal flight, as a function of speed. It serves as a basis for discussing many aspects of performance in powered flight. The *Flight* program will calculate a power curve if you, the user, first define your bird by assigning values to three morphological variables (mass, wing span and wing area) and also to gravity and the air density. These wing measurements cannot be obtained from museum specimens or ornithological "morphometrics". They have to be measured carefully, according to the aeronautical definitions in Chapter 1. Then, if you select the "Power curve" option, the programme will assign default values (which you can change) to gravity, and a number of other variables. Having set or amended the numbers in the Setup screen to your satisfaction, you can run the power curve calculation, view a summary of the results and a graph, and optionally save more comprehensive output, together with the input data, in an Excel spreadsheet or a text file. The online manual explains how to do this under "Saving Output", and also how to generate a record of the effect of varying an input variable on some variable in the output.

The graph actually shows two curves, for mechanical and chemical power, plotted against True airspeed in metres per second (Figure 3.2). The *True* airspeed is the speed at which suspended particles, such as snow flakes, stream past, relative to the bird, and it is usually different from the *Equivalent* airspeed, which determines the magnitude of aerodynamic forces (below). The mechanical power is the rate at which the flight muscles have to do work, while the chemical power is the rate at which fuel energy has to be consumed. The chemical power is typically around 4–5 times larger than the mechanical power, although this factor varies at different speeds. This is because only a fraction of the chemical energy consumed is converted into work (the rest being lost as heat), and also because of some additional "overheads" which have to be added (below and Chapter 7). The units (watts) are the same for both mechanical and chemical power. A watt is the SI unit of power, and represents a rate of doing work of 1 joule per second. A joule is the work done when a force of 1 newton moves its point of application through a distance of 1 metre. A newton is the amount of force that will impart an acceleration of 1 m s$^{-2}$ to a mass of 1 kilogram (Chapter 1).

The mechanical power in *Flight's* output applies to measurements made in unaccelerated flight, even if a steady speed and height is maintained only briefly. It is calculated from forces and speeds, and does not involve physiology. The chemical power is about the variables that are measured in physiological experiments, such as rates of consumption of fuel and oxygen, and it is only meaningful in sustained, aerobic flight, as in migration. The mechanical power directly expresses the effort required from the bird's muscles to support its weight against gravity and propel it along. *Flight* calculates it first, and then estimates the chemical power by asking at what rate fuel energy must be consumed, in order to sustain the mechanical power required from the muscles.

This has nothing to do with basal metabolism. The basal metabolic rate (BMR) is a property of a resting animal that has no direct connection with the energetics of flight, or indeed of any kind of locomotion. It is misleading to express "flight metabolism" as a multiple of BMR, as there is no connection between the two. Expressing measured rates of oxygen consumption in this way serves only to make the original

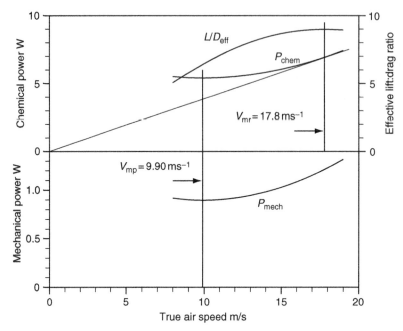

FIGURE 3.2 The *Flight* programme will calculate the mechanical power required from the flight muscles for any bird, as a function of the forward speed. This example is for the starling from the "Preset Birds" database, flying at sea level. The lower curve shows the mechanical power, which passes through a minimum at the "minimum power speed" ($V_{mp}$), which is just below 10 m s$^{-1}$ in this case. The upper graph shows the chemical power, which is the rate at which fuel energy is required in aerobic flight. According to the rules by which *Flight* derives the chemical power from the mechanical power, the minimum is at the same speed ($V_{mp}$) for both curves, but the chemical power is larger in amount because of losses in converting fuel energy into work, and also because a number of "overheads", assumed to be speed-independent, have been added. The upper diagram also shows a graph of the effective lift:drag ratio, which is not shown in *Flight*'s own graphs. This is proportional to the distance flown per unit of fuel energy consumed. It passes through a maximum at the "maximum range speed" ($V_{mr}$), which is nearly 18 m s$^{-1}$ in this case. $V_{mr}$ can also be found graphically by drawing a tangent to the chemical power curve from the origin.

power measurements difficult or impossible to extract from the published results. It is, of course, absurd to calculate "flight metabolism" as a function of the body mass alone, without taking account of the morphology of the wings, the strength of gravity or the air density.

*Flight* estimates the total mechanical power required to fly at any particular speed by calculating three components of power, the induced power (needed to support the weight), the parasite power (needed to overcome the drag of the body) and the profile power (needed to overcome the drag of the wings). The following six paragraphs are an outline of the thinking behind *Flight*'s method of calculating these three mechanical components, assembling them into a power curve, and then estimating the chemical power. The calculations themselves are described in Boxes 3.1–3.3.

BOX 3.2 **Parasite power.**

Parasite power is the rate at which work must be done to overcome the drag of the body, not including the wings. Birds (but not insects) have streamlined bodies in the aeronautical sense. A typical bird's body, in the posture for level flight, is approximately circular in cross section and elongated, with the widest cross section roughly a quarter to a third of the body length behind the front end, and the rear end tapering to a point.

**Variable definitions for this box**

| | |
|---|---|
| $A$ | Equivalent flat-plate area |
| $C_{Db}$ | Body drag coefficient |
| $D_b$ | Drag of the body |
| $m$ | Bird's all-up mass |
| $P_{par}$ | Parasite power |
| $V_{mp}$ | Minimum power speed |
| $V_t$ | True airspeed |
| $S_b$ | Body frontal area |
| $\rho$ | Air density |

The drag ($D_b$) of a streamlined body can be expressed as:

$$D_b = \frac{(\rho V_t^2 S_b C_{Db})}{2}, \tag{1}$$

where $\rho$ is the air density, $V_t$ is the True airspeed, $S_b$ is the frontal area of the body and $C_{Db}$ is the body drag coefficient. The parasite power is found by simply multiplying the body drag by the speed:

$$P_{par} = \frac{(\rho V_t^3 S_b C_{Db})}{2}. \tag{2}$$

This result comes about as follows. The factor $\frac{1}{2}\rho V_t^2$ is called the *dynamic pressure*. It is the pressure increase (above ambient) in a blind tube with its open end pointing into wind. Such a tube is called a *pitot tube* and is used in aircraft as an airspeed sensor (Chapter 14, Box 14.1) and also in albatrosses (Chapter 11, Box 11.3). We multiply the dynamic pressure by the body frontal area, which is the cross-sectional area at the widest point. A pressure times an area is a force. This particular force is a drag force, because it acts in the same direction as the incident airflow. It is a

BOX 3.2 *Continued.*

theoretical "reference" drag force that would be developed, if the incident air were brought to a halt over the whole of the frontal area. The actual drag is less than this, because of the streamlined shape of the body. The incident air divides around the body, and joins up again downstream. It is not brought to halt, but it is slowed down a bit by the obstruction in the flow. This causes some drag, but not nearly as much as the reference drag. The body drag coefficient ($C_{Db}$) is a dimensionless number less than 1, which is the ratio of the actual drag to the reference drag, and it expresses the degree of streamlining. A drag coefficient of 1 means that the body behaves like a flat plate perpendicular to the flow, while the ideal value of zero (never attained in practice) means that the body creates no drag at all.

**Practical estimates of parasite power**

Estimating parasite power comes down to measuring or estimating a particular bird's body frontal area and drag coefficient. The frontal area, of course, varies from one bird to another, whereas the drag coefficient should be much the same for different birds whose body shapes are similar. Frontal area can be measured by photographing the body from the front with a scale in the picture, or by wrapping a tape around the body at the widest part. This measurement is not very repeatable, because of the compressible nature of the feather layer, and doubt about how much it would be compressed in flight. It appears that the frontal area of many different kinds of birds can be estimated quite well from the formula

$$S_b = 0.00813m^{0.666}, \qquad (3)$$

where $S_b$ is in square metres, and $m$ is in kilograms (Pennycuick et al. 1988). The implication is that different bird bodies have roughly the same shape, so that the frontal area varies with the two-thirds power of the mass, as expected for isometric scaling. Some authors claim that passerines have higher frontal areas than Equation (3) predicts, implying that their bodies are shorter and thicker than those of other birds, but it is difficult to be sure whether the difference is real, or due to variations in technique when measuring the frontal area. The drag coefficient is even harder to measure, because experiments designed to measure it actually measure the product of the frontal area and the drag coefficient, sometimes called the *equivalent flat-plate area* ($A$), where

$$A = S_b C_{Db}, \qquad (4)$$

The body frontal area ($S_b$) has to be measured, with whatever accuracy can be achieved, before a measurement of $A$ can be converted into an estimate of $C_{Db}$. An overestimate of $S_d$ results in an underestimate of $C_{Db}$ and vice versa.

**The body drag anomaly**

It would seem to be quite straightforward to get a measurement of the drag of a bird's body by removing the wings from a dead bird, freezing the wingless body and mounting it on a drag balance in a wind tunnel. This has been done many times by different authors and, when combined with a

BOX 3.2 *Continued.*

measured frontal area, always produces anomalously high estimates of the drag coefficient, usually between 0.2 and 0.4. Such high drag coefficients would be associated with "bluff bodies" in engineering experiments, meaning bodies that are blunt rather than pointed at the downstream end. Streamlined bodies, tested at a similar scale to birds' bodies, give much lower values, below 0.1. The high drag of frozen birds' bodies is known as the "body drag anomaly", and results from separation of the airflow from the surface soon after it passes the widest point, leaving a wide, turbulent wake behind the bird, in which the pressure is lower than over the front end of the body (Figure 3.4A). The pressure difference between the front and back of the body accounts for the high measured drag. The feathers at the rear end of a frozen bird's body in a wind tunnel can be seen lifting and fluttering, making the separated boundary layer plainly visible. Close-up film of live birds, either in wind tunnels or in free flight, does not show this. It seems that the flow follows the tapering body shape and closes up downstream, leaving only a thin wake caused by viscous effects along the surface (Figure 3.4B), and much less drag. The massive flow separation seen on frozen bodies (and associated high drag) appears to be an artefact that does not occur on live birds.

Direct evidence that the body drag coefficient of live birds is much less than that of frozen bodies eventually came from measurements of the wing-beat frequencies of two different birds, a thrush nightingale and a teal, flying horizontally in a wind tunnel, over a range of different speeds (Pennycuick et al. 1996a). Although the variation of wingbeat frequency with speed was small, the measurements at each speed were very consistent, and it was possible to identify a speed at which the wingbeat frequency passed through a minimum. This speed was much higher than the value for $V_{mp}$, estimated on the basis that the body drag coefficient was in the region of 0.3–0.4 (from measurements on frozen bodies), and actually neither bird would fly as slowly as these $V_{mp}$ estimates. Adjusting the value of $C_{Db}$ downwards to 0.08 raised the estimate of $V_{mp}$ to agree with the speed at which the wingbeat frequency was a minimum, for both birds. Identifying this "minimum-frequency speed" as being the same as $V_{mp}$ was justified by a second experiment in which the wind speed was kept constant, and the tunnel was tilted by various amounts. The wingbeat frequency increased linearly with tunnel tilt, in the sense that it increased if the bird was forced to climb, i.e. if the bird had to work harder, the wingbeat frequency increased. Thus, the minimum in the frequency curve should coincide with the minimum in the power curve. See Chapter 15, Box 15.4 for more about this experiment.

## Default values in *Flight*

The default body drag coefficient in *Flight* is now 0.1. Early versions of the programme used higher default values based on drag measurements on frozen bird bodies, which are now known to be erroneous (above). There is a theoretical expectation that drag coefficients would be lower in large birds than in small ones, but this is not represented in the programme as the effect has not been observed in experiments. The current default is that

BOX 3.2 *Continued.*

$C_{Db} = 0.1$ goes for all birds, but the real value may be higher for species whose bodies are not well streamlined, such as those with long legs (storks) or big heads (pelicans). *Flight* also assumes that the bird is aerodynamically "cleaned up", especially that the legs and feet are folded and covered by the body feathers, where that is possible. Any bird has the option to lower its feet into the air stream in order to provoke flow separation, thereby increasing its drag coefficient. The webbed feet of sea birds make especially effective airbrakes, and are used in this way in steep descents.

BOX 3.3 **Profile power.**

As noted in the main text, calculating profile power by strip analysis does not lead to a simple general formula for profile power that can be applied to any bird, in the same way that induced and parasite power can be estimated (Boxes 3.1 and 3.2). However, early attempts to do this yielded one insight which can be exploited to create a simplified approach. The bird needs profile power to overcome the drag resulting from the airflow over the wings, and this flow has two components in flapping flight. The first component is the flow due to the forward speed of the whole bird. The wing profile drag arising from this comes from the whole wing, and would be expected to increase roughly with the square of the speed. The associated component of profile power would increase with the cube of the speed, like parasite power. The second component arises from the flapping motion, which rotates the wing, relative to the bird's body. This component of the flow depends on the flapping frequency and amplitude, and comes more from the distal part of the wings than from the shoulder region. The associated profile power is dominant at low speeds, where flapping frequency and amplitude are highest, and is expected to drop to a minimum round about the minimum power speed, and then build up gradually at higher speeds.

Adding the two components together, the total profile power first builds up because of the increasing forward speed, then levels off or drops slightly through the middle range of speeds as a result of decreasing wingbeat frequency and amplitude, before (conjecturally) rising again at very high speeds. This suggests an easy way to approximate the profile power. First, *Flight* restricts its performance estimates to "middle speeds", which can be defined as speeds between the minimum power speed ($V_{mp}$) and the maximum range speed ($V_{mr}$), with minor extensions at both ends of this range. It does not attempt to provide performance estimates for very low speeds, at which unsteady aerodynamic processes are likely to be important, or at very high speeds, for which there are no data. Second, *Flight* assumes that the profile power is independent of speed within this limited speed range. This means that adding profile power to the induced and parasite power does not change the minimum power speed.

If the profile power is independent of speed (within the specified limits) then how can its amount be estimated? The notion behind *Flight*'s simple

BOX 3.3 *Continued.*

method is that profile power in level flight, at medium speeds, is essentially a by-product of the bird's efforts in generating induced and parasite power, and can be found from the minimum value of the sum of these two components.

**Variable definitions for this box**

| | |
|---|---|
| $B$ | Wing span |
| $C_{db}$ | Body drag coefficient |
| $C_{pro}$ | Profile power constant |
| $g$ | Acceleration due to gravity |
| $k$ | Induced power factor |
| $m$ | All-up mass |
| $P_{am}$ | Absolute minimum power |
| $P_{mech}$ | Mechanical power required to fly |
| $P_{pro}$ | Profile power |
| $R_a$ | Aspect ratio |
| $S_b$ | Body frontal area |
| $V_{mp}$ | Minimum power speed |
| $V_t$ | True airspeed |
| $X_1$ | Profile power ratio |
| $\rho$ | Air density |

The mechanical power ($P_{mech}$) for an ideal bird (i.e. one with no profile power), flying at a True airspeed $V_t$, is first computed as the sum of the induced power from Equation (16) of Box 3.1, and the parasite power from Equation (2) of Box 3.2

$$P_{mech} = \left[ \frac{2k(mg)^2}{(V_t \pi B^2 \rho)} \right] + \left[ \frac{(\rho V_t^3 S_b C_{Db})}{2} \right], \tag{1}$$

where $m$ is the all-up mass, $g$ is the acceleration due to gravity, $\rho$ is the air density, $V_t$ is the True airspeed, $S_b$ is the body frontal area and $C_{Db}$ is the body drag coefficient. The minimum power speed ($V_{mp}$) can be found by differentiating this expression with respect to $V_t$, setting the result to zero, and solving for the speed, which gives:

$$V_{mp} = \frac{(0.807 k^{1/4} m^{1/2} g^{1/2})}{(\rho^{1/2} B^{1/2} S_b^{1/4} C_{Db}^{1/4})}. \tag{2}$$

Substituting this speed for $V_t$ in Equation (1) gives the absolute minimum power ($P_{am}$), meaning the power required for the ideal bird to fly at $V_{mp}$:

$$P_{am} = \frac{(1.05 k^{3/4} m^{3/2} g^{3/2} S_b^{1/4} C_{Db}^{1/4})}{(\rho^{1/2} B^{3/2})}. \tag{3}$$

The profile power is then set at a fixed multiple ($X_1$) of the absolute minimum power.

BOX 3.3 *Continued.*

$$P_{\text{pro}} = X_1 P_{\text{am}}. \tag{4}$$

$X_1$ is called the "profile power ratio", and it was assigned a fixed value of 1.2 in early versions of the Basic programmes from which *Flight* was developed. It was later noted that profile power would most likely be proportional to wing area, other things being equal, so an additional constant was introduced, the "profile power constant" ($C_{\text{pro}}$), with a default value of 8.4. $X_1$ is then defined as $C_{\text{pro}}$ divided by the aspect ratio $R_a$:

$$X_1 = \frac{C_{\text{pro}}}{R_a}. \tag{5}$$

The effect of this for a bird with an aspect ratio of 7 (like a pigeon), is that $X_1=1.2$, which was the original, fixed default value. For a bird with a higher aspect ratio than 7, $X_1$ is lower than this, and *vice versa*. The calculation of profile power is the only part of the power curve calculation in *Flight* that involves the wing area. The value of $C_{\text{pro}}$ can be adjusted by the user in *Flight*'s Setup screens.

## 3.2.1 INDUCED POWER

The induced power is the rate at which a bird has to do work with its muscles, continuously accelerating air downwards so as to produce an upward reaction that supports its weight. If you think of the bird as stationary, with the air streaming horizontally towards it, as in a wind tunnel, the air has zero downward velocity as it approaches the bird. The bird's wings deflect the air downwards as it flows past, adding a vertical component of velocity, the "induced velocity". The rate of change of downward momentum is the same as the upward aerodynamic force and must be equal in magnitude to the weight. It is the rate (mass/time) at which air is streaming past the wings, multiplied by the eventual downward component of velocity, which is actually twice the induced velocity (Box 3.1). The power required to impart this downward push to the air is proportional to the induced velocity. At high speeds, with plenty of air flowing past the wings, only a small induced velocity is needed, and the power required is also low. The lower the speed, the less the mass rate of flow, the higher the induced velocity, and hence the higher the induced power. At zero speed, there is no air flowing past at all, but a hovering bird or helicopter avoids the need for an infinite induced velocity (and induced power), by sucking stationary air down from above and creating its own vertical *induced wind*. The induced power is not infinite in hovering, but it is higher than at any non-zero forward speed.

A

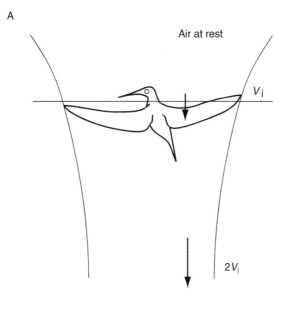

B

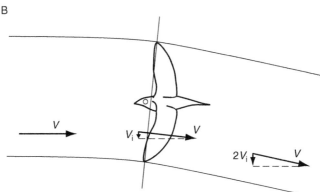

FIGURE 3.3 (A) The wing disc of a hovering bird develops an area of reduced pressure above it, which sucks stationary air down from above, accelerating it to the induced velocity $V_i$ as it passes through the wing disc. Increased pressure below the wing disc makes the air continue to accelerate downwards, to an eventual velocity of $2V_i$. The tube of air passing through the wing disc eventually narrows until its cross-sectional area is half that of the wing disc, and its diameter is reduced by a factor of $1/\sqrt{2}$. (B) In level flight the bird is flying along horizontally at an airspeed $V$, or alternatively, from the bird's point of view, it "sees" a relative wind $V$ blowing towards it from ahead. The circular tube of air that passes through the wing disc is deflected downwards, acquiring a downward component of velocity $V_i$ at the wing disc, and continuing to accelerate to an eventual downward velocity $2V_i$ behind the bird. The downwash angle is $\sin^{-1}(2V_i/V)$, and it is exaggerated in the diagram to about 11 degrees, so that it can be easily seen. For a bird flying at $V_{mp}$, the downwash angle (calculated by *Flight*) is typically between 2 and 4 degrees.

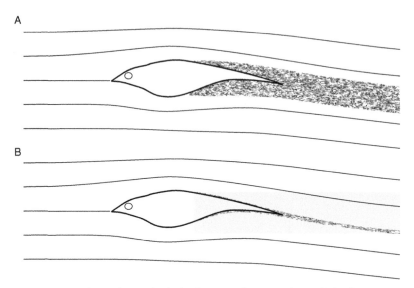

FIGURE 3.4 (A) When a frozen bird's body is tested in a wind tunnel, the flow separates from the surface at the widest part, leading to a broad area of chaotic flow downstream. This results in drag similar in amount to that behind a "bluff body" or a sphere. (B) The flow follows the surface of the same bird, when alive, leaving only a narrow turbulent wake, and creating much less drag.

## 3.2.2 PARASITE POWER AND THE MINIMUM POWER SPEED

In addition to the induced power, which is the cost of supporting the weight in air, the bird also has to do work to propel its body along, against the resistance of the air. This resistance is a drag force, meaning one that acts parallel to the incident airflow, as seen by the body. In horizontal flight (only) the body drag acts horizontally backwards. Without a forward-directed thrust force to balance it, the body drag would cause the bird to decelerate. Fixed-wing aircraft have a propeller or jet, which generates a horizontal thrust force that balances the drag, whereas birds achieve the same result in a more roundabout way by flapping the wings. The term "parasite power" is an old one from the early days of aeronautics, when structures such as the cockpit, under-carriage and wing struts, which added drag but did not contribute to supporting an aircraft's weight, were considered "parasitic". In birds, the parasite power is the drag of the body (not including the wings) times the forward speed.

If no air is flowing past the bird (zero speed), then there is no body drag, and no power is required to overcome it, but as the speed increases above zero, the curve of parasite power curves strongly upwards, following the cube of the speed (Box 3.2). The bird has to supply both induced and parasite power at the same time. When these two components are added together, and plotted against the speed, the curve first slopes downwards, because of the steeply decreasing induced power, then levels off, then increases ever more steeply at higher speeds, where the parasite power predominates. A flying bird is never "at rest" in the sense of a dog dozing on a chair, but it does have an identifiable *minimum power speed*, at which the power required to fly is less than at either slower or faster speeds. If the muscles cannot produce this level of power, then the bird cannot fly horizontally, and if the heart and lungs cannot sustain the minimum power aerobically, then it can only fly in short bursts, if at all. The minimum power required to fly is a matter of physics, and depends not only on the bird's mass and wing measurements, but also on some environmental variables, especially the strength of gravity and the air density. The maximum power available from the muscles is a matter of physiology (Chapter 7). Calculations of performance in sustained flapping flight depend on comparing the physiology with the physics.

### 3.2.3 PROFILE POWER

Like any obstruction, a wing resists the flow of air past it, and creates drag. Part of that drag results directly from generating downwash to support the weight. In a fixed wing this can be identified as the *induced drag*, a steady force which can be multiplied by the speed to find the induced power (above). However, all wings produce more drag than that which can be attributed directly to creating downwash. Any additional drag is the *profile drag* which, when multiplied by the speed, gives the profile power. Rotary and flapping wings also require profile power, but it is not possible to identify a steady profile drag force in these more complicated cases. It is possible to calculate profile power for a helicopter rotor blade by dividing it into a number of chordwise strips, estimating the speed and direction of the incident airflow (different for each strip), and then estimating the magnitude of the lift and drag force acting on each strip from wind tunnel experiments on wing cross-sectional shapes. Such a "strip analysis" was attempted for a pigeon flying in a wind tunnel (Pennycuick 1968b), but this proved unduly complicated, and involved many variables that were difficult or impossible to measure. It yielded a result for the pigeon, but did not lead to a formula like those of Boxes 3.1 and 3.2, that could readily be generalised to calculate the profile power for any bird.

A simpler method was devised for such general performance estimates by Pennycuick (1969), and this is still used by the *Flight* program, with only one modification (Box 3.3). Rather than attempting to calculate profile power directly, *Flight* assumes, in effect, that it is essentially a by-product of the bird's efforts in generating induced and parasite power. The *absolute minimum power* for an "ideal" bird, i.e. one with no profile power, would be the sum of the induced and parasite powers, when the bird is flying at the minimum power speed. *Flight* calculates this first and then multiplies it by a "profile power factor" to get the profile power. This factor was fixed in early versions of the programme, which meant that the profile power depended on the wing span, but was independent of the wing area, as the formulae for induced and parasite powers do not involve the wing area. The modification that was later introduced made the profile power factor inversely proportional to the aspect ratio, which also means that it is proportional to the wing area, other things being equal.

This method of calculating profile power lacks the direct link to classical aeronautical theory that can be claimed for induced and parasite powers. It has the merit of being simple, and involving no greater degree of guesswork than a complex strip analysis. It is not entirely satisfactory since, as Figure 3.5 shows, the profile power that it calculates is usually not a trivial fraction of the total power. There is now an indirect way to get at it experimentally, by measuring the total power from vertical accelerations of the bird's body (Chapter 14, Box 14.4), then estimating the induced and profile powers, whose theoretical basis is more robust, and finding the profile power by subtraction. Experiments on different birds along these lines would help to show whether *Flight's* method for estimating profile power gives realistic estimates. Present indications, so far as they go, are that *Flight's* estimates for the total power at medium speeds, and the fuel consumption needed to account for it, are consistent with the known performance of long-distance migrants (Chapter 8).

### 3.2.4 "INERTIAL POWER" IS NOT REQUIRED

Some authors attach great importance to a supposed further component of power, due to angular acceleration of the wing's inertia. At the beginning of each downstroke, the wing is fully extended but stationary, in terms of angular motion about the shoulder joint. The pull of the pectoralis muscle is balanced partly by the aerodynamic force on the wing, and partly by the wing's inertia. As the wing accelerates into the downstroke, it acquires a component of kinetic energy, which

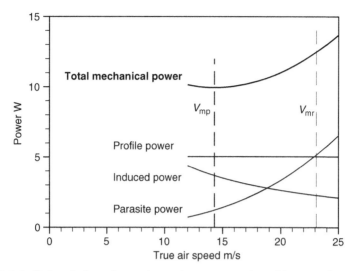

FIGURE 3.5 *Flight* calculates the mechanical power curve by adding together induced, parasite and profile power, over a range of speeds from just below $V_{mp}$ to just above $V_{mr}$. At any given speed, the values of the three thin lines at the bottom of the graph are added together to get the thick line for mechanical power above. This example is for the wigeon (*Anas penelope*) in the Preset Birds database. The power components are not available separately in the *Flight* output. The tangent construction of Figure 3.2 would not identify $V_{mr}$ correctly on this graph, because *Flight* calculates this from the chemical power.

is proportional to its moment of inertia, and to the square of its angular velocity. This energy comes from work done by the pectoralis muscle, and, so the theory goes, must be multiplied by the wingbeat frequency to get the "inertial power". This would be true if the kinetic energy gained by the wing were dissipated and lost at each wingbeat, but that is unlikely. Unlike work done against drag, which is dissipated irreversibly, the wing's kinetic energy is a high-grade form of energy which can be reconverted into work. If that work can be used to accelerate air downwards, then it contributes to the induced power, and if it accelerates air backwards, it contributes to the parasite and/or profile power. To achieve this, the wing's rotation has to be stopped at the end of the downstroke by aerodynamic forces, which themselves result from acceleration of air in the required direction.

A perfectly efficient bird would recover all of the original inertial work, and use it as part of the aerodynamic work that has to be done at each wingbeat. If the bird is not perfectly efficient, then the work that is lost appears as additional induced power in the form of an increase in the induced power factor, and/or as an increase in the body drag coefficient, and/or as an increase in the profile power ratio (see Boxes 3.1–3.3). It is not necessary, practical, or even meaningful to attempt to account for inertial power separately, and this component does not feature in *Flight*'s power calculation.

### 3.2.5 ASSEMBLING THE POWER CURVE: THE MINIMUM POWER SPEED

Adding together the three components of mechanical power (induced, parasite and profile power) gives the rate at which the flight muscles have to supply mechanical work at any particular speed (Figure 3.5). *Flight* does not begin calculating the power curve at zero speed (hovering), because the assumptions on which the calculation is based are insecure at very low speeds. Instead, it begins by estimating the minimum power speed (Box 3.4), and then moves to a slightly lower speed to start calculating the curve. It finds the induced power at that speed from Equation (16) of Box 3.1, the parasite power from Equation (2) of Box 3.2 and the profile power from Equation (4) of Box 3.3, and adds these three components together to get the total power. Then it increases the speed by $0.1 \text{ m s}^{-1}$, repeats the calculation, and continues repeating this. As the starting speed was below the minimum power speed, the power first decreases at each speed step. When the power levels off, the programme notes the speed and identifies it as $V_{mp}$, the minimum power speed. This is the speed at which the lowest rate of muscular exertion is required to fly, and at which a given amount of fuel will last for the maximum flying time.

### 3.2.6 FINDING THE CHEMICAL POWER

The chemical power, which is also listed in the output, is required for migration calculations, and is considered further in Chapters 7 and 8. At each speed, the largest component of it is found by dividing the mechanical power by the conversion efficiency, which is assumed to be constant, with a default value of 0.23. This is the direct equivalent in fuel energy of the work that is done by the flight muscles. Basal metabolism is then added as an "overhead", which is assumed to be independent of speed, but declines during migration because it is based on the declining body mass (see Chapter 8, Box 8.5). Finally, a 10% overhead is added to the total chemical power to allow for the power required to operate the lungs and circulatory system (Chapter 7). Besides the chemical power, the programme also calculates a number of other variables at each speed step, which can be inspected and

BOX 3.4 **Effective lift:drag ratio and range.**

## Variable definitions for this box

| | |
|---|---|
| $A$ | Equivalent flat-plate area of the body |
| $B$ | Wing span |
| $C_{db}$ | Drag coefficient of the body |
| $g$ | Acceleration due to gravity |
| $k$ | Induced power factor |
| $m$ | All-up mass |
| $N$ | Effective lift:drag ratio from fuel consumption |
| $N_{mech}$ | Effective lift:drag ratio from mechanical power |
| $N_{ult}$ | Maximum mechanical effective lift:drag ratio for ideal bird |
| $P_{chem}$ | Chemical power required to fly |
| $P_{ind}$ | Induced power |
| $P_{mech}$ | Mechanical power required to fly |
| $P_{mr}$ | Mechanical power required to fly at the maximum range speed |
| $P_{par}$ | Parasite power |
| $P_{pro}$ | Profile power |
| $S_b$ | Frontal area of the body |
| $S_d$ | Disc area |
| $V_t$ | True airspeed |
| $V_{mp}$ | Minimum power speed |
| $V_{mr}$ | Maximum range speed |
| $\eta$ | Conversion efficiency |
| $\rho$ | Air density |

## Effective lift:drag ratio and maximum range speed

When flight calculates a power curve, it first estimates the minimum power speed from Equation (2) of Box 3.3. Starting at a speed a little below this, it works out the mechanical power required from the muscles ($P_{mech}$) at speed intervals of $0.1\,\mathrm{m\,s^{-1}}$, by adding together the induced power ($P_{ind}$) from Equation (16) of Box 3.1, the parasite power from Equation (2) of Box 3.2 and the profile power ($P_{pro}$) from Equation (4) of Box 3.3:

$$P_{mech} = P_{ind} + P_{par} + P_{pro}. \tag{1}$$

It also works out the effective lift:drag ratio ($N_{mech}$), whose original definition, based on the mechanical power ($P_{mech}$) is

$$N_{mech} = \frac{mgV_t}{P_{mech}}. \tag{2}$$

As power is force times speed, the ratio $P_{mech}/V_t$ can be seen as the average horizontal force needed to propel the bird along, and $N_{mech}$ is then the ratio of the weight to this horizontal force. It is proportional to the ratio of distance travelled forwards to mechanical work done by the muscles. In bird migration studies, one is usually more interested in the distance flown per

BOX 3.4 *Continued.*

unit of fuel energy consumed, and *Flight* therefore calculates $N$, a slightly different version of the effective lift:drag ratio, from the chemical power ($P_{chem}$) rather than from the mechanical power:

$$N = \frac{mgV_t}{\eta P_{chem}}, \qquad (3)$$

where $\eta$ is the conversion efficiency of the flight muscles. $N$ is still the ratio of two mechanical powers, but its value is lower than that of $N_{mech}$, because $P_{chem}$ includes some metabolic "overheads", in addition to the direct fuel equivalent of the mechanical power. Under this definition, $N$ is proportional to the distance flown per unit of fuel energy consumed. *Flight* calculates $N$ at each speed step, as it computes the power curve. As the speed increases above $V_{mp}$, $N$ increases strongly at first, but eventually levels off and starts going down again. *Flight* identifies the speed at which $N$ ceases to increase as the *maximum range speed* ($V_{mr}$), and terminates the power curve calculation just above this speed. $V_{mr}$ is the speed at which the bird covers the greatest distance for each unit of fuel energy consumed. It is not necessarily the "optimum" speed for migration, because the power required is higher than at lower speeds, and may be beyond the bird's aerobic or mechanical capacity, especially if the bird is carrying a heavy load of fat (Chapter 8).

**Maximum effective lift:drag ratio**

The maximum value of $N$ occurs when the bird is flying at $V_{mp}$, but as it involves metabolic overheads, it is more practical to compute it numerically as *Flight* does, rather than trying to calculate it. The mechanical version ($N_{mech}$) can, however, be calculated for an ideal bird (one with no profile power) by finding the speed and power at which $N_{mech}$ passes through a maximum, in the same manner as $V_{mp}$ and $P_{am}$ were calculated in Box 3.3. This time, two of the morphological variables will be expressed in a slightly different form. Instead of the wing span ($B$) we use the *disc area* ($S_d$), which is the area of a circle whose diameter is equal to the wing span:

$$S_d = \frac{\pi B^2}{4}, \qquad (4)$$

and we multiply the body frontal area ($S_b$) by the body drag coefficient ($C_{Db}$), and call the result the *equivalent flat-plate area* of the body ($A$):

$$A = S_b C_{Db}. \qquad (5)$$

$A$ may be thought of as the area of a flat plate that is perpendicular to the airflow, and stops the approaching air completely, developing the same amount of drag as the larger, streamlined body. In terms of these two variables, and the others defined above, the maximum range speed is:

$$V_{mr} = \frac{(k^{1/4}m^{1/2}g^{1/2})}{(\rho^{1/2}A^{1/4}S_d^{1/4})}, \qquad (6)$$

BOX 3.4 *Continued.*

and the mechanical power ($P_{mr}$) required to fly at that speed is:

$$P_{mr} = \frac{(k^{3/4}m^{3/2}g^{3/2}A^{1/4})}{(\rho^{1/2}S_d^{3/4})}. \tag{7}$$

By substituting from Equations (6) and (7) in Equation (2), we can get the "ultimate" lift:drag ratio ($N_{ult}$) for an ideal bird with the given wing span and body frontal area.

$$N_{ult} = \frac{mgV_{mr}}{P_{mr}}$$
$$= \sqrt{(S_d/A)}. \tag{8}$$

If the ideal bird is reduced to a wing disc swept out by the wings, with an equivalent flat-plate representing the body in the middle (Figure 3.6), then the ratio of the areas of the two discs is the square of the ultimate effective lift:drag ratio. The practical maximum value of $N$ for a real bird is considerably lower, because of the profile drag of the wings, and metabolic overheads. It is computed step by step for each value of the speed, when *Flight* generates a power curve. In *Flight's* Migration calculation, the current value of $N$ is used to calculate the fuel used at each time step, and $N$ is then re-computed, taking account of changes of mass, body frontal area and speed due to the consumption of fuel.

plotted if the results are saved as an Excel workbook or a text file. These include the effective lift:drag ratio, which is closely related to the distance flown per unit of fuel energy used (Box 3.4). When the effective lift:drag ratio peaks, the programme notes the speed as $V_{mr}$, the *maximum range speed*, and terminates the calculation after a few more steps. The power required to fly at $V_{mr}$ is more than that to fly at $V_{mp}$, but there is an even greater gain of speed, so that $V_{mr}$ is the speed at which the most air distance is covered per unit amount of fuel energy consumed. $V_{mp}$ and $V_{mr}$ are the two *characteristic speeds* that can be used to characterise a particular bird's power curve. As the calculated value of $V_{mp}$ does not involve physiology, it is more robust than that of $V_{mr}$, which does (Figure 3.6).

## 3.3 ▬ SIGNIFICANCE OF THE CHARACTERISTIC SPEEDS

The minimum in the power curve means that a bird that is exerting a little more than the minimum power required for level flight can maintain either of two different speeds, one below $V_{mp}$ and the other above.

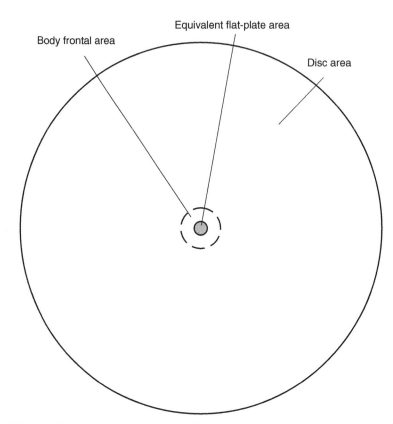

FIGURE 3.6 The same wigeon whose mechanical power curve is shown in Figure 3.5, represented by three circles, drawn to scale. The large circle is the wing disc, whose area is $S_d$, and whose diameter is equal to the wing span. The dashed circle is the body cross section at the widest part, whose area is $S_b$, the body frontal area. The small grey circle in the middle is the equivalent flat-plate, whose area $(A)$ is one tenth of the body frontal area, because the default value of the body drag coefficient in *Flight* is 0.1. If this wigeon were an "ideal bird", with no profile power, its maximum effective lift:drag ratio, from the mechanical power curve, would be $\sqrt{(S_d/A)}$, which works out to 27.9. The actual maximum effective lift:drag ratio from the chemical power curve is much lower, because it has to take account of profile power and metabolic overheads. *Flight* estimates it as 12.0, at a speed $(V_{mr})$ of 23.1 ms$^{-1}$.

However, a speed below $V_{mp}$ is inherently unstable, whereas the speed requiring the same amount of power above $V_{mp}$ is stable. This is because a bird that is exerting exactly the right amount of power to maintain a steady speed that is below $V_{mp}$ needs slightly less power than before if a gust or some other disturbance causes it to speed up slightly, because of the downward slope of this part of the power curve. Unless the bird responds very quickly by reducing its power output, it will be exerting more power than is needed to maintain the new speed, and this will cause it to speed up further, until it reaches the speed on the rising part of the power curve, above $V_{mp}$, where the power required is once again the same as the power that it is exerting (Figure 3.7). Flying around at speeds below $V_{mp}$ is possible in both birds and aircraft, but flying faster than $V_{mp}$ is easier in terms of control input, and does not necessarily require any more power. Thus, it is no surprise that airspeed measurements of wild birds flying steadily along in the field are usually slightly above $V_{mp}$ (see Chapter 15). Birds like flycatchers that require low-speed manoeuvrability for catching insect prey in the air may be regarded as specialised for flight at speeds below $V_{mp}$, and there are suggestions that most or even all insectivorous bats are also specialised for doing this.

Flying steadily at $V_{mr}$ requires more power than flying at $V_{mp}$, and a higher rate of oxygen consumption. This puts it beyond the reach of large birds like swans, and probably of many smaller birds as well, especially when they are flying high and/or heavily loaded with fat (Chapter 7). However, long-distance migrants gain both increased range and reduced flight time by flying as near to $V_{mr}$ as possible. The migration section of the *Flight* programme offers various options for ultra long-distance migrants of medium size, like knots and

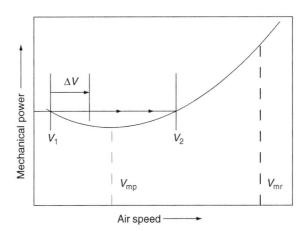

FIGURE 3.7 A speed ($V_1$) below the minimum power speed ($V_{mp}$) is unstable for a bird that is exerting just enough power to maintain that speed. If the speed increases by a small amount ($\Delta V$), the power required drops below the power that is coming from the muscles, and the bird continues to accelerate until the speed reaches the rising part of the power curve at $V_2$, which is the stable speed for the same about of power. Flying around at speeds below $V_{mp}$ is difficult, and something that birds (and fixed-wing aircraft) only do if they have a special reason, such as hawking for flying insects (after Pennycuick 1997).

godwits, representing different strategies to work the airspeed up to $V_{mp}$ as the weight declines in the early stages of a long flight (Chapter 8).

## 3.4 ⬤ EFFECT OF AIR DENSITY ON SPEED AND POWER

*Flight* calculates the power curve as a function of the True airspeed, which is the relative speed at which small particles are carried past the bird by the air. At each speed the power is a function of the air density. At higher altitudes, the air density is lower, and the whole power curve moves upwards on the graph, and expands to the right. The two characteristic speeds, $V_{mp}$ and $V_{mr}$, increase in inverse proportion to the square root of the air density. The effect of changes in the air density can be seen by running a power curve for some bird at sea level, then running it again at different heights (1000, 2000, 3000 m etc.), without changing any other variables. As you increase the height (reducing the air density), both the minimum power speed ($V_{mp}$) and the power required to fly at that speed (i.e. the minimum mechanical power $P_{min}$) vary inversely with the square root of the air density. At a height of 6700 m, where the air is only half as dense as at sea level, $V_{mp}$ is $\sqrt{2}$ times as fast as for the same bird at sea level, and $P_{min}$ is also $\sqrt{2}$ times higher. If the bird could fly at all in such thin air (unlikely), it would have to go 41% faster, and supply 41% more power from its muscles, to maintain the higher $V_{mp}$. If a bird has enough muscle power to fly a bit faster than $V_{mp}$ at sea level, then it can only just maintain $V_{mp}$ at some higher level (where $V_{mp}$ itself is faster), and at a still higher level, it no longer has enough power to maintain height.

In terms of True airspeed, a bird which is flying at $V_{mp}$ at a height of, say, 3000 m, will be flying about 5% faster than its own $V_{mp}$ at sea level (Figure 3.8). Every point on the power curve is shifted to the right by about 5% (higher speed) and also raised by the same amount (more power) at the higher altitude, although the bird is flying at the corresponding point on the curve in both cases (for example at $V_{mp}$). The airspeed indicator on an aircraft's instrument panel actually measures the dynamic pressure from a forward-pointing pitot-static tube (Chapter 14, Box 14.1). Instead of the True airspeed, it indicates the airspeed that would correspond to the measured dynamic pressure, if the air density were the same as at sea level. Pilots call this the *Equivalent* airspeed. The dynamic pressure is the basis of the aerodynamic forces acting on the wings and body, consequently corresponding points on the power curve, such as $V_{mp}$, occur at a fixed value of the dynamic

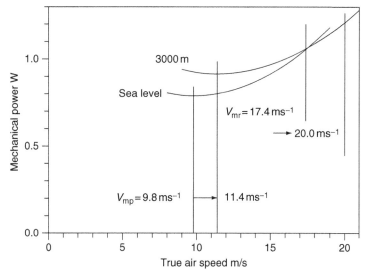

FIGURE 3.8 Mechanical power curves for the song thrush from the "Preset Birds" database, calculated by *Flight*. The effect of reducing the air density from 1.23 to 0.909 kg m$^{-3}$, corresponding to sea level and 3000 m above sea level in the International Standard Atmosphere, is to raise the whole curve and shift it to the right. $V_{mp}$ increases in proportion to the square root of the air density, and $V_{mr}$ approximately so. At low speeds, the power required is higher at the higher altitude, where the air density is lower, but the curves cross over, and at speeds above about 18 ms$^{-1}$ the reverse is the case. Since both speed and power increase by the same factor, the effective lift:drag ratio is unaffected by changes of air density (more or less).

pressure and of the Equivalent airspeed, not at a fixed True airspeed. If the prevailing air density happens to be the same as at sea level in the International Standard Atmosphere, then Equivalent and True airspeed are the same, but at higher levels, where the air density is less than at sea level, the True airspeed is higher than the Equivalent airspeed.

Care is needed when observing flight performance over a range of different heights. If you measure a bird's speed by radar, and make due allowance for the wind, you get a True airspeed, but this is not directly comparable with the True airspeeds of other birds that were measured at different heights where the air density was different. One possible solution would be to compare each bird's measured True airspeed with the value of $V_{mp}$ predicted by *Flight* for that species, but that would require a separate prediction for every observation, because the value of $V_{mp}$ depends on the air density as well as on the bird's mass and wing morphology. An easier way is to "reduce" each of the observed speeds from its original height to sea level (Chapter 15, Box 15.3), which is the same as plotting the observations against Equivalent rather than True airspeed. Then, only one power curve needs to be calculated for each species, in which the air density has been set to the sea-level value for the International Standard Atmosphere (just set the altitude to zero in *Flight*). Reducing all the observations to the same level makes them comparable with each other, so that it is valid to plot the whole data set together on the same graph. It is not valid to do this with raw tracking-radar data.

# 3.5 ⬟ ADAPTIVE SIGNIFICANCE OF MORPHOLOGY

## 3.5.1 SIGNIFICANCE OF WING MORPHOLOGY

In forward flight at any given speed, the induced power is inversely proportional to the square of the wing span (Box 3.1). In other words, long wings are the basic adaptation for economical hovering, or flight at low speeds. On the other hand, the wing *area* does not appear in any of the expressions for induced power in Box 3.1. It is an error to suppose that a low wing loading is the key to successful flight, as the Wright brothers discovered in the most frustrating way in 1901, when they increased the wing area of their unsuccessful 1900 glider, and found that the new glider performed even worse than before. The successful 1902 glider, which paved the way for powered flight, had roughly the same wing area as its predecessor, but the wing shape was different. It had a higher aspect ratio, meaning that the wing area was redistributed out to the sides, away from the centre line.

It is not the wing span by itself that determines how well a wing works, and certainly not the wing area by itself, but the *aspect ratio*, the ratio of wing span to mean chord (Chapter 1, Box 1.2). A long, narrow wing works better than a short, wide one with the same area and profile, because it develops more lift, and less induced drag at the same angle of attack. These are classical results from Prandtl's lifting-line theory (Chapter 4). On the other hand, a long, narrow wing, is more cumbersome, and provides less space in which to incorporate the necessary strength. The wings of flying animals have to compromise between having aspect ratios that are high enough to fly well, while being strong enough to allow the wing to be flapped in the air, and folded and manipulated on the ground. Alternative solutions to these practical problems have evolved in birds, bats and pterosaurs, and these are introduced in Chapters 5–6.

The significance of "wing loading" (weight divided by wing area) is that it determines the speed at which air is required to flow past the wing. This is useful as a fixed-wing notion, where the speed of the relative airflow is the same over all parts of the wing, and equal to the airspeed of the whole bird or aircraft. The airflow produces a pressure difference between the lower and upper surfaces of the wing, and this, when multiplied by the wing area, has to balance the weight in steady flight. The pressure difference is directly related to the dynamic pressure. The higher the wing loading, the more pressure is needed, and the faster the bird has to fly. The wing loading is useful in gliding calculations, where the wing is essentially fixed, for determining characteristic speeds. It also determines a bird's circling radius in gliding (not in flapping flight), which is important in thermal soaring (Chapter 10).

The wing loading has no special significance in flapping flight calculations. If the wings can be moved relative to the body, by rotation or flapping, then air can be made to flow at different speeds over different parts of the wings and body, and any simple relationship between flight performance and wing loading breaks down. Wing loading does have a connection with the circulation about the wing in flapping flight, but that is beyond the scope of the *Flight* programme (Chapter 4).

### 3.5.2 SIGNIFICANCE OF BODY SHAPE

The reason why streamlined bodies taper to a point at the rear end is to induce the flow to close up downstream, rather than separating from the surface. If the airflow separated from living birds' bodies as readily as it does from frozen bodies (Box 3.2), then there would be no adaptive advantage in the elaborate "fairings" provided by body feathers, which smoothly taper the rear end of the body to a point in flight. On the smaller scales of insect flight, the flow separates from the body anyway, regardless of whether the outline is streamlined or not. Insects, especially in the smaller sizes, usually have blunt or angular bodies, often with protruding legs or spines, as there is no biological advantage in a smoothly faired outline. The fact that birds do fair their body outlines indicates that there is a performance advantage in this, and that the high drag measured on frozen bodies is an artefact. It remains a mystery how exactly the living bird keeps the flow attached to the tapered, feathered shape, while the same body cannot do this when dead and frozen (see also Chapter 15, Box 15.4).

## 3.6 ⬤ TWO-DIMENSIONAL AEROFOIL PROPERTIES

Some of the essential terminology that is used to describe the performance of wings is derived from a huge volume of empirical experiments that were done in the twentieth century on the properties of

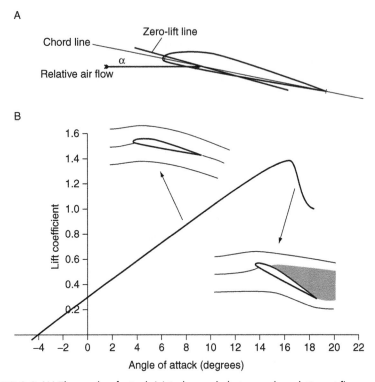

FIGURE 3.9 (A) The angle of attack (α) is the angle between the relative airflow and the chord line. A cambered wing section normally develops zero lift at a negative angle of attack, in other words, when the zero-lift line is parallel to the relative airflow. (B) The graph of lift coefficient versus angle of attack, for a wing of infinite aspect ratio, with a section such as the one shown above, is approximately linear from the zero-lift angle to the stalling angle. The break in the curve at the stall is due to separation of the airflow from the wing surface, and is accompanied by a marked rise in the drag coefficient.

"aerofoil sections", or "profiles", meaning shapes that can be used as the cross sections of wings. Aerofoil properties are often called "two-dimensional", as they refer to the cross-sectional shape of a wing only, isolated from any effects that are due to a real wing's finite span, with flow around the wing tips. When the effects of wing span are taken into account (below), the aerofoil properties refer to a wing of infinite span. This is an abstraction, but it can actually be simulated quite well in a wind tunnel, by testing a model wing whose chord and cross-sectional shape are constant, and whose span extends from one wall of the test section to the other.

Systematic measurements of the properties of aerofoil sections are traditionally based on measuring the lift, drag and pitching moment developed by the aerofoil, which of course depend on the size of the model, the wind speed and the air density. The results are invariably presented in the form of dimensionless *coefficients*, in which the effects of these variables have been eliminated, leaving a number that is primarily a function of the *angle of attack*. This latter term is often wrongly used, and further confusion arises from failure to distinguish between lift and drag forces and the corresponding coefficients (see Box 3.5). The lift coefficient is closely related to the downwash angle induced by the aerofoil, and the drag coefficient measures the extent to which the aerofoil slows down the air flowing past it.

BOX 3.5 **Properties of aerofoils and wings.**

**Variable definitions for this box**

| | |
|---|---|
| $a$ | Lift slope |
| $a_0$ | Lift slope for a wing of infinite aspect ratio |
| $c$ | Chord of a wing model |
| $C_D$ | Drag coefficient |
| $C_{Dind}$ | Induced drag coefficient |
| $C_{Dpro}$ | Profile drag coefficient |
| $C_L$ | Lift coefficient |
| $D$ | Drag |
| $D'$ | Drag per unit span |
| $k$ | Span efficiency factor |
| $L$ | Lift |
| $L'$ | Lift per unit span |
| $q$ | Dynamic pressure |
| $R_a$ | Aspect ratio |
| $S_{wing}$ | Wing area |
| $V_t$ | True airspeed |
| $\alpha$ | Angle of attack |
| $\rho$ | Air density |

**Properties of wing sections**

Vast numbers of wind-tunnel experiments were done in the twentieth century to measure the lift and drag developed by different "aerofoil sections" or "profiles", meaning shapes that might be used as wing cross sections (Abbott and von Doenhoff 1959). Although the conclusions of these experiments are not easily transferred to the jointed and infinitely adjustable wings of birds and bats, the associated terminology, which is introduced in this box, permeates all discussions of wings of every kind. Wing sections intended for low-speed flight have a rounded leading edge, a pointed trailing edge, and some *camber*, which means that the mean line is curved, convex upwards. Families of wing section shapes were created and tested by systematically varying these characteristics. The *chord line* of such a section is defined as a straight line joining the centre of curvature of the leading edge with the point of the trailing edge. The angle between the chord line and the direction of the incident airflow is the *angle of attack* (Figure 3.9).

**BOX 3.5** *Continued.*

Some biological authors have been known to use the term "angle of attack" to refer to the angle between some plane on the wing and the axis of a wind tunnel, irrespective of the direction from which the local incident airflow is coming. This is wrong and highly misleading, especially in flapping flight, where the direction of the incident airflow varies from one point to another along the wing, depending on the flapping motion, and is usually very different from the general flow direction along the tunnel axis.

**Lift and drag coefficients**
The absolute magnitudes of the lift and drag forces are not very informative by themselves, but they can be given a context by expressing them as non-dimensional coefficients, the lift coefficient ($C_L$) and the drag coefficient ($C_D$). Each coefficient is the ratio of the actual force to a reference force, which takes account of the air speed, the wing area and air density. This reference force is made up by multiplying the dynamic pressure by the wing area. The dynamic pressure ($q$) is the pressure in a blind tube pointing into the relative wind, and is given by:

$$q = \frac{\rho V_t^2}{2},\qquad(1)$$

where $\rho$ (Greek rho) is the air density, and $V_t$ is the True airspeed. The lift coefficient ($C_L$) of a real wing with a finite area ($S_{wing}$) is defined as

$$C_L = \frac{2L}{(\rho V_t^2 S_{wing})},\qquad(2)$$

where $L$ is the lift force. Alternatively, in wind-tunnel experiments that simulate a wing of infinite span (and area), the primed variable $L'$ is used to denote the lift *per unit span*. The lift coefficient in that case is

$$C_L = \frac{2L'}{(\rho V_t^2 c)},\qquad(3)$$

where $c$ is the chord of the constant-chord model. Being the ratio of two forces, $C_L$ is, of course, dimensionless. Whereas the lift *force* is a function of the air density, the speed, and the wing area, the effects of those variables are eliminated in Equation (3), so that the lift *coefficient* is primarily a function of the angle of attack. It also depends on the Reynolds number, but investigating that involves replicating whole experiments at different scales. Figure 3.8 shows a typical graph of lift coefficient versus angle of attack for a lightly cambered aerofoil section, like those traditionally used for the wings of light aircraft, at a light aircraft Reynolds number of half a million or so. Because of the camber, a section like this develops some lift when the chord line is set parallel to the incident airflow, that is, at zero angle of attack. The graph is nearly a straight line from about −4°, where the lift coefficient is zero, up to about 16°, where it is about 1.4. Further increase in the angle of attack causes the lift coefficient to drop sharply. The wing is then said to be "stalled".

Over the linear portion of the lift coefficient curve, the airflow follows the upper surface of the wing, and leaves the trailing edge with a downward component of velocity, which it did not have as it approached the wing. This downwash is responsible for the lift force, and the downwash *angle*,

BOX 3.5 *Continued.*

measured downstream of the wing, is closely related to the lift coefficient. As the angle of attack increases, the suction above the wing increases, and so does the downwash angle, until the angle of attack reaches the *stalling angle*. The wing is said to stall when the angle of attack becomes so steep that the airflow can no longer follow the upper surface of the wing, and breaks away from it, which results in an abrupt drop in the lift coefficient.

The drag coefficient ($C_D$) of a wing is defined in the same way as the lift coefficient:

$$C_D = \frac{2D}{(\rho V_t^2 S_{\text{wing}})}, \qquad (4)$$

where $D$ is the drag. For a wing model of infinite span, whose chord is $c$, the drag coefficient is calculated from the drag per unit span ($D'$):

$$C_D = \frac{2D'}{(\rho V_t^2 c)}. \qquad (5)$$

In wing sections like that of Figure 3.9, the drag coefficient is lowest at low angles of attack, and increases at higher angles of attack until the wing stalls, while the ratio of $C_L$ to $C_D$ passes through a maximum at some intermediate angle of attack. Above the stalling angle, the lift collapses, and the drag coefficient increases more steeply than before, so that the ratio of lift to drag drops catastrophically. The $C_L/C_D$ ratio is the most basic figure of merit for any wing, and it peaks at a particular value of the angle of attack, which itself corresponds to a particular value of the lift coefficient. When the wing is incorporated into an aircraft (or bird) this also defines a particular speed at which the wing works best, given specified conditions of gravity and air density.

**Wings of finite aspect ratio**
The Wright brothers and their contemporaries were intuitively aware that a wing will not work if its aspect ratio is too low, but it was not until 1919 that Ludwig Prandtl and colleagues published the "lifting line" theory that they had developed during the First World War (see also Chapter 4, Section 4.2.4). This relates the properties of a wing of given aspect ratio to the two-dimensional properties of the profile. All later aeronautical textbooks have presented this theory as part of the bedrock of classical aerodynamics, and the reader who is in search of a particularly clear modern explanation will find one in Anderson (1991). It is a fixed-wing theory, that deals with induced drag (rather than induced power as in Box 3.1), but its main results give a valid intuitive feel for the effects of increasing the aspect ratio of any wing, whether fixed, rotary or flapping. The drag coefficient of a wing or a wing profile (Equations 4 and 5) is represented as the sum of two components, the profile and induced drag coefficients:

$$C_D = C_{\text{Dpro}} + C_{\text{Dind}}. \qquad (6)$$

The profile drag coefficient ($C_{\text{Dpro}}$) is due partly to skin friction and partly to the pressure difference between the upstream and downstream sides of

BOX 3.5 *Continued.*

the wing. It is not much affected by the wing's aspect ratio, or the amount of lift it develops, and is considered to be constant in the lifting-line theory. However, the induced drag coefficient depends directly on the square of the lift coefficient ($C_L$), and inversely on the aspect ratio ($R_a$):

$$C_{Dind} = \frac{kC_L^2}{\pi R_a},\tag{7}$$

where $k$ is an "induced drag factor", greater than 1, that accounts for any deviation from an "ideal" elliptical lift distribution (Chapter 4 Box 4.2). Over the straight-line portion of Figure 3.9, $C_L$ is proportional to the angle of attack, so that $C_{Dind}$ varies with the square of the angle of attack, measured from the zero-lift angle.

The Wright brothers' 1901 glider had an aspect ratio of 3.3 and it would not fly, because the lift coefficient could not be increased enough to support a man's weight, without also raising the induced drag coefficient to a catastrophic level. Their 1902 glider, with an aspect ratio of 6.5, flew well enough to allow them to develop an effective control system, and teach themselves to fly. There is a second reason, also predicted by the lifting-line theory, for the dramatic effect on performance of increasing the aspect ratio. This concerns the "lift slope" ($a$), which is the gradient of the straight-line portion of the curve of Figure 3.9:

$$a = \frac{dC_L}{d\alpha},\tag{8}$$

where $\alpha$ (Greek alpha) is the angle of attack. If $a_0$ is the value of $a$ for a wing of infinite aspect ratio, then the lift slope for the special case of a wing of aspect ratio $R_a$ with elliptical lift distribution is:

$$a = \frac{a_0}{[1 + (a_0/\pi R_a)]}.\tag{9}$$

Delta-winged aircraft, such as Concorde and the Vulcan bomber, have low aspect ratios, and consequently a low lift slope, which means that such aircraft have to be pitched nose-up to a very high angle for take-off and landing. The Wrights' unsuccessful 1901 glider simply could not be pitched up enough to develop enough lift to fly. Published data on wing section properties are sometimes given for infinite aspect ratio, but are commonly reduced with the aid of Equation (9) to a standard aspect ratio, usually 5 or 6. The lift slope for other, more general lift distributions can also be correctly predicted from the lifting-line theory, although this is more complicated.

### Low-speed flight
Strongly cambered wing sections are often described as "high-lift" sections, but it is more accurate to describe them as having a high maximum lift

BOX 3.5 *Continued.*

*coefficient.* As an aircraft slows down when approaching to land, the lift stays the same (equal to the weight), and the lift coefficient therefore has to increase. The pilot achieves this by progressively raising the nose, so increasing the angle of attack. The lift coefficient of most aircraft wings can be increased further by trailing-edge flaps, which effectively increase the camber when deflected downwards. Airliners often have a cascade of flaps, which can be deployed at low speeds behind the main wing, with slots in between them. Such an arrangement deflects the air downwards through a larger downwash angle than would be possible by simply increasing the angle of attack with the flaps retracted, and can increase the maximum lift coefficient to 3 or even more, allowing a corresponding reduction of the landing speed. A drooped leading edge, with or without a slot, further increases the maximum angle of attack at which the wing will fly without stalling. In cruising flight, when a high lift coefficient is not required, these "high-lift devices" are retracted or "cleaned up", which minimises drag, and maximises the lift:drag ratio.

### Lift coefficients of bird wings

Because of the complicated structure of bird wings, it is not possible to test them in a wind tunnel like wing models, but it is possible to train a bird to glide in a tilting wind tunnel, and determine the minimum speed at which it can glide (Chapter 14, Section 14.4). Results indicate that both birds and bats with aspect ratios of 6–8 can glide steadily, under full control, at lift coefficients around 1.5–1.6, but that their wings are not quite fully extended. Birds can manage a slightly higher lift coefficient, perhaps 1.8, by momentarily extending the elbow and wrist joints to get maximum wing span and area, in transient low-speed manoeuvres such as landing. At very low speeds and high angles of attack, the primary feathers splay apart at the wing tips, and twist in the nose-down sense, meaning the local angle of attack at the tip is automatically reduced. This prevents stalling of the wing tips, and flattens out the sharp break in the lift coefficient curve that is seen in simple wings like the one illustrated in Figure 3.9. Bird wings have an automatic mechanism that increases the camber when the wing is fully extended (Chapter 5, Section 5.2), while bats have fore-and-aft muscles in the wing membrane that flatten the camber when a low lift coefficient is required, but relax to allow the membrane to bulge upwards in low-speed flight (Chapter 6, Box 6.1). The patagial tendon of birds pulls the leading edge of the patagium downwards when the wing is fully extended, so drooping the leading edge, and at very high angles of attack the feathers on the ventral side hinge downwards and outwards, in a manner that suggests a leading-edge flap. The leading edge of the propatagium of bats is also pulled down when the wing is fully extended, and it seems that the pteroid bone,

BOX 3.6 **Reynolds number and the effects of scale.**

The Reynolds number is a "similarity criterion", defining the scale of the flow. For two flows to be dynamically similar, both must have the same Reynolds number. One of its most common uses in engineering is for transferring the results of tests on small-scale models to full-sized aircraft. It is also an indicator of the type of flow to be expected around a body or wing of a given size, flying at a given speed.

**Variable definitions for this box**

| | |
|---|---|
| $c_m$ | Mean chord of wing |
| $d_b$ | Body diameter at widest part |
| $\ell$ | A reference length |
| $Re$ | Reynolds number |
| $Re_{wing}$ | Reynolds number based on wing mean chord |
| $Re_{body}$ | Reynolds number based on body diameter at widest part |
| $V_t$ | True airspeed |
| $\mu$ | Air viscosity |
| $v$ | Air kinematic viscosity |
| $\rho$ | Air density |

**Reynolds number**

The Reynolds number ($Re$) is defined as

$$Re = \frac{V_t \ell \rho}{\mu}, \tag{1}$$

where $V_t$ is the True airspeed, $\ell$ is a "reference length", $\rho$ is the air density and $\mu$ is the viscosity of the air. The reference length is used to compare two objects of different size but similar shape, and can to some extent be arbitrarily chosen. For wing sections, the reference length is by convention the chord length, or the mean chord for a complete wing (Chapter 1, Box 1.2). The ratio of the viscosity to the density is often called the *kinematic viscosity* and given its own symbol $v$ (Greek "nu"), where

$$v = \frac{\mu}{\rho}. \tag{2}$$

Equation (1) for the Reynolds number then becomes

$$Re = \frac{V_t l}{v}. \tag{3}$$

The Reynolds number can be seen as expressing the relative importance of inertial and viscous forces. Because of their small size and low speeds, most insects operate at Reynolds numbers of tens to hundreds, below those of most birds, and far below those of most aircraft (millions). Below a body mass of about 5 g, insects take over from vertebrates as the dominant flying animals, and one of the reasons that they look so different, and are constructed so differently, is that air, to them, is a viscous fluid. The lower the Reynolds number, the harder it is to get air to follow a curved surface. Wherever a wing or body has a convex surface, the airflow follows the surface as

BOX 3.6 *Continued.*

far as the widest point, then tends to separate from it. This results in a lot of drag, and not much lift. Special adaptations are needed to maximise lift and minimise drag at low Reynolds numbers, and they are found in insects but not in birds or bats.

Birds occupy an "intermediate" range of Reynolds numbers between about 15,000 and 500,000, where lift and drag are created by inertial forces as in aircraft, but there is more difficulty in keeping the flow attached to wings and bodies. Despite this, even small birds have bodies that are faired by a covering of feathers into smoothly streamlined shapes, which would not have any advantage unless the flow could be made to follow the shape. Some large insects (hawk moths) exhibit a degree of streamlining, but the smaller ones do not, as their Reynolds numbers are too low to have any hope of keeping the flow attached.

**Drag coefficients at different scales**

The drag of a wing section is partly "pressure drag", which is due to a pressure difference between the upstream and downstream sides, and partly "skin friction drag" which is due to viscous forces produced as the air slides over the surface. Pressure drag is due to forces that act perpendicularly to the surface, and depend on the square of the speed, whereas skin friction drag is due to tangential forces, which depend directly on the speed. Pressure drag depends on the cross-sectional area presented to the flow, whereas skin friction drag depends on the "wetted area" over which the air slides. The definition of the drag coefficient (Box 3.5) is more appropriate to pressure drag than to skin friction drag.

**Reynolds numbers in *Flight***

Separate Reynolds numbers are calculated in *Flight* for the wing and the body. They differ in the choice of a measurement for the "reference length" ($\ell$). The wing Reynolds number is defined as

$$Re_{\text{wing}} = \frac{V_t c_m}{\nu}, \tag{4}$$

where $c_m$ is the mean chord of the wing (wing area divided by the wing span). It has no connection with the arbitrary measurement known to ornithologists as the "chord", and conversely this has no significance in flight mechanics. The body Reynolds number is

$$Re_{\text{body}} = \frac{V_t d_b}{\nu}, \tag{5}$$

where $d_b$ is the diameter of a circle with the same area as the body frontal area. The programme calculates both Reynolds numbers at the minimum power speed in flapping flight, and at the best glide speed in gliding, these being the speeds at which birds are most often seen flying around in flapping and gliding flight respectively.

## 3.7 ━ SCALE AND REYNOLDS NUMBER

The notion of *scale* is expressed by the Reynolds number (Box 3.6). The lower the Reynolds number (small size, low speed), the harder it is to keep the flow attached to curved surfaces, which is the key to maximising lift and minimising drag on both bodies (above) and wings. The feathered wings and bodies of birds seem to be more effective at doing this than the wings of model aircraft, which fly in the same range of Reynolds number (Schmitz 1960; Simons 1987), but it is far from clear exactly what it is about the feathered surface that is responsible for this.

# 4

# VORTICES AND VORTEX WAKES

Vortex principles provide an alternative way to derive the same laws that underlie
the work and power calculations of Chapter 3. A flying bird leaves tracks in the
air in the form of a vortex wake, and this can be examined and used to deduce
the forces that the wings exerted on the air, and to estimate the work that they
did. Vortex principles are not explicitly used in the *Flight* programme, but are the
basis of recent and current wind-tunnel investigations on flight mechanics.

Work and power were discussed in Chapter 3 in terms of linear
motions and accelerations in the air, but in the world of real fluids this
is at best an approximation. Any process that involves work being done
on a fluid invariably involves shear (gradients of velocity) and this in
turn results in rotation. *Vortices* appear on every scale wherever work
is done on a fluid, whether the scenario is a cup of tea stirred by a
spoon, or a hurricane driven by heat from the warm ocean below.
When a bird flies by, it leaves vortices behind it in the air, which persist
for a while. The vortex wake can be observed and measured, and it
carries a record of the work that the bird's wings have done on the
air. Vortex concepts are not used explicitly in the *Flight* programme,
but some of the most interesting experiments of recent times have

approached the same mechanical questions that the programme addresses from a different direction, by looking at the vortex wake, rather than at forces on the wings and in the flight muscles.

What exactly is a vortex, and what rules do vortices obey? Those questions are easily asked, but finding answers that lead to useful predictions about real flows has taken the best efforts of some of the most famous mathematicians of the last two centuries. Leonhard Euler, Daniel Bernoulli, Jean le Rond d'Alembert, Pierre-Simon Laplace, Hermann Helmholtz, Lord Kelvin, Wilhelm Kutta, Nikolai Joukowski and Ludwig Prandtl all made major contributions. The body of theory that they and others built is not exactly simple, but it is an alternative way of looking at the flight of birds and aircraft, which has evolved in parallel with the more direct approach based on considering the impact that every air particle has on the pressure and velocity of its neighbours. In mathematicians' terms, the direct approach amounts to solving the Navier-Stokes equations, which fully describe what the fluid does, but are notoriously difficult to solve for particular cases, whereas vortex concepts are less exact but deliver practically useful results in terms of entities that can be visualised, and obey simple rules. Following the advent of huge computers, it has become practical to solve the Navier-Stokes equations numerically, but vortex concepts still provide a useful and compact way to describe many processes.

An authoritative modern account that covers both approaches can be found in Anderson (1991), and the same author's *History of Aerodynamics* is also highly recommended (Anderson 1997). My objectives in this chapter are limited to attempting to present the basic vortex concepts in an essentially pictorial form. This can be seen as an alternative view of the principles of flight covered in Chapter 3, and also as background to modern experiments on the vortex wakes of birds, and how they relate to calculations of the work done by the wings.

## 4.1 ⬤ THE CONCEPT OF THE LINE VORTEX

A *line vortex* is basically a mathematical abstraction that corresponds, under the right conditions, to a physical entity with two components, a *vortex filament*, which is a thin, rotating thread of fluid particles, surrounded by an *induced flow*, which is where the physical effects take place. The induced flow is the visible, whirling vortex, but despite that, it is said to be *irrotational*. This means that although individual particles of fluid may (or may not) circulate around closed paths, they do not rotate on their own axes. Only the fluid particles that make up the vortex filament actually rotate.

Figure 4.1A is a cross section through a line vortex. In this two-dimensional view the point in the middle is a *point vortex*, which is actually a cross section through the vortex filament. The filament can be imagined as a line passing through the point, and extending above and below the page. The filament (only) is where the air is actually rotating, and it *induces* the surrounding air to circulate around it, without the individual particles of air themselves rotating. As the radius from the filament increases, so the tangential speed of the air decreases in each cylindrical shell of air surrounding the rotating filament, in inverse proportion to the radius. If the cross section of the vortex filament were really a point (with zero radius), the speed and the angular velocity within it would have to be infinite, making what mathematicians call a "singularity", a place where the rules break down. However, the induced flow looks essentially the same if the vortex filament is replaced by a spinning core that has a finite radius (Figure 4.1B). The tangential velocity in the core (grey) is zero at the centre, and increases linearly with the radius. The outer surface of the core (at radius 2 in this case) pulls the layer of air in contact with it along, and at larger radii the tangential speed decreases in the same way as it does in the induced flow around the one-dimensional vortex filament of Figure 4.1A. Some of the most useful theoretical results of classical aerodynamics depend on the assumption that real vortices, such as those shed from the wing tips of fixed-wing aircraft, conform to the pattern shown in Figure 4.1B, with a thin core of rotating air surrounded by an irrotational induced flow.

## 4.2 ⬤ VORTEX CONCEPTS APPLIED TO FIXED WINGS

### 4.2.1 CIRCULATION AND LIFT

The speed along any of the circular paths in Figure 4.1A or B is constant around the path, and inversely proportional to the radius of the circle. If we integrate the speed around the circumference of one of the circles, the result is the same for any circle. The speed is halved for a bigger circle with twice the circumference, and therefore the integral of speed around the closed path is the same as before. This integral is called the *circulation* and the result is the same for *any* closed path, circular or not, so long as the vortex filament (or the finite core) is entirely contained within it. The circulation therefore has a fixed and measurable value for a particular vortex, and is often called the *strength* of the vortex.

If a cylinder is mounted in a wind tunnel, perpendicular to the wind (coming from the left in Figure 4.2A) the air divides symmetrically

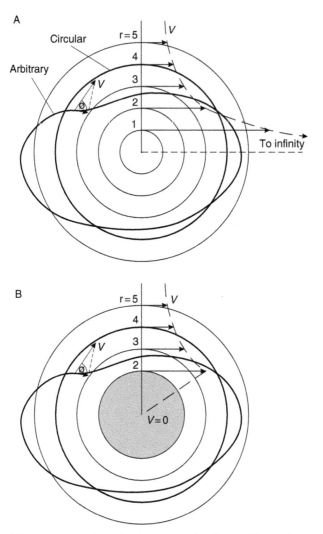

FIGURE 4.1 (A) Cross section through a line vortex. The "vortex filament", represented by the point in the middle, is the only region where particles of air actually rotate on their own axes. The "induced flow" around the filament is "irrotational", meaning that particles of air follow circular paths, but do not rotate on their own axes. Each particle moves at a tangential velocity (V) that is inversely proportional to its radial distance (r) from the filament. The "circulation" ($\Gamma$) can be found by integrating $V \cos \varphi$ (where $\varphi$ is the angle between the path and the direction of V) around any closed path that contains the filament. The result is the same whether the path is one of the concentric circles shown (where $\varphi = 0$), or any arbitrary closed path that contains the vortex filament. The value of the circulation so measured is often called the "strength" of the vortex. (B) The need for infinite angular velocity in the vortex filament can be avoided by replacing it with a core (grey) whose diameter is finite. Within the core the velocity increases linearly from zero at the centre to the core boundary. Outside the core, the velocity decreases in the same way as the induced flow around the vortex filament in (A). The circulation around any closed path that completely encloses the core is the same as that around an infinitely thin filament.

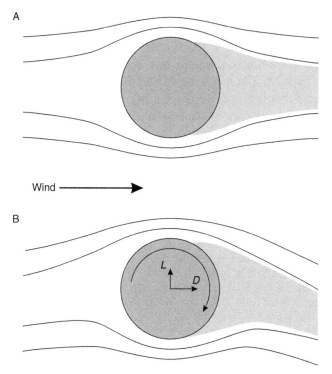

FIGURE 4.2 (A) A cylinder (dark grey) in a wind tunnel is a bluff body from which the flow separates on the downwind side to form a turbulent wake (light grey), resulting in a large amount of drag. (B) If the cylinder is spinning the induced flow of Figure 4.1B is added to the steady wind, resulting in the airflow being deflected in a direction perpendicular to the incident flow. This results in a lift force (*L*), but there is also a large drag force (*D*) due to the turbulent wake.

around it and separates from the surface, forming a turbulent wake on the downwind side (grey). This results in a lot of drag but no lift, since the airflow is slowed down by the obstruction, but not deflected upwards or downwards. If we now set the cylinder spinning as in Figure 4.2B, the effect is to add the circular induced flow of Figure 4.1B to the steady wind of Figure 4.2A. The turbulent wake is still there, along with the drag that results from it, but the induced flow from the vortex that is bound to the spinning cylinder imparts some downwash to the air leaving the cylinder on the right. This in turn produces a lift force on the cylinder (perpendicular to the incident airflow), equal to the rate at which transverse momentum is imparted to the air (Chapter 3, Section 3.1.1).

Lift results when a vortex is added to the steady flow. This is called the Magnus Effect when the vortex is due to a spinning cylinder. The amount of the lift force, per unit length of the cylinder, is directly proportional to the circulation, as defined above, and also to the wind speed, and the air density (Box 4.1). Some asymmetrical shapes develop circulation when exposed to a steady wind, without needing to spin round, and this also produces lift according to the same law. The essential characteristic of an "aerofoil" shape, that is, one that can be used as the cross section of a wing, is that a *bound vortex* develops around it when it is set at a suitable angle of attack to the incident flow, and that the lift force due to the resulting circulation is associated with a much smaller drag force.

BOX 4.1 **Lift, circulation and vorticity.**

---

### The idea of a vortex

A "vortex" is a construct with properties that may appear somewhat artificial at first sight. It consists of a "vortex filament", which is a one-dimensional line that meanders through the fluid, surrounded by a region of "induced flow". A cross section through the vortex shows the vortex filament as a point, and it is only in this infinitely small region that particles of fluid actually rotate. The surrounding induced flow is said to be "irrotational". Particles of fluid may travel in closed paths around the vortex, or they may stream past, faster on one side than the other, but they do not actually spin around on their own axes.

### Variable definitions for this box

| | |
|---|---|
| $C$ | A constant |
| $L'$ | Lift per unit span |
| $r$ | Distance from vortex filament |
| $s$ | Distance along integration path |
| $V$ | Local fluid velocity |
| $V_\infty$ | Free stream fluid velocity |
| $\Gamma$ | Circulation |
| $\varphi$ | Angle between local velocity vector and path of integration |
| $\rho_\infty$ | Free stream fluid density |

### Circulation

"Circulation" is a property that is measured in the irrotational flow around the vortex filament, not in the vortex filament itself. Figure 4.1 shows a section through an isolated vortex, in which each particle of air (outside the vortex filament) is moving in a circular path around the filament, at a speed ($V$) which is inversely proportional to the radius ($r$) from the filament. In other words,

$$V = \frac{C}{r},\tag{1}$$

where $C$ is a constant. This distribution of velocity is not as arbitrary as it looks, as it expresses the condition that the flow is irrotational. We can define the *circulation* ($\Gamma$) around any particular circular path that encloses the vortex filament as the integral of the velocity around that path:

BOX 4.1 *Continued.*

$$\Gamma = (\oint V \mathrm{d}s), \tag{2}$$

where $s$ is distance along the circular path, and the symbol "$\oint$" refers to integration around the closed path. For this case,

$$\Gamma = \left(\frac{C}{r}\right) \times (2\pi r) = 2\pi C, \tag{3}$$

in other words, the circulation is the same around any circle, irrespective of the radius. More generally, for a closed path of any shape that contains the vortex filament, the component of the local velocity along the path at any point is $V \cos \varphi$, where $\varphi$ is the angle between the local velocity vector and the path of integration, and the circulation is:

$$\Gamma = (\oint V \cos\varphi \mathrm{d}s), \tag{4}$$

The result of the integration is the same (Equation 3) for any closed path that contains the vortex filament. This is still true if the vortex filament is inside (or on the surface of) a body of arbitrary shape, such as a wing cross section, provided that the path of integration encloses the body, and the flow outside the body is irrotational. The circulation has the dimensions of length-squared/time ($L^2T^{-1}$), and it is a property of the vortex, often called its "strength".

### Bound vortex on a wing

An *aerofoil* shape is one which, when immersed in a steady flow of fluid, and set at a suitable angle of attack to the incident flow, develops a vortex around it, such that the fluid velocity on one side of the shape is higher than the free-stream velocity, and that on other side is lower. This vortex is forced into existence on a wing by the "Kutta condition" (see main text), which expresses the effect of viscous forces that equalise the speed at which the fluid leaves the upper and lower surfaces at the trailing edge. Although this type of flow does not involve particles of air moving in closed curves, but only differences in speed on the two sides of a wing, it can be seen as the combination of a free stream whose velocity is $V_\infty$ and a vortex of strength $\Gamma$. The lift on the wing, that is, the component of force at right angles to the free stream can be calculated directly from the Kutta-Joukowski theorem, which states that:

$$L' = \rho_\infty V_\infty \Gamma, \tag{5}$$

where $\rho_\infty$ is the free-stream density. The primed variable $L'$ stands for the lift per unit span of the wing. The total lift is obtained by integrating $L'$ across the span from one wing tip to the other.

### Vorticity

Unlike circulation, *vorticity* is not a property of a vortex. It is a "field variable" like pressure and density, with a continuous distribution that can be mapped in a region of fluid, and may vary with time. Vorticity is a vector quantity equal to the "curl" of the velocity, which is itself a function of the

BOX 4.1 *Continued.*

partial derivatives of velocity in the three directions of space. Vorticity can also be seen as circulation per unit area, as measured in a plane that is perpendicular to the axes of the vortices. Its dimensions are those of inverse time $(L^2T^{-1}/L^2)$. The intuitive meaning of the vorticity at any point in the measurement plane is that it is twice the angular velocity of a fluid particle at that point. One of the more arcane properties of a vortex as defined above is that the vorticity is infinite in the vortex filament, and zero everywhere else, which is another way of saying that the induced flow is irrotational.

Where there is shear, viscous effects lead to distributed vorticity that is not confined to identifiable filaments, and can be mapped. For example, where air flows over the surface of a wing, the boundary layer can be seen as a "vortex sheet", a layer of very small vortices with their axes lying along the wing span, transverse to the flow. At the trailing edge of the wing, the flow above the wing has an inward component of velocity due to the reduced pressure there, while that below the wing has an outward component of velocity. Where the two layers merge as they leave the trailing edge, this lateral motion produces a free vortex sheet with its axes aligned back along the flight path. These vortices are of course the same as the horseshoe vortices that are shed from the trailing edge (Figure 4.3B), and eventually roll up (behind a fixed wing) to form a pair of concentrated trailing vortices (Figure 4.4).

## 4.2.2 THE KUTTA CONDITION

How is a bound vortex initiated and maintained around an object like a wing, which is moving along, but not spinning around? If we consider an aircraft just starting its take-off run, the flow around a cross section of the wing first follows a pattern like that in Figure 4.3A. There is a discontinuity where the air sliding along the lower surface comes to the sharp trailing edge, and doubles back on to the upper surface. Then it doubles back again, to merge with the air coming over the upper surface as the flow leaves the wing along the same line that it followed when it approached. This pattern would persist if the viscosity of the air were zero. In the real world, the zigzag path of the air around the trailing edge results in strong local shear, which in turn produces viscous forces that force the flow to speed up over the upper surface, and slow down below. This is equivalent to adding circulation to the steady flow. The circulation around the wing builds up until the air flowing off the upper surface merges smoothly with that coming from below (Figure 4.3B). This is the "Kutta condition", and it prescribes the strength of the bound vortex that forms, and hence also the

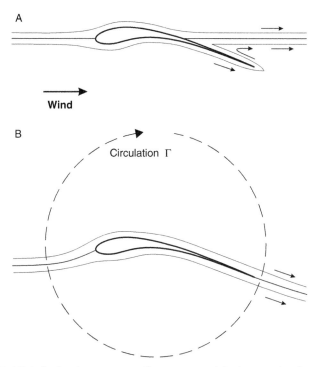

FIGURE 4.3 (A) A fluid with no viscosity flowing around the lower side of a wing would double back around the trailing edge and leave from a stagnation point displaced on to the upper surface, following the same line by which it approached. Although the surface pressure would vary on different parts of the wing, the variations would cancel one another, producing no net lift. There would also be no net drag (d'Alembert's paradox). (B) In reality, viscous forces in the zone of strong shear around the trailing edge force the flow over the upper surface to speed up and that on the lower surface to slow down, until the fluid from both surfaces leaves the trailing edge smoothly at the same speed (the Kutta condition). This amounts to forcing a circulation ($\Gamma$) to be added to the flow, which causes upwash in the air approaching the wing, and downwash as it leaves the trailing edge. This in turn gives rise to the lift. There could be neither circulation nor lift without viscosity.

amount of lift that develops. The physical meaning of the bound vortex is that the air flows faster past the top surface of the wing than past the lower surface. Air molecules do not circulate in closed paths around the wing, because the steady component of the flow is always faster than the circulating flow due to the vortex.

## 4.2.3 HELMHOLTZ'S LAWS

To see how this works in a three-dimensional wing, we need a couple of general properties of line vortices that were discovered by Hermann

Helmholtz in the nineteenth century. The first of Helmholtz's laws is that the circulation (strength) of a line vortex, defined as above, is the same at every cross section along the vortex, however long it is. The second is that a line vortex cannot end in the fluid. It can end by butting against a solid surface, or it can bend round and join on to itself as a vortex ring, but it cannot just end. Helmholtz's laws imply that a line vortex, once present, lasts for ever, and conversely, if no vortex already exists, it is not possible for one to start.

The resolution of this paradox lies in the small print. The conditions that Helmholtz assumed when he derived his laws included the assumption of an "inviscid" fluid, which is one whose density is finite, but whose viscosity is zero. In practice, Helmholtz's laws describe the behaviour of vortices in viscous fluids like air or water rather well, so long as the flow does not contain any regions of strong shear, meaning regions where the speed changes sharply over a short distance in the fluid. There is always shear wherever the fluid slides along a solid surface. The layer of fluid in contact with the surface sticks to it without slip, and viscous forces tend to hold back the layers of fluid sliding past above. Conversely, the motion of the fluid tends to pull the surface along with it. For fluids of low viscosity like air and water, these effects are confined to a thin *boundary layer* next to the surface. It is here that rotation is introduced into the fluid, in the form of a *vortex sheet*, rolling along the surface. Once a vortex is carried away from the solid surface by the flow, forces due to viscosity become negligible, and the vortex behaves (more or less) according to Helmholtz's laws.

## 4.2.4 THE THREE-DIMENSIONAL FIXED WING

Once a bound vortex has formed on a wing of finite span, it is forbidden to end in the fluid. It cannot just stop at the wing tips. It bends round to form a pair of trailing vortices, whose strength is the same as that of the bound vortex, leading back along the flight path from the wing tips, to the point on the runway where the lift developed. There they are joined together by a "starting vortex", which has the same strength as the bound vortex, but the opposite direction of rotation. A fixed-wing aircraft that flies from one airport to another actually creates an elongated, rectangular vortex ring, closed at one end by the starting vortex which is left behind on the departure runway, and at the other by a "stopping vortex", which is left on the landing runway, when the bound vortex is shed from the wing on landing.

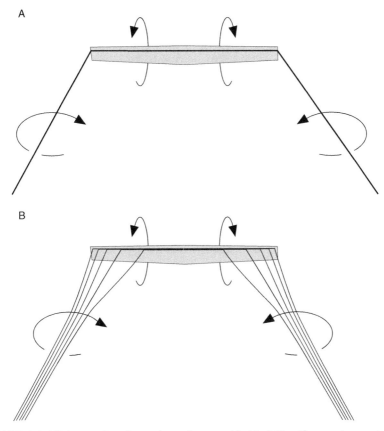

FIGURE 4.4 (A) A wing (grey) seen from above and behind. The lift per unit span is proportional to the strength of the bound vortex (black line), which cannot end at the wing tips, and bends back to form two parallel trailing vortices. (B) The number of vortex filaments making up the bound vortex is here shown as highest in the middle, decreasing as filaments are shed one by one from the trailing edge. In reality a continuous "vortex sheet" is shed from the trailing edge, but it rolls up a short distance downstream, to form a pair of trailing vortices of the same strength as in (A) but closer together.

When an aircraft is cruising along at a constant speed in level flight, its weight is balanced by the lift associated with the bound vortex on the wing. A pair of trailing vortices whose strength is the same as that of the bound vortex, stretch out behind (Figure 4.4A). The starting vortex has been left far behind, and for practical purposes the trailing vortices extend backwards to infinity. They grow continuously as the aircraft moves forwards creating new vortex at the front, and leaving the existing vortices behind where they persist, convecting slowly downwards as each vortex is carried down by the induced flow of the other and (in the case of large aircraft) creating a hazard to other air-craft that may happen to fly through them. As the trailing vortices grow, downward momentum is continuously added to the downwash zone between them (Figure 4.4A), at a rate which has to account for the lift force (Chapter 3). At the same time, work is needed to set each newly created section of the vortex pair spinning. The rate at which this work has to be done can be calculated from the strength and spacing of the vortex pair, and the forward speed. It is the same as the induced power required from the aircraft's engine, as calculated in a different way in Chapter 3, Box 3.1. The rates at which momentum and energy appear in the wake can be measured, and used to deduce the forces acting on the aircraft, and the power that is being supplied by the engine. The persisting vortex wake is a kind of "footprint" in the air. It contains a record of the forces that the aircraft (or bird) has applied to the air, and the work that has been done by the engine or flapping wings.

A single "horseshoe vortex" as shown in Figure 4.4A would imply that the circulation of the bound vortex is the same at every cross section from one wing tip to the other, which is not usually the case. Typically, the circulation is strongest in the middle, and tapers off to zero at the wing tips. As we pass outwards from the wing root towards the wing tip, the circulation around each cross section gets less, and this means that a part of the vortex bound to the wing root must have bent round and left the wing as a trailing vortex. In Figure 4.4B, the bound vortex near the wing root is made up from five horseshoe vortices, all of equal strength. Together, they make a "lifting line" along the wing, whose strength decreases in steps, as the vortex filaments are shed one by one from the trailing edge. By making the number of vortex filaments larger, and the strength of each smaller, the strength of the lifting line can be made to decline smoothly to zero at the wing tip, while the circu-lation is shed as a continuous vortex sheet from the trailing edge of the wing. Such a sheet is unstable, and soon rolls up into a concentrated, tightly wound vortex, whose circulation is the same as that due to the lifting line at its strongest point. The end result, as far as the cross-sectional view of the wake is concerned, a short distance behind the wing, is nearly the same as for a single horseshoe vortex. There are two trailing vortices, each with the same strength as the strongest part of the lifting line, but they are a little closer together than they would be if the lifting line were of constant strength (Figure 4.5). The wake of a gliding kestrel was measured and analysed by Spedding (1987a), and found to conform closely to this pattern.

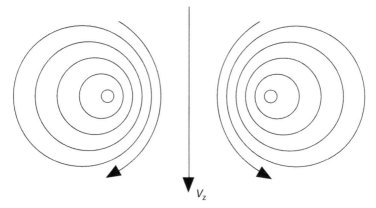

FIGURE 4.5 A cross section through the wake a short distance downstream from the wing of Figure 4.4B. Each vortex of the pair induces additional downward flow on the near side of the other, so that the downward velocity ($V_z$) in the centre is stronger than either vortex would produce on its own. Also, each vortex makes the other convect downwards at a velocity $V_z/4$. The upward velocity on the outside of each vortex is less than it would be without the other vortex. As the vortex pair grows along the direction of flight, net downward momentum is added at a rate that must balance the aircraft's weight. The rate at which work is done (induced power), must account for all of the motion irrespective of its direction.

## 4.3 ⬥ LIFTING-LINE THEORY APPLIED TO FLAPPING WINGS

### 4.3.1 VORTEX RINGS AND TRAILING VORTICES IN BIRDS' WAKES

Prandtl's lifting-line theory assumes that the vortex sheet, which forms where the air slides over the surface of the wing, can be regarded as the combined effect of a large number of vortex filaments which individually obey Helmholtz's laws, and which, once shed from the wing, roll up into the pair of trailing vortices which are actually observed. The theory gives a good account of the properties of fixed wings, and this indicates that the flow over such wings does indeed consist of vortex filaments surrounded by irrotational induced flow, as outlined above. If Helmholtz's laws are true in general, then one would expect that the theory could be extended to cover flapping wings, a notion with great creative possibilities, which has given rise to a minor industry in recent years.

Rayner (1979a,b) postulated a simple type of wake for a pair of flapping wings, in which the wings are assumed to build up a circulation during the downstroke, and to be completely unloaded (zero circulation) during the upstroke. As the wing tips separate at the beginning of the downstroke, vortices begin to stream from their tips as soon as the circulation becomes established. These join to become a single vortex connecting the wing tips, which grows in length as the wing tips move apart. At the end of the downstroke, the circulation ceases and the two ends of the vortex are shed from the wings. Being forbidden to end in the fluid, they join together to form a complete vortex ring, which is left behind as the wings are repositioned for the next downstroke. Just as an aircraft creates one long, rectangular vortex ring per flight, so a slow-flying bird creates one vortex ring per wingbeat, comprising starting and stopping vortices, connected by very short trailing vortices. The momentum of the vortex ring, divided by the wingbeat period, gives the average lift force, which must be equal to the bird's weight if the flight path is horizontal on average. Likewise, the energy of the vortex ring, divided by the wingbeat period, is the induced power.

In the first quantitative study of the wake of a flying bird, Spedding (1986) trained a jackdaw to fly very slowly through a cloud of tiny helium-filled soap bubbles, and mapped the motion of the air in three dimensions by taking stereoscopic multiple-flash photographs. He observed vortex rings which resembled Rayner's predictions, but found that he could only account for about half the momentum needed to support the jackdaw's weight (Spedding et al. 1984). The reason for this "momentum deficit" remained a mystery for another twenty years. Spedding (1987b) also observed a different type of wake, which had not been predicted, in a kestrel which had been trained to fly through a bubble cloud at a normal cruising speed. In this case, the circulation of the bound vortex did not drop to zero during the upstroke, in fact it did not change at all. As the wing tips moved up and down, they streamed a pair of continuous vortices of constant strength, as a fixed wing would do, implying that the lift *per unit span* did not change. However, the wing tips moved in during the upstroke, as the bird reduced its wing span by flexing the elbow and wrist joints, and out to full span during the downstroke. By varying its wing span in this way, the bird developed more lift during the downstroke than during the upstroke, which is necessary to produce a net forward force over the wingbeat cycle, to balance drag forces. This "concertina" wing motion is invariably seen in high-speed films of birds in fast flapping flight.

### 4.3.2 BIRDS DO NOT NEED GAITS

Spedding's observations were misinterpreted by others to imply that birds must either use a "vortex-ring gait" at low speeds, or a "constant-circulation gait" at cruising speeds, and must "shift" from one to the other as they speed up or slow down, in much the same way that a horse shifts back and forth between a walk and a trot. Although this idea has been widely repeated, there is actually no evidence for it, and there is no known reason why a bird cannot change smoothly from one type of wake to the other as it speeds up, without any discontinuous shift. In the hypothetical sequence of Figure 4.6, discrete vortex rings form at very low speeds, because the local airspeed over the wing during the upstroke is so low that the circulation cannot be maintained, and the wing has to be unloaded (Figure 4.6A). Once the bird accelerates to a modest forward speed, a small amount of circulation (and lift) can be maintained during the upstroke (Figure 4.6B). A stopping vortex forms to close the ring as the wing is partially unloaded at the end of the downstroke, but it does not contain the full amount of circulation that comes off the wing in the form of trailing vortices during the downstroke. The difference remains in the form of weaker trailing vortices that continue, closer together, during the upstroke. As the forward speed continues to increase (Figure 4.6C), the trailing vortices become stronger during the upstroke, and the "rungs" of the ladder become weaker until finally they disappear altogether, leaving a pair of trailing vortices of constant strength, but variable spacing. No gait shift is required.

## 4.4 ☞ WIND TUNNEL STUDIES OF BIRD WAKES

### 4.4.1 DPIV EXPERIMENTS

Experimental studies of bird wakes entered a new phase with a series of papers by Spedding et al. (2003a,b), Rosén et al. (2004) and Hedenström et al. (2005) on the wakes of small birds flying in a wind tunnel, observed by digital particle imaging velocimetry (DPIV). Like the helium-bubble method, this technique depends on tracking particles in the air, but in the wind tunnel the bird is stationary and the air streams past, carrying any vortex structures in the wake along with it. The particles were tiny liquid droplets introduced into the circulating air stream by a fog generator, and they were illuminated by a thin light sheet coming from a pulsed laser. The light sheet illuminated a vertical plane aligned along the direction of the air flow behind the bird, and the particles in it were photographed from the side. By statistically comparing two photographs, separated by a short time interval, variations of velocity in the plane of the light sheet could be mapped. In slow flight, when the bird was generating vortex rings, the starting and stopping vortices of each ring could be identified and measured, but the structure of the "trailing" parts of the vortex structure had to

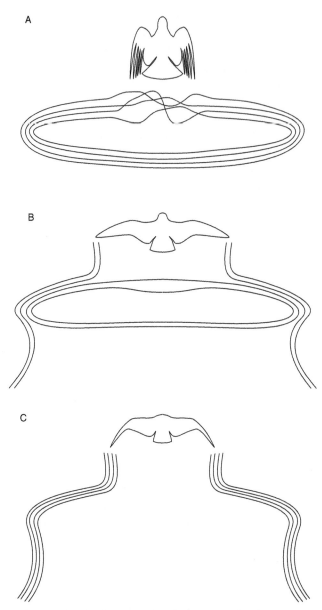

FIGURE 4.6 Schematic bird wakes, seen from behind and slightly above the bird, which is shown during the upstroke of flapping flight, at three different speeds. A (observed). In very slow flight the wing is unloaded during the upstroke, shown by separated flight feathers. The starting vortex formed during the previous downstroke, shown as a bundle of vortex filaments, is shed from the wing when the lift collapses at the end of the down-stroke, and its ends join to make a free vortex ring. B (hypothetical). At a moderate for-ward speed, some lift continues during the upstroke, and some vortex filaments continue to stream from the wing tips, while others are shed and join up to form a vortex

be inferred by comparing maps in which the illuminated plane varied from the centre line to beyond the wing tip.

The experiments of Spedding et al. (2003b) on a thrush nightingale covered a speed range from well below the thrush nightingale's estimated minimum power speed ($V_{mp}$) to about 2 m s$^{-1}$ above $V_{mp}$, and the results demonstrated a couple of important points. In the first place, there was no suggestion of any discontinuous "shift" between two or more different "gaits" at different speeds. The bird could fly at any speed between minimum and maximum, adjusting its wake structure continuously to suit the speed. Secondly, the results at intermediate speeds did not display a regular "ladder" structure as proposed in Figure 4.6B. A well-defined starting vortex could be identified at the beginning of each downstroke, but the stopping vortex was less well defined, and weaker than expected. When the results were analysed in terms of vortex filaments and irrotational flow, there was a momentum deficit as observed in the earlier helium-bubble experiments. Hedenström (2006) has described these experimental wakes, with diagrams of their inferred structure.

## 4.4.2 DISTRIBUTED VORTICITY

In the case of a vortex filament, which is the only part of a line vortex where the air actually rotates, the *vorticity* can be defined as twice the angular velocity of the filament. The induced flow surrounding the filament is irrotational, meaning that its vorticity is everywhere zero. However, vorticity can also be measured at a point in the fluid as the "curl" of the three-dimensional velocity. This is a vector quantity, derived from the local gradients of velocity in the three dimensions of space. Spedding et al. (2003a,b) and Hedenström et al. (2005) used the DPIV technique to map the magnitude and direction of the vorticity in the space immediately downstream of the bird, and found that the vorticity was not wholly confined to well-defined vortex filaments. Starting and stopping vortices could be seen at each wingbeat, but they were not of equal strength. The flow around them was not irrotational, but

---

ring. The combined effect is a pair of continuous wing tip vortices, which move out as the wings are fully extended during the downstroke, and in as they are flexed in the upstroke, together with a series of transverse vortices circulating alternately in opposite directions. C (observed). At higher speeds, the transverse starting and stopping vortices dwindle in strength and eventually disappear, leaving only the trailing vortices, undulating up and down, and moving in during the upstroke and out during the downstroke. The circulation of the bound and trailing vortices remains constant. The lift *per unit span* is therefore also constant, but the lift is reduced during the upstroke by shortening the wing span.

contained distributed vorticity that was not concentrated into thin filaments. When this distributed vorticity was taken into account, the momentum deficit disappeared.

It seems that real, flapping-flight bird wakes do not necessarily consist of structures made up of vortex filaments obeying Helmholtz's laws, as postulated by Rayner (1979a,b,c), or as shown in Figure 4.6. There are identifiable vortex cores, but the flow around them is not, in general, irrotational. This means, unfortunately, that it is not practical to calculate the momentum and energy of the wake by elaborating Prandtl's lifting-line theory, as this depends on representing the wake as an array of line vortices, which individually conform to Helmholtz's laws. Although it should be possible in principle to estimate the three main components of mechanical power (outlined in Chapter 3) from a quantitative analysis of these wakes, a full accounting of the momentum and energy will have to include distributed vorticity that is not confined to thin vortex cores. This is difficult, and is a major challenge for theorists. It may also underlie some of the performance differences between bird wings and their artificial counterparts, especially their resistance to boundary-layer separation (below).

### 4.4.3 IMPLICATIONS OF WAKE STUDIES FOR FLIGHT PERFORMANCE CALCULATIONS

The power calculation used in *Flight*, and described in Chapter 3 for flapping wings, does not explicitly take account of the structure of the vortex wake. If it were possible to do that, the effects would include changing the method of calculating the induced power. The present method depends on the much-derided concept, introduced in Chapter 3, of an "actuator disc", which is swept out by the pair of flapping wings, and imparts a downward induced velocity to the air passing through it. The actuator disc is imagined as adding a constant downward velocity to the whole of a circular tube of air passing through the disc, whose diameter is the same as the wing span. In this it resembles a fixed wing with an elliptical spanwise lift distribution which, according to Prandtl's lifting-line theory, produces a constant downwash velocity across the span, from one wing tip to the other. This case is discussed at length in every aeronautical textbook, together with the proof that this particular lift distribution results in a lower induced drag than any other. Despite its somewhat artificial appearance, the actuator disc predicts the same amount of induced power as a wake like that of Figure 4.7, coming from a fixed wing with an elliptical spanwise lift distribution. The formula for induced power, derived from the actuator disc in Chapter 3, Box 3.1, is exactly the same as the standard formula for the induced drag of a fixed wing with elliptical lift distribution, if the drag is multiplied by the speed to get the induced power (Box 4.2).

### 4.4.4 BETTER ESTIMATES OF THE INDUCED POWER FACTOR

Because an elliptical lift distribution produces the lowest induced drag that is possible with a fixed wing, corrections to induced drag or induced power calculations take the form of multiplying the calculated drag or power by a number that is somewhat greater (but not much greater) than 1, to account for losses due to deviations from the assumed constant downwash velocity across the span. In *Flight*, the default value for the induced drag factor in gliding flight is 1.1, and that for the induced power factor in flapping flight is 1.2. These values are essentially guesses based on aeronautical experience. The power curve calculation in *Flight* currently identifies two speeds that characterise a particular bird, as defined by its mass, wing span and wing area, taking account of the strength of gravity and the air density. These are the minimum power speed ($V_{mp}$) and the maximum range speed ($V_{mr}$) which define the lower and upper limits of the speed range in which

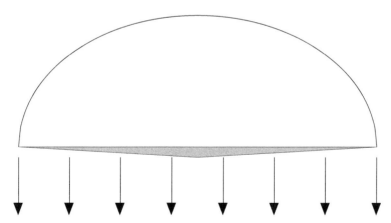

FIGURE 4.7 A fixed wing (grey) seen from behind. The strength of the bound vortex, shown by the curve above the wing, is greatest in the middle, declining to zero at the wing tips. The curve represents the local circulation, and also the lift per unit span, at each point along the span. If (and only if) this curve is half of an ellipse as shown, then the downwash velocity immediately behind the wing (arrows), caused by the shedding of a vortex sheet as in Figure 4.4B, is constant across the span. According to Prandtl's lifting-line theory, this "elliptical lift distribution" gives a lower induced drag coefficient than any other, for a given lift coefficient.

BOX 4.2 **The actuator disc versus vortex wakes.**

**Variable definitions for this box**

| | |
|---|---|
| $B$ | Wing span |
| $C_{Dind}$ | Induced drag coefficient |
| $C_L$ | Lift coefficient |
| $D_{ind}$ | Induced drag |
| $g$ | Acceleration due to gravity |
| $k$ | Induced drag (or power) factor |
| $m$ | Bird's all-up mass |
| $P_{ind}$ | Induced power in horizontal flight |
| $R_a$ | Aspect ratio |
| $V_t$ | True airspeed |
| $S_{wing}$ | Wing area |
| $\rho$ | Air density |

**Induced drag from lifting-line theory**

The induced drag of a fixed wing can be deduced from Prandtl's classical lifting-line theory, which begins by representing the circulation around the wing as a stack of horseshoe vortices as shown in Figure 4.3B. The total circulation at any particular point along the span depends on the number of vortex filaments still bound to the wing at that point. By specifying where vortex filaments are shed, the theorist can define a particular "lift distribution", in which the strength of the circulation is plotted along the wing span, from one wing tip to the other (Figure 4.6). The same graph also represents the lift-per-unit-span at each point along the span, since this is proportional to the circulation. All aeronautical textbooks (such as the excellent one by Anderson 1991) give a detailed account of the "elliptical" lift distribution, in which the graph is one half of an ellipse. This particular lift distribution is "ideal" in the sense that it produces less induced drag than any other, for a given amount of lift. It also has another (related) characteristic, that the downwash velocity immediately behind the wing, shown by the vertical arrows in Figure 4.6, is constant across the span. According to the lifting-line theory, the coefficient of induced drag ($C_{Dind}$) for this special case depends only on the lift coefficient ($C_L$) and the aspect ratio ($R_a$):

$$C_{Dind} = \frac{C_L^2}{\pi R_a}. \tag{1}$$

The aspect ratio (Chapter 1) is

$$R_a = \frac{B^2}{S_{wing}}, \tag{2}$$

where $B$ is the wing span, and $S_{wing}$ is the wing area. For a fixed-wing aircraft in level flight, the lift force must balance the weight ($mg$), and the lift coefficient is therefore

$$C_L = \frac{2mg}{(\rho V_t^2 S_{wing})}, \tag{3}$$

BOX 4.2 *Continued.*

where $\rho$ is the air density, and $V_t$ is the true airspeed. Substituting this expression for $C_L$ in Equation (1) makes the induced drag coefficient

$$C_{Dind} = \frac{4(mg)^2}{(\pi\rho^2 B^2 V_t^4 S_{wing})}. \tag{4}$$

The induced drag itself ($D_{ind}$) is

$$D_{ind} = \frac{C_{Dind}\rho V_t^2 S_{wing}}{2}$$

$$= \frac{2(mg)^2}{(\pi\rho B^2 V_t^2)} \tag{5}$$

Since the induced drag for an elliptical lift distribution, given by Equation (5), is the lowest that is possible, the induced drag of a real wing has to be multiplied by an *induced drag factor* ($k$), which is a number that is never less than 1, to account for any deviations from the ideal lift distribution, and from constant downwash velocity along the span:

$$D_{ind} = \frac{2k(mg)^2}{(\pi\rho B^2 V_t^2)}. \tag{6}$$

This formula is used when calculating a bird's glide polar in *Flight* (Chapter 10, Box 10.1) to find the induced drag. The default value assigned to $k$ in gliding is 1.1.

### Induced power from the actuator disc

The starting point for calculating the induced *power* in flapping flight comes from helicopter theory rather than fixed-wing theory. A helicopter rotor sweeps out a circular disc, which can be approximated as an "actuator disc", whose theoretical property is that the air pressure increases in a stepwise manner as the air passes through the disc. This pressure step imparts a downward induced velocity to a tube of air that flows through the disc. The cross-sectional area of this tube is assumed to be that of a circle, whose diameter is the same as the rotor diameter in the case of a helicopter, or the wing span in the case of a bird (even though the wings do not sweep out the full area of the disc). The derivation of the induced power ($P_{ind}$) needed to support the weight ($mg$) is given in Chapter 3, Box 3.1, and the result is

$$P_{ind} = \frac{2(mg)^2}{(V_t\pi B^2 \rho)}. \tag{7}$$

This is for an "ideal" actuator disc which produces a constant induced velocity from edge to edge of its circular area, which is also the condition for minimum induced power to support a given weight. In practice, variations of induced velocity across the disc require an "induced power factor" ($k$), which is a number whose value is never less than 1. The default value of $k$ in *Flight* is 1.2 for flapping flight, and the formula used for induced power in the power curve calculation is

BOX 4.2 *Continued.*

$$P_{\text{ind}} = \frac{2k(mg)^2}{(V_t \pi B^2 \rho)}. \tag{8}$$

The reader who believes that it is more "realistic" to calculate induced power from the vortex wake than from an actuator disc should notice that the induced drag for a fixed wing with an elliptical lift distribution (from Equation 5), when multiplied by the speed to get the corresponding power, gives exactly the same result as the induced power from Equation (8). The lifting line and actuator disc calculations are actually different views of the same theory. The basic assumption in both cases is that the downwash velocity is constant across the wing span.

**The " bow-tie" fallacy**

The advantage of the actuator disc approach for flapping flight is that it requires no information about the wings except the wing span. It does not require that the airspeed "seen" by a point on the wing is the same as the forward speed of the whole aircraft or bird, or that the local airspeed has to be constant across the span, and this permits a massive simplification in considering rotary or flapping wings. The calculation does not even require the angle through which the wings are flapped to be specified, although it does require the "disc area", through which air passes as it is accelerated downwards. If a bird or insect flaps its wings through an angle that is less than the full 180° available to each wing, then the wings sweep out a double sector shaped like a bow tie, and some authors have assumed that this "swept area" should be used instead of the area of the full circle, when calculating the induced power. If this were so, the argument of Chapter 3 Box 3.1 would require that the induced velocity would have to be inversely proportional to the swept area. Extending this line of thought to a fixed wing, in which no area is swept at all, the induced velocity would have to be infinite. However, we know from the lifting-line theory, which has been part of the bedrock of theory for generations of aeronautical engineers, that the induced velocity for a fixed (non-flapping) wing is not infinite, but the same as that for a circular actuator disc, whose diameter is the same as the wing span. Common sense is misleading in this case. The "bow-tie" concept is wrong.

most birds fly most of the time, and for which the calculations are reasonably robust. Speeds lower than $V_{\text{mp}}$ are reserved for unsteady activities, such as landing, taking off and hawking for flying insects, while speeds higher than $V_{\text{mr}}$ may be used, if they are used at all, by predatory birds such as falcons and skuas for pursuing other birds.

Improved understanding of the vortex wakes may make it possible to identify these characteristic speeds with particular types of wake structure, which can themselves be associated with particular values of a variable such as the reduced frequency (Box 4.3) that summarises the wing motion in a single number. If this proves to be feasible, then it may be possible to go further, and predict a value for the induced power factor ($k$) from the reduced frequency. When *Flight* calculates a power curve, it would then calculate the reduced frequency at each speed, and recalculate $k$, instead of using a fixed value for $k$ at all speeds, as it does at present.

BOX 4.3 **Reduced frequency and Strouhal number.**

Wind-tunnel experimenters tend to describe different wake geometries in terms of the forward speed at which they are observed, but obviously the same wake pattern, if it is seen in different species, is not likely to occur at the same speed. To achieve a more general description, wake patterns will need to be connected with some variable that involves the ratio of the wingbeat frequency to the speed. To make such a variable dimensionless, a reference length is also required. Two dimensionless variables, the reduced frequency and the Strouhal number, which involve different reference lengths, have been introduced by theorists for various purposes.

**Variable definitions for this box**

$A$    Wingtip amplitude
$c_m$   Mean chord
$f$    Wingbeat frequency
$f_{red}$  Reduced frequency
$St$   Strouhal number
$V_t$   True airspeed

The "reduced frequency" ($f_{red}$) is defined by Spedding (1992) as

$$f_{red} = \frac{\pi f c_m}{V_t}, \qquad (1)$$

where $f$ is the wingbeat frequency, $c_m$ is the mean chord (ratio of wing area to wing span) and $V_t$ is the true airspeed. It characterises the wake geometry, being equal to the ratio of the distance that the bird travels forwards in one wingbeat cycle to the mean chord. High values of the reduced frequency (high frequency, low speed) indicate rapid changes of flow geometry through the wingbeat cycle, and the likelihood that unsteady aerodynamic effects will need to be considered, whereas a low reduced frequency (low frequency, high speed) indicates that quasi-steady aerodynamics may give a satisfactory account of the flow.

"Strouhal number" ($St$) is a related dimensionless variable. A version of it used by Nudds et al. (2004) is:

$$St = \frac{fA}{V_t}, \qquad (2)$$

BOX 4.3 *Continued.*

where $A$ is the wingtip amplitude, defined as the vertical linear excursion of the wing tip above and below its position with wings level. This is the same as Equation (1), except that $A$ is substituted for $\pi c_m$. Some care is needed with the term "amplitude". The traditional mathematical usage refers to a sine wave, in which the value of some quantity swings in each cycle from zero to $+A$, then down to $-A$, before returning to zero. $A$ is the amplitude, and on this definition, the difference between the positive and negative peaks (the "peak-to-peak swing") is $2A$. However, some authors define the "amplitude" as the peak-to-peak swing, and others neglect to mention what exactly they mean by the term.

The Strouhal number gives an indication of the angle with which the wing tip moves up and down, relative to the flight path, whereas the reduced frequency does not, and this could be seen as an advantage for describing wake geometries. The practical difficulty with the Strouhal number is that the wingtip amplitude is difficult to measure. Bird wing tips are thin and pointed, and are apt to disappear in photographs when seen edge-on. Measuring wingtip amplitude is challenging in the wind tunnel, and impractical in the field, where the observer has no control over the camera geometry. The reader should not be unduly impressed by papers in which large numbers of Strouhal numbers have been calculated from published observations, whose original authors were not paying special attention to measuring wingtip amplitudes, and were not measuring them in a standard way. The reduced frequency is less susceptible to such uncertainties, as it depends on the wingbeat frequency, which is easy to measure from video, and on the standard morphological variables defined in Chapter 1.

## 4.5 ⬭ FEATHERED WINGS

There are two possible reasons why bird wakes might have different characteristics from those of fixed wings. The differences might be inherent in the flapping motion, and in that case one would expect bats to show similar wakes, with distributed vorticity and ill-defined vortex structures. Another possibility is that the feathered surface of birds' wings and bodies is responsible, and if that were the case, the wakes of birds and bats would differ. It has been noted elsewhere in this book that the boundary layer appears to remain attached far more tenaciously over a living bird's body, than over the same body at the same Reynolds number when dead and frozen, or over a smooth-surfaced model of similar shape (Chapter 15, Box 15.4). It appears from the DPIV studies that the wings of small birds are also more resistant to flow separation than model aircraft wings at a similar scale, but not much is known about the proneness or otherwise of bat wings and bodies to boundary-layer separation. Hedenström et al. (2007) have recently studied the wakes of a small nectar-feeding bat by the DPIV technique (above), but their measurements were limited to speeds from the minimum power speed downwards, which were said to be the normal speed range for the species. It may be that Microchiroptera generally are specialised for manoeuvrable flight in the unstable speed range below $V_{mp}$, and that they differ in that respect from the majority of birds. This would complicate comparisons between bird and bat wakes, but be that as it may, two interesting possibilities have been raised. Birds (but not bats) may be able to suppress boundary-layer separation by exploiting some property of the feathered surface, or alternatively, both birds and bats may be able to do this in some way that does not depend on feathers. Either way, new aerodynamic principles are likely to be involved.

# 5

# THE FEATHERED WINGS OF BIRDS

A bird's flight feathers are elaborate structures that could only have evolved to meet the need to resist bending and torsion in a wing of moderate to high aspect ratio, and to transmit aerodynamic forces to the arm skeleton. The shafts of the flight feathers form a distributed spar, arranged in such a way that the bird can vary the span and area of its wings in flight, with automatic adjustment of the cross-sectional shape. The whole wing can be instantly deployed or folded away. The legs are independent of the wing structure.

The physical principles of wings, which were introduced in Chapters 3–4, determine some structural characteristics that must be shared by all wings, whether they belong to animals or machines. Building or evolving a wing is not just a matter of increasing the surface area, as some authors suppose. A structure with a large area may behave as a wing, or as a parachute, or as something in between. It will only work as a wing if the cross-sectional shape is able to create a bound vortex, and if the area is arranged in the right way, that is, sticking out to the sides. This is because the efficiency of a wing is expressed by its lift:drag ratio, and that depends strongly on its *aspect ratio*. The aspect ratio is the ratio of the wing span to the mean chord (Chapter 1, Box 1.2). An efficient wing has to be long (large wing span) and also narrow (not too much chord). The evolution of animal wings, at least in the early stages, is a matter of modifying an existing structure to one that combines an aerofoil cross-sectional shape with an adequate aspect ratio (Chapter 16). Some modern gliders have aspect ratios exceeding 35, which results in extreme performance, especially at low speeds, at the expense of being somewhat unwieldy. The majority of bird and bat wings have aspect ratios in the range 5–12.

# 5.1 ⬤ GENERAL STRUCTURAL REQUIREMENTS

## 5.1.1 WINGS HAVE TO RESIST BENDING AND TORSION

In level flight, a wing develops a lift force that acts upwards, some distance out from the side of the body (Figure 5.1) and therefore exerts an upward bending moment about the wing root. The need for bending strength to resist this determines the maximum aspect ratio that can be achieved with given structural materials and a given type of construction, because making the wing longer and narrower not only increases the moment arm, but also reduces the depth of the wing root, which has to incorporate the structural strength that resists the bending moment. The aspect ratios of bird wings begin at around 5 in the short, stubby (and inefficient) wings of small passerines, and reach a maximum of about 16 in the largest albatrosses.

In addition to creating a bending moment as shown in Figure 5.1, the aerodynamic forces also tend to twist the wing in the nose-down sense. The basic structural requirement of any wing is that it is an elongated structure supported only at the root end, which has to resist large bending and torsional (twisting) moments. The arm skeletons of bats and pterosaurs, which are described in Chapters 6 and 7, look very different from those of non-flying animals, because they are modified to form a "spar", meaning a structure that resists bending and torsion, but is not required to resist compression (as in the legs of ungulates) or tension (as in the arms of apes). The arm skeleton of birds also acts as a spar in the inner part of the wing, but distally the shafts of the flight feathers take over this function. They are made of a more elastic material (keratin rather than bone), and act as a "distributed spar", an arrangement unique to birds (Figure 5.2).

## 5.1.2 THIN-WALLED TUBES AND EULER BUCKLING

Box 5.1 introduces the characteristics needed by a structure that is to resist bending and torsion, in particular the notion of the second moment of area of the cross section, which expresses the way in which

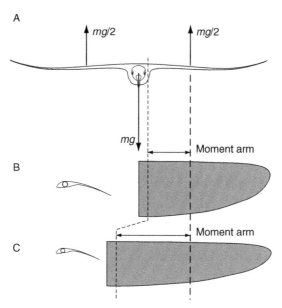

FIGURE 5.1 (A) In steady gliding flight, each wing provides a lift force that balances half the weight, but is displaced outwards from the body, typically by about 20% of the wing span (i.e. 40% of the semi-span, as shown). The moment arm of the lift force about the shoulder joint is the distance between its line of action (thick dashed line) and the axis about which the shoulder joint rotates (thin dashed line). Wing Planform (C) has the same area as Planform (B), but its span is greater by a factor of 1.25, and its chord is less by a factor of 0.80. This amounts to increasing the aspect ratio by a factor of 1.56, which would have a major effect on flight performance. The lift force stays the same, but the moment arm (and also the wing-root bending moment) increases by a factor of 1.25. At the same time the depth of the wing skeleton decreases by a factor of 0.80 (left). The second moment of area of the bone cross section, which has to resist the bending moment, being proportional to the square of the depth, decreases by a factor of 0.64. Reconciling this by allometric changes of wing proportions is the basic difficulty in making high aspect ratio wings.

the material is distributed in a cross-sectional view. Practical methods for making these and other measurements on wings are given in Box 5.2. The swan humerus shown in Figure 5.5 is essentially a bony tube with a thin wall, which means that its second moment of area is large, in relation to the mass of bone that it contains. The wider the tube, the less material is needed to resist given bending and torsional loads. On the other hand, if the wall is too thin, a bending moment may cause the wall to buckle on the compression side. This is called "Euler buckling". It is a common mode of failure of thin-walled structures, and it is due to instability of the wall shape, rather than to any lack of strength in the wall itself. The hollow humerus of Figure 5.5 contains bony struts ("trabeculae") whose

BOX 5.1 **Strength and stiffness of beams.**

A wing is a beam that is supported at one end (the wing root), and loaded with bending and torsional moments, but not with significant amounts of compression or tension. The bending of beams is a classical topic in engineering, which has been presented in a biological context by Wainwright et al. (1976). The strength and stiffness of a beam are two different quantities, both of which depend partly on the geometry of the structure, and partly on the properties of the material from which the beam is made. This box introduces the concepts of strength and stiffness, while Box 5.2 gives practical methods of making the measurements.

**Variable definitions for this box**

| | |
|---|---|
| $A$ | Area of beam cross section |
| $\Delta A$ | Area of thin strip of cross section |
| $E$ | Modulus of elasticity |
| $F_t$ | Tension force |
| $\Delta F_t$ | Tension on strip of cross section |
| $G$ | Shear modulus |
| $H$ | First moment of area |
| $H_{na}$ | First moment of area about neutral axis |
| $I$ | Second moment of area |
| $I_{xx}$ | Second moment of area in the $X$ direction |
| $I_{yy}$ | Second moment of area in the $Y$ direction |
| $J$ | Polar second moment of area |
| $M$ | Applied moment |
| $R$ | Radius of curvature |
| $y$ | Distance above arbitrary baseline |
| $y_{na}$ | Distance of neutral axis above arbitrary baseline |
| $\varepsilon$ | Strain |
| $\sigma$ | Stress |

**Stress, strain and elastic modulus**

*Stress* is force applied per unit area. It has the same dimensions (force/area) as pressure, and the same SI unit, the pascal (Pa), is used to measure it. One pascal is equal to one newton per square metre, and it is an inconveniently small unit. Discussions of mechanical stress usually involve megapascals (MPa) or even gigapascals (GPa). When a stress ($\sigma$) is applied to a sample of some material, the resulting *strain* ($\varepsilon$) is a dimensionless measure of the distortion that the stress causes. The strain is the ratio of the change in the length of the sample to the original length. Thus, if the sample shortens (or lengthens) by 1% when the stress is applied, the strain is 0.01. The ratio of the stress to the strain is the *elastic modulus* ($E$) of the material

$$E = \frac{\sigma}{\varepsilon} \tag{1}$$

Since strain is dimensionless, $E$ has the same dimensions as stress. A "stiff" material is one with a high elastic modulus. A linear or "Hookean" material is one in which the strain is directly proportional to the stress, and $E$ is constant, over some range of stress that is of practical interest. The *strength* of a material

BOX 5.1 *Continued.*

may be variously defined as the stress at which the linear stress/strain relationship breaks down, or at which the material is permanently deformed, or breaks.

### Stress distribution in a bent beam

We consider a beam whose cross-sectional shape is arbitrary, although it remains constant along the length of the beam (Figure 5.2A). One end of the beam is fixed, and a force is applied to the other end, perpendicular to the axis of the beam. The beam is straight when unloaded, but the moment due to the force applied to the tip causes it to bend. This bending moment is greatest at the root, because the moment arm is greatest there, dwindling to nothing at the tip. At some intermediate point, the moment is $M$, and it bends the beam to a radius of curvature $R$ at the same point. The curvature means that the top edge of the beam has lengthened by a small amount and the bottom edge has shortened, but not necessarily by the same amount. The length is unchanged on a two-dimensional surface part-way down the beam, which appears as the *neutral axis* in a cross-sectional view, that is, as a line across the section. Above the neutral axis the material of the beam is subjected to a tensile stress, and below it to a compressive stress.

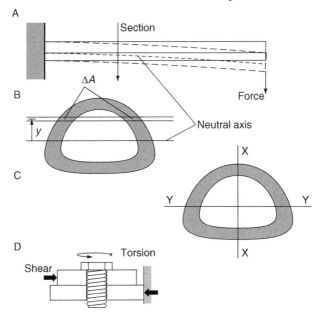

FIGURE 5.2  (A) A beam that is fixed at one end bends when a force is applied to the free end. The upper part of the beam is stretched by tensile stress, and the lower part is compressed by compressive stress. Somewhere in between is a plane where the stress and distortion are zero. (B) In a cross section view this plane is seen as a line, the "neutral axis". The tensile force applied by elements of area above the neutral axis has to balance the compressive force applied by elements of area below it. (C) A neutral axis (YY) drawn horizontally through the section refers to bending in the dorso-ventral ($y$) direction. A second neutral axis (XX) perpendicular to this refers to bending from side to side (the $x$ direction). (D) Shear stress and torsion.

BOX 5.1 *Continued.*

If $y$ is the distance of a point in the beam above the neutral axis (negative if it is below), then the strain at that point due to the curvature is

$$\varepsilon = \frac{y}{R},$$  (2)

and if the elastic modulus of the material is $E$ (and constant), then the stress is

$$\sigma = \frac{Ey}{R}$$  (3)

In the cross-sectional view, the point at position $y$ is seen as part of a thin strip (Figure 5.2B), parallel to the neutral axis (i.e. constant $y$), whose area is $\Delta A$, so that the tension force ($\Delta F_t$) exerted by that strip, being the stress times the area, is

$$\Delta F_t = \frac{Ey\Delta A}{R},$$  (4)

and the total tension ($F_t$) is

$$F_t = \left(\frac{E}{R}\right) \int y \mathrm{d}A,$$  (5)

where the integral is taken from the neutral axis to the top of the beam on the tension side. Likewise, the total compression force on the compression side is the same integral, taken from the bottom of the beam up to the neutral axis. If the beam is in equilibrium, then the tension and compression forces must be equal, and this requirement can be used to locate the position of the neutral axis.

### First and second moments of area

The integral $\int y \mathrm{d}A$ in Equation (5) is called the "first moment of area" of the cross section, denoted here by $H$. If $y$ is measured from an arbitrary baseline, and the height ($y_{na}$) of the neutral axis above this baseline is initially unknown, it can be found by dividing the first moment of area, which has the dimensions of volume, by the total cross-sectional area ($A$) of the cross section:

$$y_{na} = \frac{(\int y \mathrm{d}A)}{A}$$  (6)

This equation can be approximated by a summation, to find the neutral axis on a drawing of a beam cross section (Box 5.2). If the first moment of area is calculated about the neutral axis ($H_{na}$), the result is of course zero, because the contributions from the elements of area above the neutral axis (positive $y$) balance those below the neutral axis (negative $y$). In other words:

$$H_{na} = \int (y - y_{na})\mathrm{d}A = 0$$  (7)

Calculations of beam stiffness require the "second moment of area" ($I$), which is

$$I = \int (y - y_{na})^2 \mathrm{d}A$$  (8)

BOX 5.1 *Continued.*

The position of the neutral axis has to be known before *I* can be calculated. The integral is taken from the bottom to the top of the beam. *I* is always positive because the deviations from the neutral axis are squared. If the cross section were cut from sheet metal of constant thickness and density, *I* would be its moment of inertia about the neutral axis. This is the reason why it is traditionally denoted by the symbol *I*, and sometimes (confusingly) referred to as the "moment of inertia" of the cross section. Actually, the definition of the second moment of area [Equation (8)] does not involve mass or inertia, and it is best to think of it as a purely geometrical quantity.

**The beam equation**

The second moment of area of the cross section appears in the "beam equation", which determines the radius of curvature (*R*) to which the beam will be bent by an applied moment (*M*):

$$R = \frac{EI}{M} \tag{9}$$

The bigger the moment (*M*), the smaller the radius of curvature (*R*), i.e. the more the beam is bent. The product *EI* is called the "flexural stiffness" of the beam. Increasing the flexural stiffness results in a stiffer beam (less deflection for a given moment), and this can be achieved either by making it from a stiffer material (higher *E*), or by redistributing the material so as to increase *I*. The value of *I* depends both on the amount of material in the cross section (which determines its mass per unit length) and also on the way the material is distributed, in terms of its distance from the neutral axis. A beam shaped as a hollow box or tube weighs the same as a solid bar with the same amount of material in its cross section, but is stiffer, because the material is further from the neutral axis, making *I* larger.

**Bending in different directions and twisting**

For a beam with a circular cross section, any diameter of the circle will serve as a neutral axis, and the second moment of area is the same, regardless of the direction in which the beam is bent. This is not true of beams in general, with no restrictions on the cross-sectional shape. In Figure 5.2C, the same cross section as in Figure 5.2B is shown with a second neutral axis, at right angles to the first. The two neutral axes (and any others that may be calculated) intersect at a point called the *centroid* of the cross section, which would also be its centre of mass if the shape were cut from sheet metal. The choice of directions is, of course, arbitrary, but in the case of wing spars (or bird humeri) there is a "primary" direction for bending, which is normally assumed to be the dorsoventral direction. If this is assumed to be upwards in the figure, then the corresponding neutral axis can be marked "YY" to indicate that it refers to bending in the Y direction, and a second one is marked "XX" for bending in the X direction. The corresponding second moments of area about these two axes will generally be different, and can be denoted by $I_{xx}$ and $I_{yy}$ respectively.

While a bending moment applied to a beam results in compression in one part of the cross section and tension in another, twisting results in "shear". This is a different type of stress (Figure 5.2D). If two plates are held together by a screw passing through a hole in both plates, then shear is the

BOX 5.1 *Continued.*

type of stress applied to the screw, by trying to force one plate to slide past the other. In this case the shear stress has the same magnitude and direction over the whole cross section of the screw. If a spanner is used to apply a twisting moment (torque) to the screw, this also results in shear stress, but varying in magnitude and direction over the cross section. The stress is everywhere tangential (perpendicular to the local radius), and increases linearly from the centre of the circle, where the stress is zero.

The stiffness of a beam when subjected to a torsional moment depends on the *polar second moment of area* ($J$), which is simply the sum of the second moments of area, measured in any two directions mutually at right angles, such $I_{xx}$ and $I_{yy}$ above:

$$J = I_{xx} + I_{yy}. \tag{10}$$

The "torsional stiffness" $JG$ determines the angle through which a beam will twist when subjected to a twisting moment or "torque", and is the product of the polar second moment of area and the *shear modulus* ($G$). The shear modulus is the ratio of shear stress to shear strain in the material, and it is analogous to the flexural stiffness $EI$, which determines the response to a bending moment [Equation (9)]. As it refers to a different type of stress, $G$ is not expected to be numerically the same as the elastic modulus ($E$) as defined in Equation (1).

function is to hold the wall in shape, and prevent it from buckling under load. The trabeculae are densest near the ends of the bone, where muscle attachments apply strong local forces that tend to distort the wall, and are sparser in the middle part of the tube (Figures 5.3 and 5.4).

Flight feather shafts are responsible for the bending and torsional stiffness of the outer part of the wing. They are made of keratin, which is a more elastic material than bone, but not necessarily any less strong. They too are cylindrical tubes in cross section, but only at their bases. Further out, where the vanes of the feather begin, the shaft cross section changes to a box shape, in which dorso-ventral bending is resisted by slabs of keratin that form the top and bottom of the box (Figure 5.6). The sides of the box serve to hold these "spar booms" apart, so as to maintain the second moment of area of the whole cross section. The upper spar boom, whose function is to resist compression, has ridges that project into the cavity of the feather and stiffen it against buckling, while the interior of the shaft is filled with a light-weight keratin foam or "parenchyma", which holds the walls in shape without itself carrying significant loads. The side and bottom walls are grooved in a way that allows the shaft some freedom to twist, while resisting bending (Figures 5.5– 5.7).

BOX 5.2 **Structural measurements on wings.**

This box introduces practical methods for measuring the quantities introduced in Box 5.1. Unlike the basic morphological measurements described in Chapter 1, these structural measurements involve cutting up the wings of dead birds. Quantities which are defined as integrals, like moment of inertia and second moment of area, can be approximated in practice by summation, dividing the wing into strips. Kirkpatrick (1990, 1994) surveyed a number of such quantities in birds and bats, and some of his regression equations can be found in Chapter 13, Table 13.3.

**Variable definitions for this box**

$a_{lin}$   Linear acceleration
$a_{ang}$   Angular acceleration
$A_i$    Area of strip $i$
$F$    Force
$H$    First moment of area
$I$    Second moment of area
$I_{xx}$   Second moment of area in the $X$ direction
$I_{yy}$   Second moment of area in the $Y$ direction
$I_w$    Moment of inertia of wing
$J$    Polar second moment of area
$M$    Moment about shoulder joint
$m$    Mass
$m_i$    Mass of strip $i$
$r_i$    Radius of strip $i$ about shoulder joint
$r_m$    Mean radius of lift
$y_i$    Distance of strip $i$ above datum
$y_{na}$   Distance of neutral axis above datum

**Spanwise centre of lift**

The bending moment which the lift force on a wing exerts about a bird's shoulder joint is made up of contributions from elements of area, at different distances from the axis of the joint. This can be approximated by ruling a tracing of the wing (fully extended) into chordwise strips, parallel to the axis of rotation of the joint (Figure 5.3). If the strips are narrow, and the bird is gliding, the lift on each strip can be assumed to act through the quarter-chord point, that is, a quarter of the local chord, measured back from the leading edge. If strip $i$ has an area $A_i$, and the pressure difference across the wing is the same all over, then the strip applies a moment proportional to $A_i r_i$ about the shoulder joint, where $r_i$ is the radius of the strip about the shoulder joint. The total moment $M$ is then proportional to the sum of the contributions of all the strips:

$$M \propto \sum A_i r_i, \tag{1}$$

and the mean radius ($r_m$) can be estimated by dividing this by the total area:

$$r_m = \frac{\sum A_i r_i}{\sum A_i} \tag{2}$$

BOX 5.2 *Continued.*

In the downstroke of flapping flight, the local air speed varies along the wing, because the whole wing has a component of forward speed, and this is combined with a component due to flapping, which increases with the radius. The area of each strip can be weighted by a factor that involves the square of the local airspeed. The effect of this is to move the spanwise centre of lift outwards. The amount of the adjustment depends on the postulated combination of forward speed and angular velocity of the wing,

**Moment of inertia**
According to Newton's Second Law of Motion, a linear force $F$, when applied to a body of mass $m$ results in a linear acceleration $a_{\text{lin}}$ (rate of increase of velocity), where

$$a_{\text{lin}} = \frac{F}{m} \tag{3}$$

The analogous relationship for rotation says that the rate of change of *angular* velocity ($a_{\text{ang}}$) is the ratio of the applied moment ($M$) to the body's *moment of inertia* ($I_{\text{w}}$ for a wing).

$$a_{\text{ang}} = \frac{M}{I_{\text{w}}} \tag{4}$$

The dimensions of $I_{\text{w}}$ are those of force times distance ($\mathbf{M\,L^2\,T^{-2}}$), divided by those of angular acceleration ($\mathbf{T^{-2}}$), that is, $\mathbf{M\,L^2}$. If the body is made up of particles at different distances (radii) from the axis of rotation, each particle's contribution to the moment of inertia is its mass, multiplied by the square of its radius.

When a bird's wing starts on the downstroke, the axis of rotation passes through the head of the humerus. To measure the wing's moment of inertia, a tracing of the wing is first marked into chordwise strips as in Figure 5.3. Each strip contributes to the moment of inertia an amount equal to the mass of the strip times the square of its radius. Measuring the mass of each strip ($m_i$) is a somewhat messy operation, involving a sharp knife or paper guillotine, and a saw if the bird is large. The wing is stretched fully out, and carefully cut into strips, starting at the tip. After each cut, the material that was in the strip, including any fragments of feathers, is collected and weighed. The moment of inertia is then estimated as

$$I_{\text{w}} = \sum m_i r_i^2 \tag{5}$$

**Second moment of area of bone and feather cross sections**
To measure the second moment of area (Box 5.1) of a bird's humerus or of a flight feather shaft, the first step is to cut cleanly through the shaft, and photograph the cut surface, remembering to include a length scale in the picture. The neutral axis can be located on a drawing of a cross section by first superimposing a grid on the cross section as in Figure 5.4. The scale at the top shows that this particular grid has 1 mm×1 mm squares. An arbitrary baseline is drawn below the cross section. For each strip (such as strip number $i$), the vertical distance from the baseline to the middle of the strip

**BOX** 5.2 *Continued.*

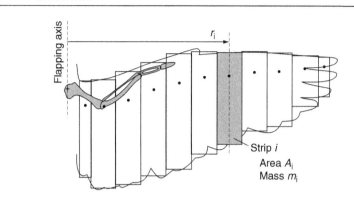

FIGURE 5.3 The spanwise centre of area of a bird's wing can be found by ruling a tracing of the wing into chordwise strips, and measuring the area of each strip ($A_i$) and its radius ($r_i$) about the shoulder joint. The first moment of area is found by adding up contributions equal to the area of each strip times its radius. Dividing this by the total area gives the mean radius. The same diagram can be used to find the moment of inertia about the shoulder joint, but in this case the wing has to be physically cut into strips. Strip number $i$ makes a contribution to the total moment of inertia equal to its mass ($m_i$) times the square of its radius ($r_i$) about the shoulder joint.

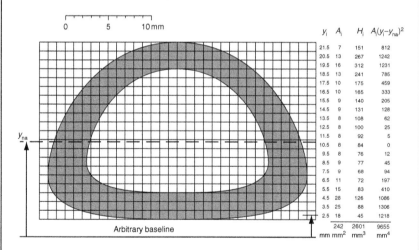

| $y_i$ | $A_i$ | $H_i$ | $A_i(y_i-y_{na})^2$ |
|---|---|---|---|
| 21.5 | 7 | 151 | 812 |
| 20.5 | 13 | 267 | 1242 |
| 19.5 | 16 | 312 | 1231 |
| 18.5 | 13 | 241 | 785 |
| 17.5 | 10 | 175 | 459 |
| 16.5 | 10 | 165 | 333 |
| 15.5 | 9 | 140 | 205 |
| 14.5 | 9 | 131 | 128 |
| 13.5 | 8 | 108 | 62 |
| 12.5 | 8 | 100 | 25 |
| 11.5 | 8 | 92 | 5 |
| 10.5 | 8 | 84 | 0 |
| 9.5 | 8 | 76 | 12 |
| 8.5 | 9 | 77 | 45 |
| 7.5 | 9 | 68 | 94 |
| 6.5 | 11 | 72 | 197 |
| 5.5 | 15 | 83 | 410 |
| 4.5 | 28 | 126 | 1086 |
| 3.5 | 25 | 88 | 1306 |
| 2.5 | 18 | 45 | 1218 |
| | 242 | 2601 | 9655 |
| | mm mm$^2$ | mm$^3$ | mm$^4$ |

FIGURE 5.4 The cross section of Figure 5.2 overlaid by a 1 mm × 1 mm grid. The first column at right is the ordinate ($y_i$) of the middle of each row of squares, above the arbitrary baseline drawn in below. The next column is the area ($A_i$) in each row that is occupied by structural material (grey). The third column is the contribution of each row to the first moment of area relative to the baseline, obtained by multiplying the first column by the second. The total of this column (2601 mm$^3$) when divided by the total area (242 mm$^2$) gives the ordinate of the neutral axis ($y_{na}$) above the baseline. The last column is the contribution of each row to the second moment of area about the neutral axis. The largest contributions come from the rows at the top and bottom of the section, because they are furthest from the neutral axis, and these distances are squared.

BOX 5.2 *Continued.*

$(y_i)$ is measured, and the area $(A_i)$ is measured by counting squares along the strip that are occupied by structural material. The strip then makes a contribution $y_iA_i$ to the first moment of area. Each contribution can be thought of as a volume, obtained by multiplying an area by a length. In Figure 5.4 the sum of the contributions for all the strips is 2592 mm$^3$ (bottom right). This is the first moment of area, about the baseline. The "average $y$" for all the strips can be found by dividing this by the total area (242 mm$^2$), which gives 10.7 mm as the height of the neutral axis above the baseline. To summarise, the first moment of area $(H)$ is

$$H = \sum A_i y_i, \tag{6}$$

and the height of the neutral axis $(y_{na})$ above the arbitrary baseline is

$$y_{na} = \frac{\sum A_i y_i}{\sum A_i} \tag{7}$$

The neutral axis can now be drawn in. The second moment of area, for bending about the neutral axis in the $Y$ direction, can now be calculated from the existing measurements as:

$$I_{yy} = \sum [A_i(y_i - y_{na})^2] \tag{8}$$

If required, another neutral axis, at right angles to the first, can be found and drawn in, as in Figure 5.2C, and a separate second moment of area $(I_{xx})$ can be calculated about that axis for bending in the $X$ direction. The polar second moment of area $(J)$, for torsion, is the sum of the second moments of area in the $X$ and $Y$ directions.

$$J = I_{xx} + I_{yy} \tag{9}$$

**Scaling the second moment of area**

Each of the contributions that makes up the second moment of area in Equation (8) is made up of an area, multiplied by a length squared. Thus, its dimensions are those of length raised to the fourth power ($\mathbf{L}^4$), and the SI unit is the m$^4$. In Figure 5.4 heights above the baseline are measured in mm, and areas in mm$^2$. The estimate for the second moment of area (bottom right) is therefore 9655 mm$^4$. It is best to convert this to the basic SI unit (m$^4$) before attempting to do any calculations with it. The factor to convert mm to m is $10^{-3}$. Therefore, the factor to convert mm$^4$ to m$^4$ is $10^{-12}$, and the estimate for the second moment of area is $9.655 \times 10^{-9}$ m$^4$.

If the cross section can be made into a digital drawing, in which the structural material is shown as black and everything else (including holes in the material) is white, then it is a simple matter to program the above calculations, using a pixel as the unit of area, and the pixel interval (the distance between neighbouring pixels) as the unit of distance. In that case, the second moment of area comes out in units of pixel interval raised to the fourth power. To find the factor to convert these units into m$^4$, first determine the

BOX 5.2 *Continued.*

linear factor to convert pixel intervals to metres *on the original bone*, by referring to the scale that was photographed with the cross section, and included in the drawing. Then raise this linear factor to the fourth power.

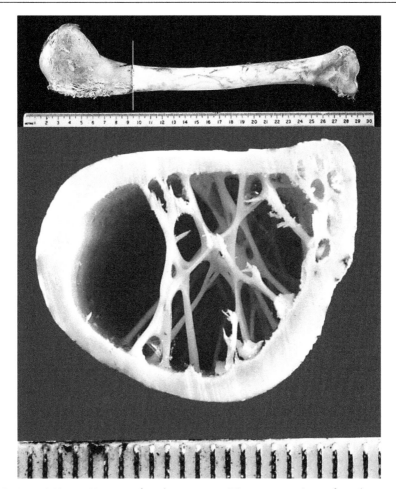

FIGURE 5.5 Upper: Humerus of a whooper swan (*Cygnus cygnus*) seen from the ventral side, with the head (proximal end) to the left, and the anterior side of the bone downwards. The vertical white line marks the outer end of the pectoralis muscle insertion, which runs along the anterior edge of the proximal end of the bone. The centre of rotation is difficult to locate because of the rolling action of the shoulder joint, but is near the left edge of the smooth convex bulge at the left end of the picture. Lower: Cross section through the bone as marked by the white line in the upper picture, looking distally. This is the point of maximum bending and torsional moments carried by the bone in flight. The cavity inside the bone is connected to the inter-clavicular air sac, and is air-filled. Bony trabeculae in the cavity brace the outer wall, and prevent it from buckling under load. This structure is not a truss. The loads are carried by the wall, not by the trabeculae. Photos by C.J. Pennycuick.

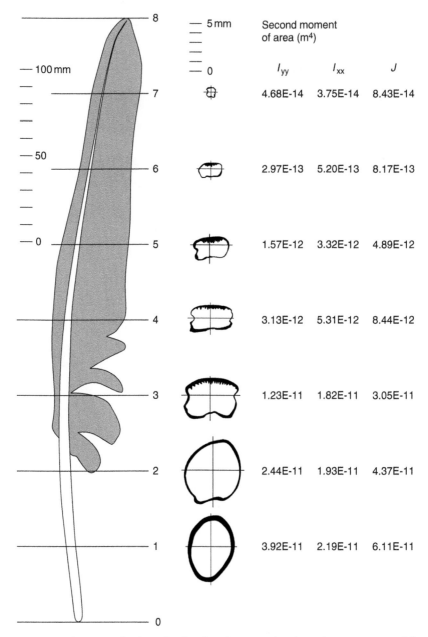

FIGURE 5.6 A primary feather of a Greylag Goose, with enlarged cross sections of the shaft at stations 1–7 shown on the right. Different length scales are shown for the feather as a whole (left) and for the cross sections (right). The solid keratin walls are shown black in the cross sections. The two mutually perpendicular straight lines across each section are the neutral axes about which the second moments of area were calculated (listed at right). $I_{yy}$ is the second moment of area for bending in the dorso-ventral ($y$) direction,

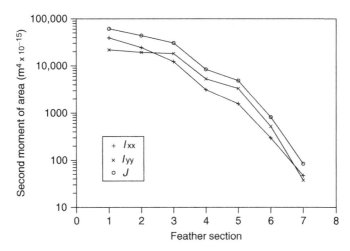

FIGURE 5.7 The three second moments of area (tabulated at the right side of Figure 5.6) decrease as the feather tapers from root to tip. This is because the shaft at any point has to resist the moments applied by the feather area distal to that point. Further out along the feather, the remaining area dwindles, and its mean moment arm shortens. All three second moments of area are, of course, zero at stations 0 and 8, but this cannot be shown on the graph, as a logarithmic $y$ scale has been used.

## 5.2 ⬤ MECHANICS OF THE BIRD WING

### 5.2.1 ADJUSTABLE WING SPAN AND AREA

A bird's wing looks very strange to anyone accustomed to aircraft wings, because the spar has three joints in it (elbow, wrist and metacarpal joint), which are used to adjust both the span and the area of the wing. The elbow and wrist joints are hinges, which allow movement only in one plane. In a gliding bird, the effect of flexing the elbow and carpal joints is to make the planform of the wing into an "M" shape (Figure 5.8). Each flight feather slides over its neighbour on the distal side, and under its neighbour on the proximal side, so that the

---

$I_{xx}$ is for the $x$ direction, and $J$ is the polar second moment of area, for torsion. The hollow interior of the shaft is filled with a foam or "parenchyma", made of thin keratin sheets, whose function is to stabilise the shape of the load-bearing walls, and prevent them from buckling when the feather is subjected to bending and torsional moments. Except at the base of the shaft, which is oval in cross section, the section shape is a modified box, which resists dorso-ventral bending, but allows some twisting. The upper wall of the box is subjected to compressive stress, and has longitudinal ridges, which increase its depth and its resistance to Euler buckling. The posterior vane (right) is wider than the anterior one, which causes the feather to twist in the nose-down sense when loaded, pressing the posterior vane of each feather against the underside of the anterior vane of its neighbour.

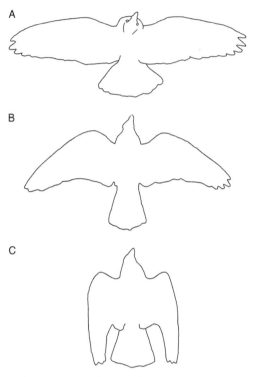

FIGURE 5.8 Tracings from photographs taken by a camera looking perpendicular to the (tilted) air stream, of a pigeon gliding at different speeds in a tilting wind tunnel (from Pennycuick 1968a). (A) Equivalent air speed 8.6 ms$^{-1}$, wing span 0.65 m (near maximum). (B) 12.4 ms$^{-1}$, wing span 0.57 m. (C) 22.1 ms$^{-1}$, wing span 0.25 m. At low and medium speeds the wing area is reduced by approximately the same factor as the wing span, so that the mean chord remains roughly constant, but at very high gliding speeds (C), the hand-wing rotates to be parallel to the air flow, and the mean chord increases. After Pennycuick (1968a).

wing surface closes like a fan. The wing span can be reduced to less than half its full extent, without any effect on the structural stiffness or strength of the wing. As the joints flex, the wing area is reduced by approximately the same factor as the span, meaning that the mean chord stays much the same. The ability to adjust the wing span and area in this way gives birds major performance advantages in both gliding and flapping flight, which may possibly have been shared by pterosaurs, but are not shared by bats (Chapter 6).

A gliding bird can achieve better gliding performance, by adjusting its wing span and area, over a wider range of speed than would otherwise be possible (Chapter 10). In flapping flight, the elbow and wrist joints are flexed and extended during every wingbeat cycle, extending the wing to its maximum span and area during the downstroke, and reducing it during the upstroke. This shortening of the wings during the upstroke is too fast to be obvious to the naked eye, but is plainly visible in television films of flying birds of any kind, where the motion is usually slowed down to make it look steadier. Wing span variation during the wingbeat cycle is implied by the notion of a "constant-circulation wake". It has been argued that this minimises or eliminates energy expenditure in creating transverse vortices at each wing beat, and thus is an adaptation to economical long-distance cruising (Chapter 4). On the other hand, the wing has to continue developing some lift during the upstroke to achieve this, and this means that the pectoralis has to be forcibly extended while still exerting tension, rather than relaxing completely and allowing the lift force to raise the wing passively. This would involve some expenditure of fuel energy, but not enough is known to compare the cost of maintaining a constant-circulation wake with the gain due to eliminating transverse vortices. Observed values of the "span ratio", meaning the ratio of the wing span during the upstroke to that in the downstroke, are compatible with a constant-circulation wake in cormorants, but falcons in level flight reduce the wing span too much in the upstroke (Pennycuick 1989; Pennycuick et al. 1994). Corvids also reduce their wing span during the upstroke to an extreme degree that makes their wing motion read-ily recognisable to the eye, especially jays, which appear to close their wings completely at each wingbeat. This may have some connection with the "bounding" flight seen in smaller species of the same order (Passeriformes), in which a burst of several wingbeats alternates with a ballistic phase with the wings fully closed (Chapter 9, Box 9.1).

Unlike a bat's wing membranes, which are an encumbrance when folded around the body, a bird's wings can be folded instantly by fully flexing the elbow and wrist joints. They fit snugly against the sides of the body, and can be extended just as quickly, ready for instant flight. Pilots of sailplanes and hang-gliders, who are accustomed to rigging and de-rigging their unwieldy wings, are perhaps the only ones who fully appreciate what a remarkable adaptation this is.

## 5.2.2 MECHANICS OF THE WING SKELETON

The wing skeleton starts at the head of the humerus, and extends only about half way to the wing tip. The most proximal element is the humerus, which angles back from the leading edge at the wing root. It is a thin-walled, air-filled tube of bone, a shape that is very effective for resisting bending and torsional moments with a minimum amount of material (Box 5.1), especially as the bone structure is adapted to resist torsion at the submicroscopic level (de Margerie et al. 2005). Its articulation at the shoulder joint has far more freedom of movement than any aircraft wing root. It can rotate through large angles up and down, forward and back, and nose-up and nose-down about its own axis. In flight, the root bending moment is balanced by the downward pull of the pectoralis muscle, inserting on the underside of the deltoid crest, which projects forwards from the proximal end of the shaft (Figure 5.5). The forward position of the insertion also results in a nose-down torsional moment, which balances the nose-up moment caused by the fact that most of the wing's area lies ahead of the axis of the humerus (Figure 5.9).

The bending and torsional moments carried by the humerus are transferred in their entirety to its outer end from the radio-ulna, through the elbow joint. The ulna, which is the main structural element of the "forearm" section of the wing, is also a thin-walled tube of bone, although it is not filled with air. Its characteristic curved shape, with a thin, straight radius between its ends, is the basis of the unique bird system for the

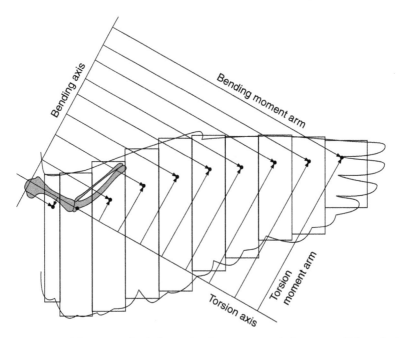

FIGURE 5.9 A gliding wing divided into chordwise strips, with the centre of lift marked at the quarter-chord point of each strip. Nearly all the lift contributions exert a nose-up moment about the torsion axis of the humerus. This moment has to be balanced by the downward pull of the pectoralis muscle, which inserts ahead of the humerus axis. If similar axes were drawn for the radio-ulna, the torsional moment would be predominantly nose-down. The bending moment at the base of the radio-ulna is transmitted through the elbow joint as a torsional moment applied to the outer end of the humerus. After Pennycuick (1967).

control of the wing's profile shape (below). The ulna angles forwards from the elbow joint, and runs obliquely to the leading edge of the wing at the wrist. Depending on the degree of flexure of the elbow joint, the ulna is often roughly perpendicular to the humerus in flight, although the joint typically opens to an angle of about 150 degrees when fully extended. Unlike the shoulder joint, which can move in any direction, the elbow is a hinge joint, which is only free to move in flexion and extension. The radio-ulna cannot rotate up and down, relative to the humerus. Consequently, the bending moment applied by the outer part of the wing to the ulna becomes a torsional moment when it is transferred to the outer end of the humerus. Most of the area of the wing lies behind the extended axis of the radio-ulna, which therefore has to carry a nose-down torsional moment, in addition to the bending moment. These moments come partly from the "hand" wing through the wrist joint, and partly from the secondary flight feathers, which are attached to the rear side of the ulna.

The carpal (wrist) joint connects the outer ends of the radius and ulna with the next section of the bony spar, which consists of two curved meta-carpal bones, fused together at their ends to form a single unit, with a space between them in the middle. When the outer part of the wing is producing an upward force, the carpal joint behaves as a hinge joint, allowing rotation in only one plane. The hinge axis is not parallel to that of the elbow joint, but tilted so that the hand-wing swings in a plane that is inclined nose-up relative to the plane of the humerus and radio-ulna. The joint unlocks if the wrist joint is flexed, and the outer (hand) part of the wing can drop if there is no aerodynamic force to hold it up. Beyond the metacarpals there are two small bony elements, representing an inde-terminate number of fused phalanges. These have only a small amount of fore-and-aft movement relative to the metacarpals.

The outer end of the second of these bones is the end of the bony skele-ton of the wing, but it is nowhere near the wing tip. In most birds, the skele-ton extends only about half way from the shoulder to the wing tip. The shafts of the flight feathers radiate outwards and backwards from the wing skeleton, forming a "distributed spar", which collects the aerodynamic forces from the whole wing surface, and transfers it in the form of bending and torsional moments to the wing skeleton. This arrangement is unique to birds. Nothing like it is found in any other animal or machine.

### 5.2.3 MECHANICS OF FLIGHT FEATHERS

Most of the area of the wing is supported by primary flight feathers, which radiate from the hand skeleton, and secondary flight feathers, which are attached to the rear side of the ulna. Each flight feather has a shaft or "rhachis", which acts as an independent spar, made of keratin rather than bone. The lifting area consists of two vanes on either side of the rhachis, which are made from closely packed "barbs", each of which is a thin, tapered cantilever beam, supported at its base by the rhachis. Feathers are dead structures made of keratin. The base of each feather is surrounded by a follicle, which is a pocket in the skin containing the chemical machinery that creates a new feather. When the feather is fully formed, its base remains in the living follicle, which then provides the mechanical anchorage that holds the feather in the skin. In the case of flight feathers, the follicle is reinforced with connective tissue to form a robust socket, whose function is to transfer the bending and torsional moments at the base of the feather to the underlying skeleton. Each feather is connected to the wing skeleton at one point, the follicle at its base, unlike the wing membrane of a bat, which has to be stretched between two skeletal supports.

The function of the feather shaft is to resist deformation due to the bending and torsional moments set up by the aerodynamic force acting on the vane, and to transmit these moments to the skeleton. The cross section of the shaft at any particular point has to resist the moments generated by forces acting *further out* along the feather. At a point near the tip, such as Section 7 in Figure 5.6, the cross section is very small, because there is only a small amount of vane area further out, and the forces acting on it have only a short moment arm about that point on the shaft. On the other hand Section 1 has to resist the entire force acting on the whole of the feather, and the bending moment arm, averaged for the whole of the vane, is a substantial fraction of the length of the feather. Still nearer the base, where the shaft enters the follicle, the moments are progressively transferred to the skeleton, and the cross section dwindles, eventually to nothing. The second moments of area (Box 5.2) are shown in Figure 5.6, and plotted on a logarithmic scale in Figure 5.7. They decrease gradually at first, and then ever more steeply, through three orders of magnitude from Section 1 to Section 7 (Figure 5.10).

## 5.2.4 RELATION OF PRIMARY FEATHERS TO THE SKELETON

When a bird's wing is extended, the primaries and secondaries spread out like a fan, and appear to form one continuous series, but actually these two sets of flight feathers transmit forces and moments to the skeleton in different ways (Figure 5.10). The follicles of the innermost primaries are bound by robust connective tissue to the metacarpal unit

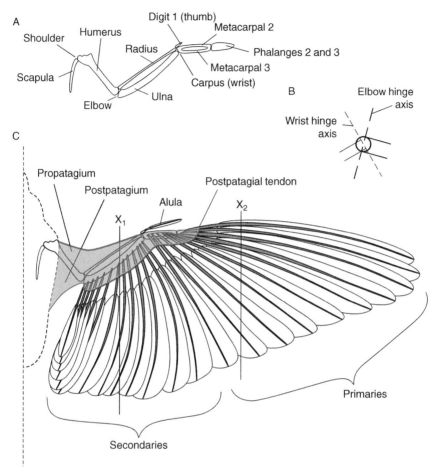

FIGURE 5.10 (A) A bird's wing skeleton (also shown in C) collects all of the aerodynamic force generated by the wing, although it extends less than half the distance from the shoulder joint to the wing tip. (B) Looking along the axis of the ulna (thick circle), the hinge axis of the wrist joint is inclined at an angle of 40–45 degrees relative to the hinge axis of the elbow joint, in the nose-down sense. (C) The area coloured pink is occupied by patagia (skin membranes), which are supported by the skeleton. Patagia constitute the entire wing area of bats and pterosaurs, but provide only a small fraction of the wing area in birds. Most of the wing area is made up of "flight feathers" (remiges), which are divided into "primary" feathers attached to the hand skeleton, and "secondary" feathers attached to the ulna. The gap between the innermost secondary feather and the body is faired by "scapular" feathers (not shown). Although the scapulars are similar in structure to flight feathers, their follicles are not attached to the skeleton, but are embedded in the postpatagial membrane. The postpatagial tendon (red line) runs along the upper side of the primary and secondary follicles, connecting them together, and controlling the fanwise spreading of the flight feathers when the elbow and wrist joints are extended. Because of the curvature of the ulna, this tendon also increases the camber of the wing by forcing the secondaries downwards, when the joints are fully extended. The cross sections $X_1$ and $X_2$ are shown in Figure 5.11.

(two fused bones), while those further towards the wing tip are similarly bound to one or other of the two phalanges. In flight, each primary feather tries to bend upwards, and to twist in the nose-down sense, and it transmits a bending and a twisting moment via its follicle to the underlying bone. The moments contributed by all the primaries are collected at the base of the metacarpal unit, and thence transmitted via the wrist joint to the outer end of the radio-ulna. None of the primary feathers have any freedom of movement relative to the bones to which they are attached. The joints between the metacarpals and the phalanges allow a small amount of movement, but this is constrained to fanwise spreading, in the plane of the wing. Birds do not have any means of moving primary feathers up or down relative to the bones, or of moving individual primaries in any direction relative to their neighbours.

## 5.2.5 RELATION OF SECONDARY FEATHERS TO THE SKELETON

Secondary flight feathers are mechanically similar to primaries, but the way in which they transmit moments to the wing skeleton is different. Like primaries, their bases are supported by reinforced follicles, but instead of being rigidly attached to the ulna, these are bound by connective tissue to the rear side of the ulna. In many birds there are regularly spaced "tubercles" (bumps) on the ulna, marking the points to which the secondary follicles are attached. The attachment is strong but flexible, so that the secondaries can hinge up and down, relative to the ulna. To transmit their bending moments, they depend on a basal tendon, which connects the outer ends of all the primary and secondary follicles together (Figure 5.10). The tendon is anchored at its inner end to the ulna at the elbow joint, and at the other end to the outer phalanx bone. Its action depends on the unique curved shape of the bird ulna, to which the bases of the secondary follicles are attached in an arc. As the wrist joint is extended, the outer end of the tendon is pulled outwards, so straightening the tendon as it curves around behind the ulna. The hinge axis of the wrist joint is at a "nose-up" angle relative to that of the elbow, so that the basal tendon pulls the secondary follicles downwards. This arrangement automatically adjusts the "camber" of the inner part of the wing, between the elbow and wrist joints, meaning the curvature of the wing cross section. When the elbow and wrist joints are fully extended, the flight feathers not only spread out to their maximum area, but the secondaries are also pulled downwards by the basal tendon, so increasing the camber. This is a "high-lift" configu-ration, seen during the downstroke of flapping flight at low speeds, and when gliding at minimum speed or landing. At higher gliding speeds, or when lift is reduced in the upstroke of flapping flight, the elbow and wrist joints are flexed, so reducing the wing span and area, and also allowing the cross section to take up a flatter (less cambered) shape (Figure 5.11).

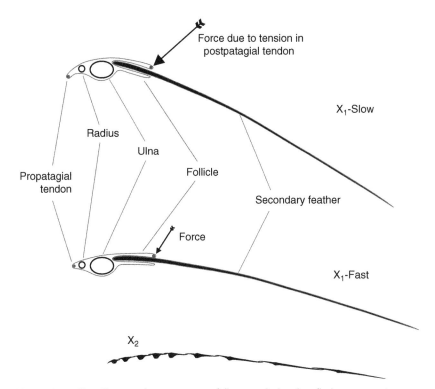

FIGURE 5.11 The elbow and wrist joints are fully extended in slow flight, causing the propa-tagial tendon to pull downwards, extending and drooping the leading edge in Section $X_1$ (from Figure 5.10). The same movement tightens the postpatagial tendon resulting in a down-ward force on the outer ends of the secondary feather follicles. This droops the follicles, which are flexibly attached to the ulna, and increases the camber of the inner part of the wing. In fast flight, flexing of the elbow and wrist joints relaxes the tension in the postpatagial tendon, and allows the secondary follicles to hinge upwards, so flattening the camber. The profile of the hand wing (Section $X_2$ from Figure 5.10) consists of a single layer of primary feathers, whose follicles are rigidly attached to the metacarpals and phalanges. The feather shafts form a distributed spar. Stiff barbs attached to each side of each shaft form the anterior and posterior vanes of the each feather. Each vane overlaps the neighbouring feathers on either side, and can slide over them as the joints are flexed and extended, so varying the wing area.

### 5.2.6 PATAGIAL MEMBRANES

The cross section of the inner part of the wing is also partly determined by the small triangular patagium (wing membrane) ahead of the elbow joint. The leading edge of this is supported by the patagial tendon, which has an elastic section in the middle, and is attached to the radius near the wrist at its outer end, while its inner end is tensioned by a slip of the deltoid muscle, coming off the shoulder. When the elbow joint is fully extended, this "pro-patagium" is pulled into a curved shape, concave on the underside. This effectively droops the leading edge of the wing ahead of the elbow joint, complementing the downward deflection of the secondaries to the rear, caused by tightening of the basal tendon.

There is also a "post-patagium", which bridges the angle between the elbow joint and the side of the body, behind the humerus. This supports a number of large feathers that appear mechanically similar to flight feathers, but they are actually contour feathers, not flight feathers, because their follicles are embedded in the skin of the post-patagium, without any direct mechanical connection to any part of the skeleton. When it is extended, the post-patagium, and the feathers it supports, form a "wing-root fairing" which connects the wing surface aerodynamically to the side of the body.

## 5.3 ◦ FLAPPING THE WINGS

### 5.3.1 THE DOWNSTROKE IN FLAPPING FLIGHT

A bird's shoulder joint is very different from any aircraft's wing root, in that the head of the humerus (corresponding to the root end of the main spar) can rotate through large angles up or down, and forward or back, as well as being able to rotate either way around its own axis. The lift force on a gliding bird's wing acts upwards, far out from the body, creating an upward bending moment about the shoulder joint, which has to be resisted in order to prevent the wings from "clapping hands" above the animal's back. In gliding flight, this lift moment is balanced by a downward moment, produced by a steady tension force in the pectoralis muscle, pulling downwards on its attachment on the humerus, a short way out from the shoulder joint. Using a muscle rather than a passive structure to maintain a steady force or moment requires some expenditure of metabolic power, even though the muscle does no work so long as it maintains a constant length (Chapter 7, Box 7.2). If the muscle shortens, it does work. In that case the muscle becomes an "engine" that converts some fraction of the fuel energy consumed into work. In level flight, the downstroke is the power stroke. The amount of work done in each wingbeat, by the right and left pectoralis muscles together, must account for the aerodynamic work done in the same period against the drag of the wings and body. The mechanical *power* is the work done in each downstroke by both pectoralis muscles, multiplied by the wingbeat frequency, and this has to equal or exceed the aerodynamic power, calculated in Chapter 3. This in turn is the main component of the metabolic (or chemical) power in prolonged flight, which is the rate of consumption of fuel energy.

The pectoralis is a tapered muscle. At its "origin" or inner end, the muscle fibres exert a force which is distributed over a large area of bone, mostly on the expanded sternum, but also on the coracoid and clavicle. The fibres do not run the full length of the muscle, but progressively transfer their force to connective tissue elements within the muscle. The muscle "attachment" on the humerus consists entirely of tendinous connective tissue, and covers a much smaller area than the origin, because the stress that can be transmitted by a passive tendon is about 200 times greater than the isometric stress that a muscle fibre can exert. Even so, the insertion is spread along the bone so that its outer end is typically twice as far (or more) from the centre of rotation of the shoulder joint than the inner end, and therefore the insertion at that end moves twice as far when the humerus rotates through a given angle.

## 5.3.2 THE UPSTROKE IN FLAPPING FLIGHT

The wing is raised by the supracoracoideus muscle, which is much smaller than the pectoralis in most birds, and arranged in a manner that is unique to birds. The muscle originates along the sternum, where it is completely covered by the pectoralis. Its tendon runs up beside the coracoid, through the *foramen triosseum* where the coracoid, clavicle and scapula meet, to its insertion on the top side of the humerus. In most flight regimes the upstroke is a recovery stroke, whose function is to position the wing, ready for the next downstroke. In fast flight, the aerodynamic force on the wing raises it passively, but some work may be required from the supracoracoideus in very slow flight, below the minimum power speed. Hummingbirds differ from other birds in that they are specialised for hovering, which they do with the body in an upright posture, sweeping the wings forward and back, rather than up and down. During the "upstroke" (which should perhaps be more accurately named a "backstroke"), the wings are rotated at the shoulder so that what is normally the upper surface faces downwards, and the wings develop inverted lift, directed upwards. The upstroke and downstroke are both power strokes in a hovering hummingbird, whereas in other birds that are able to hover (flycatchers, kingfishers) the upstroke is a complex "flick" movement, whose function is to extract the wing from the vortex ring that has been created by the downstroke, and position it for the next downstroke. The supracoracoideus is relatively much larger in hummingbirds than in other birds, although still not as large as the pectoralis.

## 5.4 ▬ THE REST OF THE SKELETON

### 5.4.1 THE AXIAL SKELETON

The bird skeleton (Figure 5.12) is based on that of bipedal saurischian dinosaurs, but apart from the wings, it has some other unique modifications, which set birds apart from other vertebrates. Dinosaurs stood and walked in an oblique head-up posture, with a long tail that approximately balanced the body weight about the hip joint. The visible tail of birds is a fan of feathers which weigh very little, and although birds stand in a posture that is much the same as that of dinosaurs, their tail skeleton is reduced to a size that is far too small to act as a counterweight. There is no way that birds can balance their weight about their hips, because the hip joint is too far back, and the body leans forwards from it. The head-down moment about the hip joint is instead balanced by postural muscles that pull downwards on the rear end of the pelvis, so levering the front end up.

The pelvic lever is the *synsacrum*. The saurischian pelvis is a ring made up of three bones on each side. The ilium, and behind it the ischium, articulate with the sacral vertebrae, while the two pubic bones join together ventrally, as they do in our own skeletons. In birds, the ilium and ischium are fused to the vertebrae rather than articulating with them, and they are hugely expanded. The ilium is a long plate extending forwards as far as the base of the neck, with lumbar and some thoracic vertebrae fused to it, while the ischium is an expanded curved plate behind the hips, also fused to the vertebrae, and to the ilium. The two pubes do not join ventrally, but are elongate elements fused to the rest of the synsacrum, and lying below the edge of the ischium. Although the whole structure is thin and light, its curvature gives it a large second moment of area in a section through the hip region, making it a stiff longitudinal lever (Box 5.1). It is also a curved plate with the concave side downwards, and in a cross section through the bird's body, it faces the oppositely curved dorsal plate of the sternum (Figure 5.12C). Intercostal muscles flex and extend the joints between the ribs and ventral ribs, so moving the sternum up and down

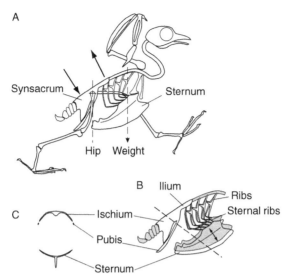

FIGURE 5.12 (A) The weight of a standing or running bird's body is supported at the hip joint, but as there is no balancing tail, the body's centre of mass is well ahead of the hip. The bird has to hold the front end of the body up, by pulling downwards on the rear end of the synsacrum (arrows). The synsacrum is shaped to resist longitudinal bending, and acts as a "pelvic lever". (B) The synsacrum consists of the three pelvic bones (ilium, ischium and pubis) together with the sacral and lumbar vertebrae and some thoracic vertebrae, all fused together into a single rigid unit. It is connected to the sternum by the ribs and sacral ribs. Air is pumped in and out of the respiratory system by intercostal muscles, which flex and extend the joints between the ribs and the sternal ribs, so moving the sternum up and down relative to the synsacrum. (C) Section along the dashed line in B. The synsacrum and sternum operate as the two halves of a bellows in ventilation. The pubic bones of the two sides do not join ventrally, as they do in most other tetrapods. The synsacrum is a thin sheet of bone that is curled downwards at the sides, forming a light channel girder that is very resistant to longitudinal bending.

relative to the synsacrum. A bird breathes by this bellows action, not with a diaphragm as mammals do.

## 5.4.2 THE KEELED STERNUM

Birds differ from bats and pterosaurs in having a prominent keel on the sternum, on which both the pectoralis and supracoracoideus muscles originate. This serves no obvious function as regards the mechanics of flapping the wings. Bats have no such keel. Their pectoralis muscles pull against one another in the ventral midline, with no bony septum between them, and since both always contract together, this causes no problems. There is no way that a bird or a bat can flap one wing and not the other. That would make the body roll, without doing any work on the air. The function of the keel must be something that applies to birds but not to bats. The clue is in the unusual arrangement of the avian respiratory system, which is radically different from that of any mammal, including bats. More details about this are in Chapter 7, Box 7.6. In birds (only), a system of thin-walled, non-respiratory "air sacs" draws air right through the actual lung, which is a small and compact structure. The inter-clavicular air sac is the one that sends branches into the cavities inside the hollow humeri of birds, and it also sends channels inside the keel of the sternum. These in turn send branches into the pectoralis muscle itself, lying parallel to the muscle fibres (Figure 5.13).

Everybody knows that birds have "pneumatic bones", but it is not so widely known that their pectoralis muscles are also pneumatic. The air channels can sometimes be demonstrated in dissections in which coloured latex is vacuum-injected through the trachaea of a pigeon, although the flexibility of the surrounding muscle makes the penetration difficult to achieve. When the pectoralis is dissected away from its origin on the keel in an ordinary dissection, the cut surface has an

FIGURE 5.13 Dissection of a wandering albatross (*Diomedea exulans*), seen from the ventral side. The keel of the sternum runs across the bottom of the picture, and the bird's head is to the left. The ventral edge of the pectoralis muscle has been detached from its origin on the side of the keel, and reflected out to the bird's left side, revealing conspicuous cavities inside the muscle. It is sometimes possible to demonstrate corresponding cavities in a latex-injected pigeon dissection, and they are then seen to branch off the inter-clavicular air sac. Photo by C.J. Pennycuick.

obvious rough texture, not seen in transverse cuts of other muscles. In some large birds such as albatrosses, the cut surface contains plainly visible channels (Figure 5.13). The most likely function of these air channels is as sites of evaporative cooling, allowing heat to be extracted directly from the muscle fibres during flapping flight. Without the keeled sternum to hold the muscles away from the ventral side of the body, the air channels would close up when the muscles contract, so preventing water vapour from escaping from the interior of the muscle. No such arrangement is needed in bats, because they can dispose of large amounts of heat by passing blood through the wing membrane, which presents a vast area to the air stream, and makes a highly effective convective cooler. Birds have a much smaller area that can be used for convective cooling of the blood, limited to the undersides of the forearms, and the sides of the body that are covered when the wings are folded.

## 5.5 ⬤ ADAPTATIONS FOR GLIDING

"Gliding" is flight on outstretched wings, without flapping, something that most birds do for some of the time. Some glide nearly all the time, resorting to flapping flight only at moments of dire necessity. A gliding bird loses height or speed or both, relative to the air, but these losses can be offset by "soaring" behaviour, in which the bird extracts useful energy from motions of the atmosphere (Chapter 10). A bird that glides for most of the time has to support its weight by exerting a steady downward moment on its humeri with its pectoralis muscles. There are two ways to reduce the metabolic power required to maintain this moment, first by moving the muscle insertion further out from the shoulder joint, which reduces the force required, second by using a "slow" muscle, meaning one with a low maximum strain rate. The metabolic power required to maintain a given force is proportional to the maximum strain rate, which is a property of the biochemistry of the particular muscle (Chapter 7, Box 7.2). The pectoralis muscles of Old World vultures have two distinct parts, a large superficial part which is deep red in colour, and a smaller deep part which is pale pink. The colour difference indicates that the superficial part is adapted to maintain a higher level of aerobic specific power than the deep part. The likely interpretation is that the superficial part is a "fast" muscle whose maximum strain rate is matched to flapping the wings, while the deep part is a slower muscle, used to maintain tension economically in gliding flight. The ultimate "slow muscle" is a tendon, which cannot shorten at all. Having a maximum strain rate of zero, a tendon can maintain tension indefinitely at no metabolic cost. Albatrosses are the only birds known to exploit this, with a wing lock based on a tendinous sheet in parallel with the pectoralis muscle. This is arranged in such a way that it prevents the wing from rising above the horizontal position when the humerus is protracted fully forward, but unlocks to allow flapping flight when the humerus is retracted (swung back) by a small amount.

As to the wing morphology of soaring birds, it is an interesting oddity that only albatrosses and frigatebirds show any resemblance to the planform shapes of sailplanes, and even they have modest aspect ratios of 14 to 16. Most birds that are specialised for soaring in thermals, such as vultures, eagles, cranes, storks and pelicans, have almost rectangular wings with aspect ratios of 8 to 10, but they also have characteristic slotted wing tips, due to "emarginated" distal primary feathers. This means that, instead of tapering smoothly like the goose primary feather of Figure 5.6, the distal primaries of these soaring birds show an abrupt narrowing of both the anterior and posterior vanes, some distance from the tip, which results in the primaries separating into a series of "winglets" at the wing tip. The flexibilities of the feather shafts are graduated, so that the first feather bends in gliding flight until its tip points almost straight up, and subsequent feathers bend less, resulting in a cascade of up to 6 feathers around the wing tip, each one of which is in the downwash of the feather in front of it. It would appear that this arrangement modifies the flow around the wing tip, presumably by moving the trailing vortices outwards, so making the wing perform as though its span were greater than it really is. The low aspect ratios of wings of this type are likely to be an adaptation for take-off performance and manoeuvrability, rather than for gliding performance.

# 6

# THE MEMBRANE WINGS OF BATS AND PTEROSAURS

Unlike a wing made of feathers, one that evolved by extending a lateral patagium, like that of flying squirrels, has to be tensioned between two or more skeletal members. The diversity of bats is much less than that of birds, because of their less versatile wings, whose structure also constrains the evolution of the legs. The wings of pterosaurs also involved the legs, but may have had an elastic membrane that allowed control of span and area, with a degree of versatility nearer to birds than to bats.

This book is primarily about birds, but any animal that flies has to overcome the same mechanical problems, in the process of transforming its ancestral limb structure into a pair of wings. The other two groups of flying vertebrates, bats and pterosaurs, started from the same basic tetrapod limb structure as birds, but evolved wings in which the surface area is provided by a *patagium*. This is essentially a double layer of skin, which has no bending stiffness in itself, and has to be stretched out like a hang glider's sail by a skeletal frame, rather than being supported at one end only, as sailplane wings and flight feathers are. The diversity of both groups has been restricted in comparison

Modelling the Flying Bird

**135**

with that of birds, as a direct result of the mechanical basis of the patagium, in ways that can be observed in bats, and inferred in pterosaurs. Beyond the limitations inherent in patagial wings, the two groups are very different.

## 6.1 ⬭ BATS

Bats (Chiroptera) are a widespread and highly successful order of mammals, with more living species (over 1000) than any other order. The earliest known bat fossils are from the Eocene period, some millions of years after the last pterosaur died at the end of the Cretaceous. The few survivors of the catastrophe that ended the Mesozoic Era included the ancestors of modern birds and mammals, and most of the modern orders of both groups are first known from Eocene fossils. Bats are similar in size and mass to small and medium-sized birds, but there are no goose-sized or larger bats. Like birds, they originated from ancestors that did not fly, and modified their original anatomy so as to fulfil the requirements for flight. They have wing spans and aspect ratios in the same range as birds of similar size, and the *Flight* programme, which only requires that information, will calculate their flight performance without distinguishing between them and birds.

The physical problems of evolving wings are the same for birds and bats (Chapter 3) but the solutions that the two groups have evolved are different in almost every respect. Bats are excluded from a vast range of ecological niches in which birds use their legs for walking, perching, running, swimming and catching prey, because the leg is a primary element of the wing structure in bats, whereas in birds it is not. Most bats find their way around and locate their prey by echolocation rather than vision, which makes them pre-eminent as nocturnal aerial insectivores, but not so good at other forms of predation. There are bats that eat other bats, and one bat species (*Nyctalus lasiopterus*) is believed to prey on nocturnally migrating songbirds, on the basis of feathers in its droppings at migration time. On the other hand two entire orders of birds (raptors and owls) are specialised as predators, including many raptor species which catch birds and bats in flight. Likewise, whole orders of birds are specialised for living and hunting in the water (Chapter 12), whereas there are no true water bats. Caribbean fishing bats can detect ripples on the water surface caused by a fish swimming just below, and catch the fish by dipping their hooked hind claws in the water, but no bat can swim around under water like a cormorant in pursuit of fish, or plunge-dive like a kingfisher. The mammalian method of reproduction requires bat mothers to carry

FIGURE 6.1 A female *Rousettus aegyptiacus* carrying a baby in flight. This is a small fruit bat (Megachiroptera) with a mass of about 120 g and a wing span of about 0.5 m. The plagiopatagial muscles can be seen on the left wing. The soles of the feet point forwards because of rotation of the leg at the hip joint. The ankle joints are deflected downwards to produce a downward curl at the trailing edge of the plagiopatagium. Unusually for a fruit bat, this species uses a primitive form of echolocation for obstacle avoidance. The lips are drawn back to emit clicks that are audible to the human ear. Photo by C.J. Pennycuick.

embryos and babies in flight (Figure 6.1), rather than laying eggs in a nest, and the limitations of mammalian lungs exclude bats from high-altitude flight (Chapter 7, Box 7.7). Birds are seen over the polar ice fields, but bats are not. Some bats migrate over land in short stages of a few tens or hundreds of kilometres, but no bat flies non-stop for thousands of kilometres over ocean or desert, as many bird species do. The abilities and limitations of bats begin with the mechanical principles of their wings.

## 6.1.1 MECHANICS OF THE BAT WING

Whereas birds have a pure cantilever wing, in which a stiff structure delivers all the bending and torsional loads to the shoulder joint, the wing membrane of a bat is flexible, with no resistance to bending or torsion. The only type of stress that the membrane can resist is tension. It has to be stretched between two stiff bony supports, which pull outwards at opposite edges. If the membrane were flat, it would only pull on the bony framework in the plane of the membrane, and would not

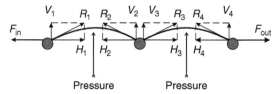

FIGURE 6.2 A bat's wing membrane can only resist tension, not bending or torsion. It has to be stretched between two or more skeletal supports (grey). The membrane bulges towards the low-pressure side of the wing, so that it pulls on the supports at an angle to the plane of the wing. The middle support is pulled by a force $R_2$ by the membrane on its left, and by a force $R_3$ by the membrane on its right. The horizontal components of these two forces ($H_2$ and $H_3$) cancel, while the vertical components ($V_2$ and $V_3$) add together, and contribute to the aerodynamic force on the wing. The vertical components on the outer supports ($V_1$ and $V_4$) also add to the aerodynamic force, while the horizontal components $H_1$ and $H_4$ have to be balanced by forces $F_{in}$ and $F_{out}$, applied by the supports. These outward forces are necessary to "tension" the wing.

exert any force perpendicular to that plane (lift). It works as a wing because the membrane bulges when excess air pressure is applied to one side of it (Figure 6.2). At every point around the edges of the membrane, where it attaches to the skeleton, it exerts a large component of force pulling inwards, which is balanced by an opposing force at the opposite edge, and a smaller component, which is unbalanced, perpendicular to the wing surface. The sum of these unbalanced components makes the aerodynamic force on the wing, which is then resolved into drag (parallel to the incident air flow) and lift (perpendicular to the incident air flow). As always, the measure of the wing's efficiency is the ratio of lift to drag. The skeletal supports have to resist the unbalanced forces that translate into the aerodynamic force on the wing, as they do in a bird's wing, and in addition, they have to provide the tension in the membrane by pulling against one another.

Figure 6.3 shows the main structural components of a bat's skeleton, and the nomenclature of different parts of the wing membrane from Norberg's (1972a) account of *Rousettus aegyptiacus*. This is a small member of the suborder Megachiroptera (fruit bats), but the same description of the main wing components also applies to the other suborder (Microchiroptera, insectivorous bats), which includes the majority of bat species. From the shoulder to the wrist, a bat's wing skeleton is similar to that of a bird, except that, as usual in mammals, the radius rather than the ulna is the main structural element of the forearm.

The way in which aerodynamic forces are developed by the different panels of a bat's wing, and transferred to the skeleton, was analysed by

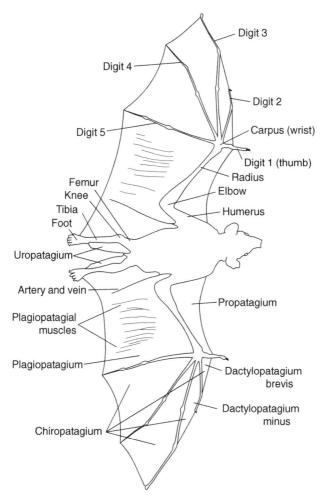

FIGURE 6.3 Nomenclature of the parts of a fruit-bat's wings, according to Norberg (1972a).

Norberg (1972b). Outboard of the wrist, where the bones of a bird's hand skeleton are reduced and thickened, those of a bat are hugely elongated and slender. The five elongated metacarpals radiate from the wrist joint, and each digit continues with three or four similarly elongated phalanges. Digit 1 (the thumb) points forwards and supports a drooped leading edge in flight, as well as being used for clambering, while Digits 2–5 support the wing surface. Digit 3 runs to the wing tip, and is augmented by the shorter Digit 2, ahead of it, to make the rhomboidal "Norberg panel", described by Norberg (1969) and shown in Figure 6.4. This is a characteristic feature of the wings of all bats, that

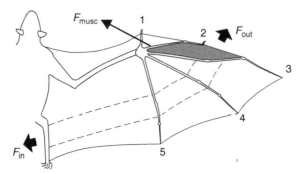

FIGURE 6.4 A bats's wing is tensioned by the *musculus extensor carpi radialis longus*, which exerts a force ($F_{musc}$) on the anterior side of the base of the second metacarpal. The Norberg panel (grey) is the rhomboid-shaped unit formed by the second and third digits, and the dactylopatagium minus between them. This is stiff in the plane of the wing, and transmits the pull to the membrane between Digits 3 and 4. Norberg (1969) explains in detail how this works. The forward pull due to the muscle rotating Digits 2 and 3 is eventually balanced by an inward pull exerted by the leg. The short arrows correspond to the forces $F_{in}$ and $F_{out}$ in Figure 6.2 The tension path between them (dashed lines) changes direction as it pass through Digits 4 and 5, which are held in compression by the pull of the membrane.

is light but stiff in the plane of the wing. It allows the hand-wing to be pulled forwards against a strong drag moment, but does not require the phalanges to be thick and heavy, as they are in a pterosaur's wing-finger (below). A hypothetical Norberg panel probably also formed an essential part of the wing of the ancestors of birds, up to and including *Archaeopteryx*, although its function has been taken over in modern birds by the fused metacarpals (Chapter 16).

The last two fingers (Digits 4 and 5) run through the membrane from the wrist to the trailing edge, and perform two distinct functions. The first is to resist the bending moment caused by the pull of the membrane as it bulges towards the low-pressure side on both sides of the finger. The bending moment in the finger is much the same as that in a flight feather shaft, but it originates differently, from the upward component of tension in the membranes attached to each side of the finger skeleton, rather than from the attachment of the cantilever bases of the stiff barbs to the sides of the feather rhachis. Besides tending to bend the finger, the tension in the membrane also tends to compress the finger towards the wrist. In resisting this compression, each finger allows the tension path in the membrane to turn. Working outwards from the body, the tension paths turn a corner as they pass Digit 5, and another at Digit 4. As a result, Digit 3 can pull forwards on the outer edge of the wing, directly opposing the leg, while pulling in an almost perpendicular direction. This allows the wings of some bats, especially Molossids, to be tensioned straight out from the body, in a rather narrow, pointed shape. Because pterosaurs lacked fingers through the membrane, they would not have been able to bend the tension path in this way, and must have depended instead on backward curvature of the wing finger to tension the membrane (below).

## 6.1.2 THE LEG AS WING SUPPORT IN BATS

Besides tensioning the patagium at the inner edge, the leg also controls its camber. The knees of non-flying mammals, such as ourselves, bend the wrong way for this. Flexing knees like ours would camber the trailing edge of the wing upwards instead of downwards. The two stereoscopic pairs of photographs in Figure 6.5 show a *Rousettus* fruit bat gliding in a wind tunnel, seen from above. Both pictures show that the hip joints allow the femurs to rotate outwards to such an extreme degree that the knees project outwards and *upwards* in flight, a position which is not ideal for walking on the ground. The feet are rotated around so that the toes curl downwards, with Digit 1 (the big toe) on the outside and Digit 5 towards the centreline. The ankle joint can flex so as to curl the trailing edge of the plagiopatagium sharply downwards, as seen in Figure 6.1 in the downstroke of flapping flight. Most bats have a limited ability to walk quadrupedally on their wrists and feet, with the thighs splayed wide apart. Some (especially vampires) are surprisingly agile on the ground, and can even jump, but they cannot stand or walk upright on their hind legs, because the hip joint has to be very far back, in order for the leg to support the posterior part of the wing membrane. Bats' toes are armed with sharp, hooked claws, and they typically roost hanging head downwards from their feet, with the wing membranes wrapped around the body. Unlike the versatile feet of birds, this simple bat foot is not readily adaptable to functions other than hanging up, or clambering about in branches (Figure 6.6).

## 6.1.3 CONTROL OF PLANFORM AND PROFILE SHAPE IN BATS

Like a hang-glider's sail, a bat's wing has to be tensioned, meaning that a steady tension force has to be applied to the outer part of the membrane, and balanced by an inward pull, where the membrane attaches to the leg skeleton, and to the side of the body. This means that if a bat reduces its wing span by sweeping back the hand wing, in the way that birds do, the membrane has to contract, which reduces the tension in its internal elastic fibres, so that the sail billows upwards (Figure 6.7).

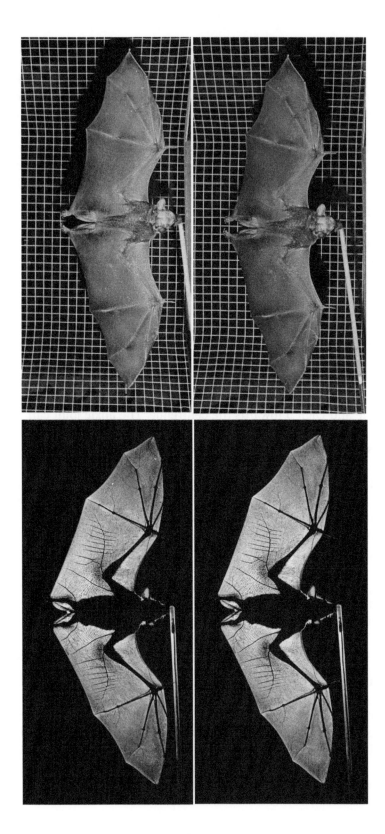

FIGURE 6.6 A typical bat foot, belonging to the fruit bat *Rousettus aegyptiacus*. All five toes are similar, with hooked claws, used for hanging inverted while roosting. Fishing bats hook fish by trailing the claws in the water. Photo by C.J. Pennycuick.

While birds drastically reduce both the span and area of their wings at every upstroke of flapping flight, without impairing the wing's ability to resist bending and twisting moments (Chapter 5), bats can only do this to a minor extent, and not without affecting the strength of the wing.

---

FIGURE 6.5 Stereoscopic photographs of a small fruit bat (*Rousettus aegyptiacus*) flying in a wind tunnel (from Pennycuick 1971). The air stream was inclined upwards by tilting the wind tunnel, so that the bat was able to glide. It was trained to maintain a constant position by feeding it with banana pulp, supplied through the tube on the right. The camera was aligned perpendicular to the air flow, and the upper stereo pair was taken by reflected light, from a flashgun mounted above the tunnel. The lower pair was taken by transmitted light, by mounting the flashgun below the bat, so that the light shone directly towards the camera, through the wing membranes. The stereoscopic effect can be seen by diverging the eyes, so that the left eye looks at the left picture, and the right eye at the right picture. The viewer will then see three images, the centre one being three-dimensional, formed by fusing the two pictures. This is easiest to achieve by holding the page in bright light, perpendicular to the line of sight, and starting with the upper (reflected light) image. When fusion is achieved, the wire netting will recede below the bat, and the central image will become solid. Viewers who are new to this may find it helpful to start by looking over the top of the page, and fixating on an object a few metres away, and then transferring attention to the bat images and fusing them. Photos by C.J. Pennycuick.

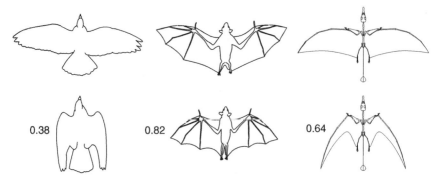

FIGURE 6.7 A pigeon (left) trained to glide in a tilting wind tunnel could reduce its wing span by a factor of 0.38 when the wind speed was increased from 8.6 ms$^{-1}$ to 22 ms$^{-1}$ (Pennycuick 1968a) while a fruit bat (centre) trained to glide in the same wind tunnel had a narrower speed range from 5.5 ms$^{-1}$ to 10 ms$^{-1}$, and could only reduce its wing span by a factor of 0.82 (Pennycuick 1971). If pterosaurs' wings (right) worked as postulated in Figure 6.10, with an elastic membrane, they would have been better able than bats to vary their wing span and area, and perhaps comparable with birds in this respect. These planform changes also occur between the downstroke and upstroke of every wingbeat, and may be responsible for the superiority of birds over bats in long-distance migration, in which case pterosaurs' flight performance may have been more comparable to that of birds than to that of bats.

If the conjecture in Chapter 4 is correct, that birds use this planform variation to obtain an energetically efficient vortex wake, then the inability of bats to do the same thing might be one reason why their migrations seem to be confined to much shorter distances than those of birds. On the other hand, Digits 4 and 5 give a bat a much greater degree of control of the cross-sectional shape of the hand-wing than is possible in a bird (Norberg 1972b), and this is the basis of the incredible agility at low speeds for which bats are famous, for instance when catching flying insects. Bats can also control the camber of the plagio-patagium to a limited extent, by shortening a set of plagiopatagial muscles that run fore-and-aft in the membrane, behind the ulna, without attaching to the skeleton at either end. In gliding flight, these muscles flatten the cross section at higher speeds, and relax to allow the membrane to bulge upwards into a more cambered shape at low speeds. A similar arrangement in hang gliders is called "variable billow". Figures 6.8 and 6.9 show contour maps of the wings of a gliding bat, at speeds near the minimum and maximum at which it would fly in a wind tunnel. Changes of profile shape and angle of attack at different speeds can be seen (Box 6.1). These maps were made by photogrammetry from stereoscopic photographs like those of Figure 6.5.

BOX 6.1 **Bat contoured plots.**

Figures 6.8 and 6.9 were made from two stereo pairs like the lower pair in Figure 6.5, taken by transmitted light (Pennycuick 1971, 1973). Enlarged positive transparencies were made from the original negatives and placed

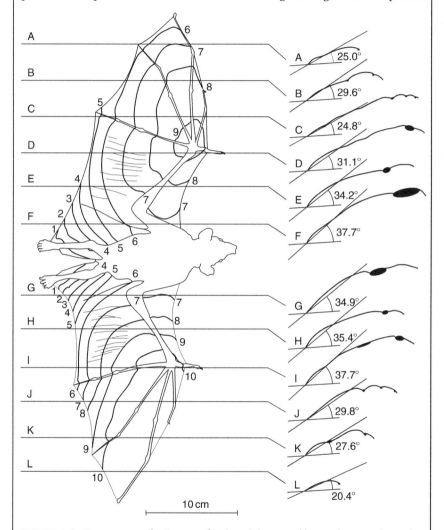

FIGURE 6.8 Contour map of a *Rousettus* fruit bat gliding steadily in a tilting wind tunnel at an equivalent air speed of 5.5 ms$^{-1}$, from a stereo pair of photographs taken by transmitted light, as in Figure 6.5. The thick contour lines are numbered with the height in centimetres above a datum plane just below the bat's feet. The thin lines marked A–L are the positions of profiles whose upper surface is shown on the right, with the cross-sectional shapes of the bones filled in approximately. The zero-lift line through the trailing edge of each profile was calculated from thin aerofoil theory according to the method of Pankhurst (1944), and as this bat was gliding, the angle of attack was measured relative to the axis of the wind tunnel. Data for the bat are in Table 6.1. From Pennycuick (1973).

BOX 6.1 *Continued.*

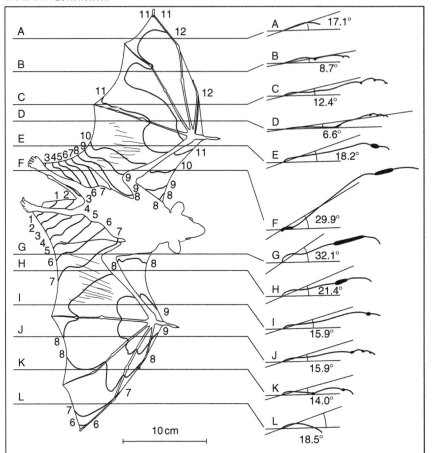

FIGURE 6.9 Contour map as in Figure 6.8, but at a higher speed (9.0 ms$^{-1}$). The bat reduces its wing span and area slightly, flattens its profile by tightening the plagiopatagial muscles, and reduces its angle of attack. From Pennycuick (1973).

in a map-making machine, normally used for making contour maps from pairs of vertical aerial photographs. A spot that appeared to the operator to float at a constant (but adjustable) height was steered by hand along the three-dimensional surface, while the machine reproduced its track on a drawing. The resulting contours (thick lines) are numbered with the height of the membrane in centimetres, above a datum level just below the bat's feet. In Figure 6.8 the highest level is Contour 10 on the outer part of the right wing, which is 10 cm above the datum level, that is, nearer to the camera.

Besides drawing the contours, transects were taken along each of the 12 chord lines A–L, and plotted as wing profiles on the right of each figure. The cross-sectional shapes of the bones are approximate, as these are seen in silhouette by transmitted light. The profiles from the upper surface can

BOX 6.1 *Continued.*

be considered according to the classical theory of thin wing sections as explained by Abbott and von Doenhoff (1959). A practical numerical method due to Pankhurst (1944) allows the zero-lift angle of attack to be calculated (see Chapter 3, Box 3.5). This is the angle between the chord line and the direction of the incident air flow, which would make the lift coefficient zero if this were a rigid profile made of, say, sheet metal. The chord line is not shown in the profiles in Figures 6.8 and 6.9 (it joins the leading and trailing edges), but they show an estimated zero-lift line drawn through the trailing edge, and a line which is parallel to the axis of the wind tunnel, and assumed (as the bat was gliding) to represent the incident air flow. The angle between these two lines is an estimate of the local angle of attack.

TABLE 6.1 Data for *Rousettus* contour maps.

|  | Figure 6.8 | Figure 6.9 |
|---|---|---|
| Body mass (kg) | 0.120 | 0.120 |
| Lift (N) | 1.14 | 1.16 |
| Wing span (m) | 0.523 | 0.500 |
| Wing area (m$^2$) | 0.0495 | 0.0484 |
| Aspect ratio | 5.52 | 5.17 |
| Lift coefficient | 1.27 | 0.485 |
| Equivalent air speed (ms$^{-1}$)[a] | 5.45 | 9.00 |
| Tunnel tilt (°) | 13.50 | 9.00 |
| Downwash angle (°) | 4.18 | 1.71 |
| Reynolds number (mean chord) | 34,000 | 57,000 |

[a] Reduced to sea level air density 1.22 kg m$^{-3}$.

## 6.1.4 FLIGHT MUSCLES OF BATS

The downstroke in flapping flight is powered by the paired pectoralis muscles, which originate over a wide area of the ribs and sternum, much like those of birds. The sternum of bats does not have an expanded curved dorsal plate like that of birds, because bats breathe with a diaphragm, not with a sternal bellows. Bats also lack the prominent ventral keel of the bird sternum, having only short bony sections at the forward and aft ends of the sternum, with a median ligamentous sheet stretched longitudinally between them. A bat's pectoral muscles originate on either side of this sheet of connective tissue, through which the left and right muscles pull directly against each other. No bony keel is needed for the muscles to flap the wings. The keel of the bird sternum serves a different function that does not apply to bats, allowing evaporative cooling directly from cavities in the pectoralis muscles (Chapter 5). As in birds, the pectoralis of bats inserts on the underside of a ridge that projects forwards from the head of the humerus, so applying a nose-down moment to the wing, which is necessary for the same reason as in birds (Figure 5.5). Bats elevate the wing with the deltoid group of muscles, which originate on the side of the vertebral column, and pull upwards on the dorsal side of the humerus (Norberg 1970, 1972a).

### 6.1.5 THERMOREGULATION AND RESPIRATION IN BATS

When a bat's wing is not tensioned, the sail hangs loose with little contraction of its area. It does not fold in the fanwise manner of a bird's wing, or contract in the manner seen in pterosaur fossils (below). Bats cannot stand upright on their back legs, and they roost by hanging head-down from the feet, with the sail wrapped around the body. The sail has a vast surface area and a copious blood supply, which can be controlled in flight to dispose of heat by convection, provided that the air temperature is below that of the bat's blood. In sunlight, the wing collects heat if the blood supply is turned on, and this may be the main reason why most bats are nocturnal, or at least crepuscular. Bats can also dispose of heat to a limited degree by fluttering the sail when roosting, but they have no system for evaporative cooling, either internal like the air sacs of birds, or external like the sweat glands of many other mammals. Their last resort in a thermal emergency is to lick their chests, and use saliva for cooling.

The lungs of bats are like those of other mammals, but very different from those of birds (see Chapter 7, Box 7.7). Oxygen diffuses into the blood from the gas in the blind cavities (alveoli) that line the wall of the lung, and carbon dioxide diffuses out. The lungs are ventilated by contraction of a muscular diaphragm which closes the posterior end of the thoracic cavity, as in other mammals. Bats' lungs are no more effective than those of mountaineers at high altitudes, unlike the lungs of birds, which can maintain blood oxygen levels sufficient for strenuous activity, at lower atmospheric pressures. Some bird species routinely migrate at heights above 6000 m ASL, whereas bats are confined to more modest altitudes, perhaps 2000 m.

### 6.1.6 TAKE OFF AND LANDING IN BATS

Most bats roost in places like trees or the roofs of caves, where they can take off by dropping into a clear space, although a few (vampires) can take off by jumping upwards from a level surface. Landing involves attaching the claws to a suitable toe-hold, and rotating the body from the flight attitude to the head-down roosting position. Small bats can land on a vertical or inverted surface, either by hooking the thumb claws on to the surface and swinging the feet up, or by rotating in the air and attaching the feet directly. Fruit bats have a somewhat different technique for landing on branches. The bat approaches slowly above the branch, with its feet trailing, and hooks the branch with its downward-curving claws, then swings over forwards into the head-down posture, furling its wings as it does so.

## 6.2 ⬭ PTEROSAURS

Pterosaurs are an extinct order of reptiles. They belonged to the archosaur branch of the Class Reptilia, which comprises birds, crocodiles and the two orders of dinosaurs, Saurischia (which were closely related to birds) and Ornithischia which were somewhat different. The archosaurs may be considered a sub-Class, or a super-Order, depending on how you look at it. The relationship between the different archosaur orders is that they all sprang from a common ancestor. That was a long time ago, but not so long ago as the still earlier ancestor that the archosaurs as a whole shared with other branches of the reptiles, such as turtles, lizards and the synapsid line that eventually led to mammals (including bats and ourselves). The common ancestor of birds and pterosaurs was not a flying animal. Birds and pterosaurs each evolved flight separately, in different ways, from an ancestor that did not fly (Chapter 16). Neither group inherited any flight adaptations from the other, or from a common ancestor.

The first pterosaur fossils are the most ancient known flying vertebrates, dating from Triassic times. Wellnhofer (1991) has written an authoritative account of their history, with sketches of all known genera drawn to the same scale. In terms of general shape, pterosaurs were like frigatebirds, with large wings relative to the size of the body, not like swans or guillemots. Early pterosaurs, characterised by a long, bony tail with a paddle on the end, are assigned to the suborder Rhamphorhynchoidea, which survived until late in the Jurassic. Some of the best-preserved rhamphorhynch specimens were found in the famous upper-Jurassic Solnhofen limestone formation of south Germany, alongside the first members of the other pterosaur suborder (Pterodactyloidea) which differed in having very short tails that could not have been used to balance the body weight about the hips. The loss of the balancing tail typical of dinosaurs was not accompanied by any drastic modification and expansion of the pelvis, like that seen in birds (Chapter 5), presumably because rhamphorhynchs had given up bipedal walking long before, when they modified the legs to support the inner end of the wing (below). The Solnhofen formation also yielded several specimens of *Archaeopteryx*, the first bird known to have had a wing that (more or less) fitted the description in Chapter 5, although the rest of its skeleton did not yet show the characteristic modifications of the limb girdles and tail that distinguish birds from dinosaurs. Pterodactyls flourished until the last days of the Cretaceous, when they disappeared along with the dinosaurs and many other groups of animals. Birds later reappeared and prospered, but pterosaurs sadly did not.

## 6.2.1 MECHANICS OF THE PTEROSAUR WING

Pterosaurs are known only from their fossilised skeletons, and from surface impressions of the wing membranes in the relaxed (dead) state. As there is no prospect of observing pterosaurs in flight, still less of flying one in a wind tunnel, the way that their wings worked has to be inferred from the similarities and differences between their wings and those of birds and bats. The pterosaur skeleton was basically dinosaur-like, and to that extent it resembled a bird more closely than a bat. However, while birds retained the bipedal stance of their dinosaur ancestors, the legs of pterosaurs were modified like those of bats to support the inner end of a flexible sail, with only a limited capacity for walking.

The wing skeleton of pterosaurs differed from those of both birds and bats, in that there was a single, jointed bony spar, running all the way to the wing tip (Figure 6.10). Wellnhofer (1991) illustrates a sectioned pterodactyl humerus, which is a thin-walled tube very similar to the swan humerus of Figure 5.5, complete with internal trabeculae. The cavity may have been connected to the respiratory system and filled with air, as in birds. The radio-ulna was quite similar to that of bats, but instead of dividing into five digits at the carpal joint as in bats, the spar continued with four tightly bundled and partially fused metacarpals. These are thought to represent Digits 1–4, while Digit 5 is presumed to have been lost at an early stage of pterosaur evolution. The metacarpal unit was short in rhamphorhynchs, but in the later pterodactyls it was longer, and formed a prominent section of the spar. Three short, clawed digits (1–3) projected forwards from the outer end of the metacarpal unit, while the spar continued along the leading edge of the wing to the tip as a

---

whales (Figure 6.14), becoming prominent when the membrane is fully contracted. (D) *Rhamphorhynchus* foot after Wellnhofer (1991). If the feet were simply rotated back in C, the soles would be upwards, and Digit 1 would be on the inside. Outward rotation at the hip brings the dorsal side of the foot upwards, with Digit 1 on the outside. Digit 5 still supports the trailing edge tendon, as in the unrotated ancestor, and therefore has to be modified so that the tendon can pass over Digits 1–4 to the outside.

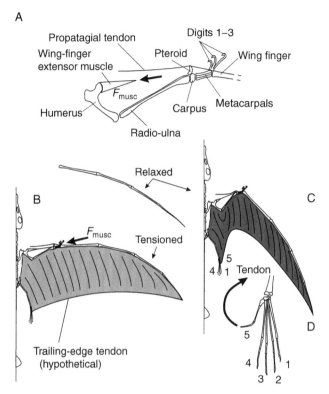

FIGURE 6.10 Pterosaur reconstruction based on *Rhamphorhynchus muensteri*, a small, tailed pterosaur from the Upper Jurassic Solnhofen limestone of southern Germany, as described by Wellnhofer (1975). The hypothetical elements of this reconstruction are from Pennycuick (1988b). (A) Arm skeleton (enlarged from (B) below). The distinctive hammer-headed humerus articulates with a straight radio-ulna (not curved like the ulna of birds). The pteroid, projecting forward from the carpus, is a bone that is peculiar to pterosaurs. The metacarpals are bound together to form a single structural unit. Beyond them the phalanges of Digits 1–3 form fingers with hooked claws, while Digit 4 is the hugely elongated "wing finger". (B) To spread the wing, the elbow joint would have been fully extended, and the wing finger fully protracted. It is proposed that a muscle originating on the head of the humerus exerted a force $F_{musc}$ to pull the wing-finger forwards, against the pull of elastic fibres in the membrane. A hypothetical trailing-edge tendon connects the fifth toe with the tip of the wing finger. The joints between the wing-finger's phalanges are assumed to be bound by elastic material, so that the finger as a whole would flex like a bow when the wing was tensioned. The isolated wing-finger above the diagram is copied from (C), where the tension is partially relaxed, allowing the joints to straighten. (C) When the pull of the extensor muscle was relaxed, the elastic membrane would have been free to contract, pulling the wing finger back, reducing the wing's span and area. The fully relaxed wing would contract so that its planform would be similar to that seen in the dead wings of fossils. The contraction would cause wrinkles to appear on the surface (thin black lines), which have been interpreted as structural "fibres", although they are strictly surface features seen in casts of dead wings. More probably they are analogous to the "pleats" seen in the throat pouches of rorqual

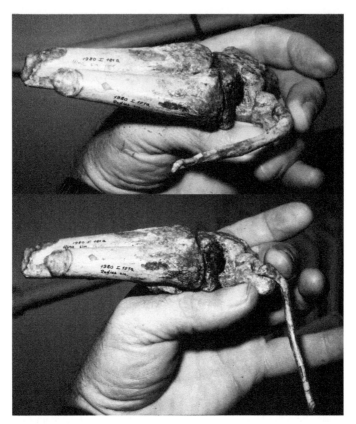

FIGURE 6.11 Two photographs of the carpus, and the outer end of the radio-ulna, of a Cretaceous pterodactyl *Santanadactylus spixi*, held by Prof. Peter Wellnhofer to show that the pteroid can be articulated with the carpus in two alternative positions. It is proposed here that extension of the wing caused the pteroid to "snap" from the upper to the lower position, so deploying the propatagium as a drooped leading edge. Photos by C.J. Pennycuick.

single, vastly elongated "wing-finger" with four phalanges, believed to be Digit 4. A small *pteroid* bone, peculiar to pterodactyls, projected from the wrist, usually pointing inwards in fossils, towards the shoulder. Its function is uncertain, but it most probably controlled the leading edge of a propatagium that stretched from the shoulder to the inner end of the wing finger (Figures 6.11 and 6.12).

The nature of the wing membrane is known from a few fossils in which surface impressions of dead wings have been preserved, especially a number of famous late-Jurassic specimens of both rhamphorhynchs and small pterodactyls from the fine-grained Solnhofen limestone. These show the outer part of the wing contracted into a narrow, sharply pointed shape which some authors (not very imaginatively) have assumed was also their

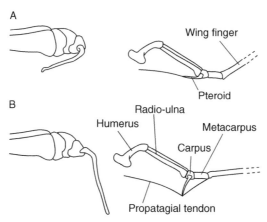

FIGURE 6.12 (A) Pteroid pointing inwards towards the shoulder, in the position normally seen in the contracted wings of fossils. (B) Pteroid in the "down" position proposed for the extended wing, deploying the leading-edge tendon to droop the propatagium.

shape in flight. In a few specimens, especially the famous "Zittel wing" now in Munich (Figure 6.13) a pattern of fine surface ridges can be seen, which were first described in 1882 as "Fasern" (fibres), and have been widely assumed ever since to be stiff structural elements made of keratin. The pattern of these "fibres" is vaguely reminiscent of the fan-like arrangement of flight feathers of a bird's wing, but there are far more of them, and they are much thinner—far too thin to be spars like feather shafts. They are also closely packed side by side, and as they radiate towards the trailing edge of the wing, new ones are interpolated between those that start further forward. At the forward end, they peter out, and there is no sign of any mechanical attachment to the wing bones. In the inner part of the Zittel wing, the "fibres" wrap around the elbow joint, appearing soft and flexible at that point, which suggests that they might have been soft and flexible over the rest of the wing as well.

It has been claimed that the "fibres" must represent solid structures, because they are so regular and sharply defined. However, elsewhere in this same Solnhofen limestone, fossilised medusae have been found, showing patterns of wrinkles where the surface contracted as the animal died in hypertonic brine. Such a soft creature would have to be preserved in a two-stage process, whereby some encrusting microorganisms such as blue-green algae first deposited a hard, negative "mould" on the surface, and mud particles were later compacted into the mould, after the organic remains had decayed away. The preservation of surface detail implies nothing at all about the mechanical strength of the original jellyfish, or about that of the Zittel wing's membrane. These pterosaur fossils were revealed when a slab was split from its counter-slab.

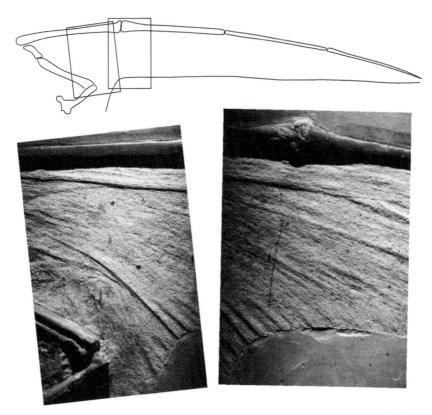

FIGURE 6.13 The Zittel wing from which the supposed "fibres" were first described. The sketch shows the contracted outline of the wing, with two rectangles corresponding to the photographs below. The ridges are even and regular in the right-hand photograph, but in the left-hand one they fold around the elbow joint. Additional ridges are interpolated as the wing widens towards the trailing edge. There is no separation or fraying of the ridges at the trailing edge, as might be expected if they were stiff fibres. Photos by C.J. Pennycuick.

Bones may be preserved in one slab or the other, but no internal structure is preserved in the wing membranes. They are strictly surface impressions, positive in one slab and negative in the other.

## 6.2.2 TENSIONING THE PTEROSAUR WING MEMBRANE

If we doubt the assumption that the surface ridges represent "fibres", then we may ask whether anything that resembles them is known in living animals. There is actually a striking resemblance (albeit on a larger scale), with the throat pouches of rorqual whales, the group that includes Blue, Fin and Humpback whales, whose feeding methods have been described by Minasian et al. (1984). When relaxed, a rorqual's throat pouch has prominent, closely packed, parallel ridges, running in a fore-and-aft direction. Whale biologists refer to the pattern of ridges and grooves as "pleats". Despite their robust appearance, the pleats do not contain longitudinal stiffening elements of any kind, and their function has nothing to do with resisting bending forces. They are a by-product of the internal structure of the wall of the pouch, which is highly elastic in the direction transverse to the pleats, but not in the longitudinal direction. The whale feeds by taking in a huge volume of water through its mouth, so expanding its throat pouch into an enormous balloon (Figure 6.14). The pouch then slowly contracts, expelling the water through the baleen plates along the sides of the mouth, while any fish, squid or krill that it contained go down the whale's throat. The pleats flatten out as the pouch expands, and reappear as it contracts. If this was also the basis of the ridges on pterosaur wing membranes, then the implication is that the membrane (unlike a bat's wing) was highly elastic, in a direction transverse to the ridges, and that the ridges (or pleats) appeared on the surface when the wing was relaxed, allowing the elastic membrane to contract. Of course, all the fossil wings are relaxed.

The outer part of the relaxed, dead wing of a pterodactyl fossil has much the same narrow, sharply pointed shape as the outer part of the wing of a dead bird, or of a living one in fast gliding flight, or during the upstroke of flapping flight. The corrugated surface and narrow planform shape of the relaxed (dead) membrane suggest that it contained much stronger elastic fibres than are present in a bat wing, and was expanded in flight by the outward pull of the wing finger, which was much thicker than the fingers of bats, and raises the possibility that this expansion and contraction might have taken place during each wingbeat cycle, as it does in birds (Figure 6.7). A bird can expand its wing to its full span and area without exerting any large forces, but a pterosaur, constructed as suggested, would have had to do work against the elastic fibres when expanding the wing at the beginning of the downstroke. However, this work would have been temporarily stored in the elastic fibres, and could in principle have been converted into aerodynamic work, when the wing was allowed to contract at the end of the downstroke. In that case, pterosaurs would have been able to vary their wing span and area in flapping flight in the same manner as birds, which is something that bats cannot do, or only to a small extent. If the implication of this kind of motion for long-distance migration, as suggested in Chapter 5 is correct, then it is possible that some pterosaurs could have been long-distance migrants, with all the adaptive opportunities that migration opens up for birds but not for bats (Chapter 8).

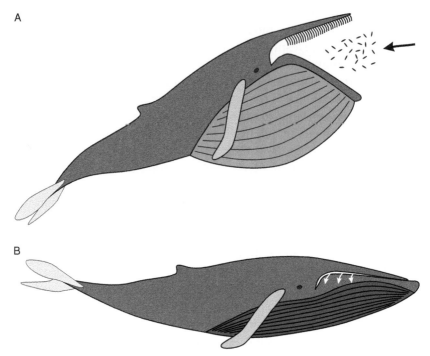

FIGURE 6.14 (A) Rorqual whales (Balaenopteridae) feed by engulfing prey in a highly distensible throat pouch, then closing the mouth and contracting the pouch, so that the water flows out through the array of baleen plates along the sides of the upper jaw. The name of the sub-order to which rorquals belong (Mysticeti) refers to the baleen plates (Greek *mystax*: moustache). (B) The contracted throat pouch fairs into the streamlined shape of the whale's body, and the contracted membrane surface then shows prominent longitudinal "pleats". It is argued here that these are directly analogous to the parallel "fibres" seen on the surface of the contracted (dead) wing membranes of some pterosaur fossils, implying that the membranes were stretched in flight to a much larger area than that seen in the fossils. See also Figures 6.10 and 6.13.

## 6.2.3 THE TRAILING-EDGE TENDON AND THE FIFTH TOE

The reconstruction shown in Figure 6.10, based on an elastic membrane, requires a tendon that runs from the foot to the tip of the wing finger, and pulls the trailing edge of the elastic patagium back when the wing finger is protracted. No such tendon is visible in any of the fossils, but that is not a compelling argument against its existence, as no other tendons are preserved in these fossils either. Pterosaur feet have an unusual feature, which at first sight appears to conflict with existence of a trailing-edge tendon. Digits 1–4 of the pterosaur foot are slender like the toes of bats, with hooked claws that look suitable for hanging up, but for little else, whereas Digit 5 is different, more robust than the other toes, and with a bend at the joint between the first and second phalanges. This looks like the anchorage for the inner end of the trailing-edge tendon, but there is a difficulty. Because of the rotation of the thighs to make the knees project dorsally, as they do in bats, the fifth toe lies on the *inside* of the foot in flight, towards the centre-line. Some authors have argued that the fifth toe supported a "uropatagium" between the legs and the tail, but have not explained why such a sturdy support would be needed for this, even if a uropatagium existed. A more likely interpretation is that the tendon originated at an early stage of evolution, at a stage when the pterosaur ancestor's fifth toe was still on the outside of the foot, as it is in flying squirrels. Subsequent rotation of the leg in the course of pterosaur evolution meant that the tendon crossed from the fifth toe above the other toes, so requiring the toe skeleton to be modified to control it.

## 6.2.4 MECHANICS OF THE WING FINGER

The pterosaur patagium was a single expanse of membrane, without any bony supports running through it, as Digits 4 and 5 do in a bat's wing. These two digits are loaded in compression, and serve to turn the direction of the tension paths in the membrane (Figure 6.4). Most reconstructions of pterosaur skeletons show the wing finger sticking straight out from the body, but this overestimates the wing span, as the wing could not have been tensioned in this position. The tension paths in a pterosaur's wing would have had to run directly from the inner edge of the membrane to the wing finger, without any corners, and this would mean that the wing finger had to bend back when the wing was fully extended. The wing finger was made up of four phalanges, each of which had oblique and slightly expanded end plates at both ends (except at the wing tip). The phalanges were connected by butt joints where the end plates met. If these joints were bound together by elastic ligaments, the finger as a whole would bend like a bow when tensioned, and this is shown in the greater curvature of the wing finger in Figure 6.10B than in C.

## 6.2.5 LARGE AND GIANT PTEROSAURS

Wellnhofer (1991) gives wing span estimates for a number of pterosaur species throughout the history of the group, and these include large pterodactyls with estimated spans between 5 and 6.2 m, throughout the Cretaceous. These estimates are based on the assumption that the wing finger ran straight out to the wing tip. The span would be less if the wing finger were bowed as in Figure 6.10A, but even so it seems that the largest Cretaceous pterodactyls had functional wing spans which were greater than those of living vultures and albatrosses, both of which reach about 3 m in the largest species. Some tertiary fossil birds such as *Teratornis* and *Argentavis* may have had larger wing spans but this depends on extrapolating from the skeleton. This is unreliable in birds, because much of the span is made up by primary feathers, which have not been preserved in these fossils.

At the extreme end of the Cretaceous, something seems to have changed with the brief appearance of the giant pterodactyl *Quetzalcoatlus northropi*. The enormous size of this animal may be judged by comparing its humerus (Figure 6.15) with that of the little rhamphorhynch in Figure 6.10. Both humeri have the same distinctive, hammer-headed shape, but the one in Figure 6.10 is only about 4 cm long, a convenient size to handle with tweezers. The *Quetzalcoatlus* remains are fragmentary, but Chatterjee and Templin (2004) estimate from the size of the known bones that the mass of *Q. northropi* was 70 kg, its wing span

FIGURE 6.15 Prof. Peter Wellnhofer, Director of the Bavarian Museum of Palaeontology, where many of the most famous pterosaur specimens from Solnhofen are kept, holding a cast of a humerus of the giant end-Cretaceous pterodactyl *Quetzalcoatlus northropi*. The hammer-head shape of the humerus is similar to that of the little *Rhamphorhynchus* illustrated in Figure 6.10, but that humerus is about 4 cm long, and if it were free from the matrix, it could be conveniently handled with tweezers. Photo by C.J. Pennycuick.

was 10.4 m, and its aspect ratio 11.3, i.e. around twice the linear size of the "standard" large Cretaceous pterodactyls.

If Chatterjee and Templin's numbers are input to *Flight's* power curve calculation, together with a flight muscle fraction of 0.15 and today's sea-level air density and gravity, its maximum rate of climb would be negative, meaning that it would not be able to maintain height when flying at its minimum power speed and exerting full power. This is a mechanical argument, not a physiological one. It makes no assumptions about the availability of oxygen, but assumes that the sea-level *density* of the air was much the same as in modern times. However, it seems likely that the atmosphere was denser throughout Mesozoic times than it is now (Budyko et al., 1985; Dudley, 1998), and there may also have been an episode of extremely high air density right at the end of the Cretaceous, when *Quetzalcoatlus* lived, caused by outgassing associated with the prolonged and massive volcanic eruptions that created the Deccan Traps (Officer and Drake, 1985). Increasing the air density reduces the minimum power speed, and also the power needed to fly at that speed, in inverse proportion to the square root of the air density, whereas the power available from the flight muscles is proportional to the wingbeat frequency, which varies in inverse proportion to the 3/8 power of the air density. These two graphs are shown in Figure 6.16B, representing nine power-curve runs, in which the air density was increased in steps of 0.5 kg m$^{-3}$ from 1 to 5 kg m$^{-3}$, while everything else was held constant. The maximum rate of climb (Figure 6.16A) is initially about $-0.1$ m s$^{-1}$, but increases through zero when the air density is just below 4 kg m$^{-3}$. This is 3.25 times the sea-level air density in the International Standard Atmosphere, and would correspond to an altitude of 14,000 m *below* sea level today. It is not inconceivable that Earth could retain such a dense atmosphere, considering that Venus currently retains an atmosphere whose surface density is more than 90 times ours, even though its gravity is weaker, and its surface temperature is much higher. So long as sufficient oxygen is still present to support the reduced level of metabolic activity needed to fly, any gas that is not actually toxic or corrosive will serve to increase the air density (see also Chapter 2, Box 2.4).

## 6.2.6 WATER PTEROSAURS?

It is a common idea that many of the larger pterosaurs were fish-eaters, although no known pterosaur shows a body form like that of wing-swimming birds such as auks. If any pterosaur could swim with its bat-like legs, then one would expect some bats to be able to swim too, but they do not. Pterosaurs did, however, fly with toes 1–4 of each

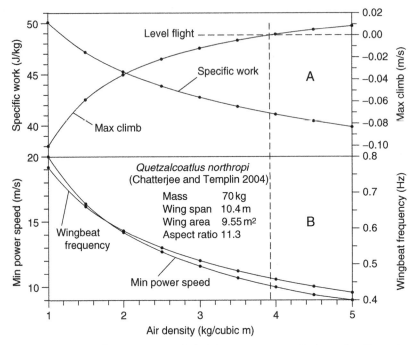

FIGURE 6.16 Output from nine runs of *Flight*'s power curve calculation for *Quetzalcoatlus northropi*, using estimated mass and wing measurements from Chatterjee and Templin (2004), with the flight muscle fraction set to 0.15, gravity to 9.81 m s$^{-2}$, and all other input variables set to default values, including the isometric stress for the myofibrils at 560 kN m$^{-2}$ (Chapter 7). (A) Rate of climb according to the calculation of Chapter 7, Box 7.5 rises above zero when the air density is just below 4 kg m$^{-3}$. As the air density increases from 1 to 4 kg m$^{-3}$, the specific work decreases from around 50 J kg$^{-1}$ (too high), to about 42 J kg$^{-1}$, which is only a little over the value for delivering maximum power (Chapter 7). (B) The minimum power speed (and with it the power required to fly) drops by a factor of 2.2 when the air density increases from 1 to 4 kg m$^{-3}$, whereas the wingbeat frequency, which determines the power available, drops by a factor of only 1.7. This is the reason for the increasing power margin, which permits the maximum rate of climb to increase.

foot curled downwards, which raises the possibility that they could have caught fish that were swimming just below the surface, by trailing their claws in the water like the fishing bat *Noctilio leporinus* (Novick and Leen, 1969). Some raptors such as ospreys and fish eagles, which snatch fish in their talons without actually entering the water, are fish-eaters without being true water birds, and the same may have been true of some pterosaurs. Such a lifestyle would be sufficient to explain the occurrence of fish remains, apparently in the body cavities of pterosaur fossils, without necessarily implying that any pterosaur could actually swim.

# 7

# MUSCLES AS ENGINES

A muscle does work by exerting tension and shortening. The work that can be done by unit mass of muscle in one contraction is essentially fixed, and the power is found by multiplying this by the contraction frequency. This is the same as the wingbeat frequency in a flying animal, and is lower in large animals than in related smaller ones. An animal's maximum rate of climb is estimated and output by the *Flight* programme, and must be at least zero if the animal is to be capable of level flight in a given environment.

The work required for powered flight comes from muscles in all flying animals. The theory of engines from aeronautics is not a great deal of help in biology, since nearly all aircraft engines are heat engines, which first convert fuel energy into heat, then convert some fraction of the heat into work, and dispose of the rest. Muscles are not heat engines, and they do not share the basic requirement of all heat engines, that the largest possible difference is needed between the temperature at which work is done (for instance in a cylinder or turbine), and that at which waste heat is dumped to the outside world. Muscles are isothermal engines, which convert chemical energy into work without first converting it into heat, and there is no thermodynamical requirement

Modelling the Flying Bird
© 2008 C.J. Pennycuick. Published by Elsevier Inc. All rights reserved.

for any part of the system to operate at a high temperature. If a muscle were 100% efficient, that is, if it could convert all of the fuel energy that it consumes into work, then it would not heat up at all. In practice, there is some heating because only a part of the fuel energy is converted into work, and the remainder appears as waste heat, which has to be disposed of. A flying animal's capacity to dispose of waste heat determines the temperature rise in its flight muscles, which may amount to a few degrees during maximal exertion, but not hundreds or thousands of degrees as in jet or rocket engines. Even so, all flying animals except possibly the smallest insects require adaptations to dispose of waste heat, which would otherwise cause overheating. On the other hand, keeping warm is seldom if ever a problem in powered flight, even at sub-zero air temperatures.

This chapter is about those properties of vertebrate skeletal muscle that an engineer-god would need to know before attempting to design a muscle-powered flying animal. The approach has its origins in a famous paper by A.V. Hill (1950) in which he generalised his own earlier studies of the mechanical properties of isolated muscles, so as to predict the limits of performance in different types of locomotion, in animals of different size. Muscles differ from most artificial engines in that the only kind of force they can produce is tension, and they only do work by shortening. Skeletal levers or hydraulic converters are required to convert a muscle's inherent "pull" force into a push force or a torque. Some "tonic" muscles are specialised for maintaining tension, without doing work, but most skeletal muscles are specialised to function as engines, whose primary function is to convert fuel energy into work. A muscle does this intermittently, producing a certain amount of work when it shortens against a load, after which it has to be lengthened passively, usually by an antagonistic muscle, before it can do some more work in another contraction. This intermittent action, with a work stroke alternating with a recovery stroke, is somewhat analogous to that of a reciprocating engine, at the level of the individual piston. The actual mechanism is basically the same in different animals, with no broad divisions such as that between piston and turbine engines. The differences between vertebrate and insect flight muscles, due to the higher contraction frequencies at which the latter operate, amount to variations of the same basic mechanism rather than radically different types of engine. Within vertebrates, it is possible to generalise about the structure and mechanical properties of skeletal muscles, because there is a remarkable degree of uniformity between different vertebrates, large and small.

# 7.1 ⬡ GENERAL REQUIREMENTS

## 7.1.1 ENGINE AND SUPPORT SYSTEMS

Like any artificial engine, muscles need to be supplied with fuel and oxygen, while carbon dioxide and heat, the main waste products of their operation, need to be removed and disposed of. Animals cannot simply suck in air and ignite the fuel as aircraft do, because the oxidation process takes place in an aqueous medium at body temperature. Oxygen first has to diffuse from the air into a liquid medium, the blood, which then carries it around the body and delivers it to tissues that require it. The "support systems" for the muscles are the lungs, where oxygen is extracted from the air and carbon dioxide is disposed of, and the blood system which provides the internal transport. The blood (at least in bats) also carries waste heat from tissues that generate it to sites where it can be disposed of, either by convection or evaporation.

In prolonged exertion, as in long-distance migration, flight is said to be *aerobic*, meaning that the maximum power available from the muscles is limited by the capacity of the support systems to keep pace with their requirements. The support systems themselves consume mechanical power, to pump air in and out in the case of the respiratory system, and for the heart to pump the blood around a closed system of blood vessels. *Flight* accounts for this when calculating the chemical power from the mechanical power, by multiplying the power calculated for the flight muscles (Chapter 3) by a "respiration and circulation factor" whose default value is 1.1. This is somewhat crude, but there is currently no theory that would provide a better estimate of the amount of power needed to support a given level of power in the flight muscles. In a short burst of maximal exertion, for instance during take-off, the power output of the muscles can exceed the capacity of the support systems by using reserves of fuel stored locally in the muscle fibres, and oxidising it anaerobically. This results in an oxygen debt which eventually has to be repaid, by oxidising the products of the initial energy-yielding reaction. The upper limit of power output in such sprint activities is determined by the mechanical properties of the engine itself, not by the support systems. The maximum power output may greatly exceed the capacity of the support systems in some animals, for instance in ambush predators like crocodiles, whose lives consist of long periods of inactivity punctuated by infrequent explosions of violent activity. Many large birds, such as condors and albatrosses, avoid the need for sustained aerobic activity by soaring, and are generally believed to be incapable of continuous flapping flight. Spiders, one may note, can only operate anaerobically, because their bodies are so arranged that they can either run or breathe, but cannot do both at the same time—hardly, perhaps, an example of intelligent design!

## 7.1.2 FUELS FOR MUSCLES

Fuels for muscular contraction fall into three broad groups, fat, carbohydrate (glycogen) and protein. Glycogen and protein are stored in hydrated form, meaning that a quantity of water is bound to the actual combustible molecule, so reducing the energy density of the stored fuel and increasing the mass of fuel that has to be carried, for a given amount of usable energy. Mainly because it does not require water of hydration, the energy density of stored fat is far higher than that of other fuels (Table 7.1). Aerobic oxidation of fat is the only practical option as the primary energy source for long-distance migration for this reason. However, it is not possible, for biochemical reasons, to burn fat only. The metabolic pathways for metabolising fat involve the consumption of some protein, and it seems that this protein accounts for around 5% of the total energy released, in a bird that is primarily consuming fat (Jenni and Jenni-Eiermann, 1998). As it happens, a long-distance migrant can meet more than half of this requirement by consuming protein from its flight muscles, in effect burning part of the engine as well as the fuel as it gets lighter. This has some interesting consequences which are considered further in Chapter 8. Anaerobic oxidation of carbohydrate can be activated more quickly than fat metabolism, and is better suited to sprint activities involving short bursts of maximum power, for instance at take-off.

## 7.2 ◕ THE SLIDING FILAMENT ENGINE

### 7.2.1 THE ACTO-MYOSIN ARRAY

In transmission electron micrographs, sections through skeletal muscle tissue show a regular array of protein filaments, which are of two types, thick and thin (White and Thorson, 1975) (Figure 7.1). In transverse

TABLE 7.1 Energy density of biological fuels.

| Fuel | Percent water | Energy density J kg$^{-1}$ |
|------|---------------|----------------------------|
| Fat | 0 | $3.9 \times 10^7$ |
| Glycogen | 73 | $4.6 \times 10^6$ |
| Protein | 69 | $5.7 \times 10^6$ |

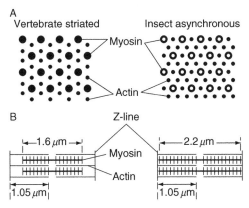

FIGURE 7.1 (A) In a transmission electron micrograph, a cross-section of vertebrate skeletal (striated) muscle (left) shows a regular hexagonal array of thick filaments made of the protein myosin, each of which is surrounded by a hexagonal ring of thinner filaments made of another protein actin. In the asynchronous flight muscles of insects (right), the array has a 3:1 ratio of actin to myosin filaments, instead of 2:1 in vertebrates, and the myosin filaments have a core made of a different material. (B) In longitudinal section, the actin filaments are about the same length in both vertebrates (left) and insects (right), but the myosin filaments are longer in insects, almost reaching the Z-lines at either end of the sarcomere, even when the muscle is at its extended length. Skeletal constraints typically allow 26% strain in vertebrates, but only about 2% in insects. After White and Thorson (1975).

section, the thick (myosin) filaments are arranged in a hexagonal array, while the thin (actin) filaments form a denser array, with a hexagon of actin filaments surrounding each myosin filament. The ratio of actin to myosin filaments is 2:1 in vertebrate skeletal muscle, but 3:1 in insects, where the geometry of the array is slightly different. In longitudinal section, the array is divided into "sarcomeres", which are the contractile units of the muscle. Each sarcomere is separated from its neighbours at either end by transverse membranes (Z-lines). The actin filaments are attached to each side of the Z-line, and run inwards towards the centre of the sarcomere. Neither the actin nor the myosin filaments themselves shorten when the muscle contracts. The sarcomere shortens because the actin filaments slide past the myosin filaments, pulling the Z-lines at either end towards the centre. The sarcomeres in vertebrate skeletal muscle fibres are around 2 $\mu$m long, and can be observed with a light microscope as striations, whose spacing can be measured in living muscles. The contractile filaments themselves can only be resolved by an electron microscope, which requires fixed preparations.

At the molecular level, each thick filament is built up of bundles of myosin molecules, each of which has a long "stem", and a pair of short "heads" that project from one end at an angle, like the bowl of an old-fashioned clay pipe. The heads or "cross-bridges" project from the side of the filament at intervals of 14.5 nm, and connect with attachment sites on the nearest actin filament. There is a short segment without cross-bridges in the middle of each myosin filament. The cross-bridges on either side of this gap pull against one another, and shorten the sarcomere by pulling opposing sets of actin filaments inwards towards the centre. To cut a very long story short, when the muscle contracts, each myosin cross-bridge exerts a force on the actin filament by bending, then reattaching to another site further along the actin filament. The effect is that the myosin filaments walk like centipedes along the actin filaments.

## 7.2.2 THE FORCE–LENGTH RELATIONSHIP

Any muscle varies in length, but in vertebrate skeletal muscle, the maximum and minimum lengths of a muscle in the intact animal are, in most cases, constrained by the skeleton. Figure 7.2 refers to a frog muscle that was removed from the animal and maintained in a saline solution that allowed it to work normally, in a classic experiment by Gordon et al. (1966). One end of the muscle was connected to an apparatus that held its length constant, at a value that could be varied over a range that extended beyond the maximum length in the intact animal. The other end was attached to a force transducer, which measured the tension that the muscle developed when it was stimulated by a continuous series of electrical impulses, but not allowed to shorten. The scale above the graph is the length of a sarcomere in micrometres ($\mu$m), observed in the living muscle with a microscope. This is correlated in the lower part of the diagram with the relative positions of the actin and myosin filaments, as determined from electron micrographs. When the sarcomere length is about 3.7 $\mu$m (right), the tips of the actin filaments have been pulled right to the tips of the myosin filaments, so that no cross-bridges are connected, and there is no tension apart from that due to stretching of the cell membrane. As the muscle is allowed to shorten, the actin filaments slide in and make contact with more cross-bridges. The force builds up until all the cross-bridges are connected at a sarcomere length of about 2.25 $\mu$m (point A). The tension remains constant as the muscle shortens to point B (2.00 $\mu$m), where the ends of the actin filaments coming in from opposite directions meet. The muscle continues to shorten with a small decrease of force, as the ends of opposing actin filaments overlap, until the ends of the myosin filaments arrive at the Z-line membranes at either end (point C, 1.67 $\mu$m). The isolated muscle will shorten even further, crumpling the ends of the myosin filaments until the tension finally

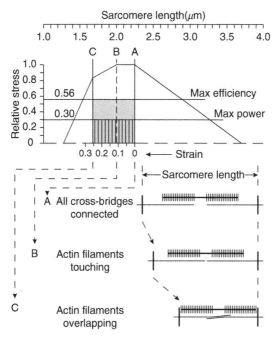

FIGURE 7.2 Graph of the isometric tension developed by an isolated frog semitendinosus muscle at different lengths, as measured by observing the length of a sarcomere through a microscope (after Gordon et al., 1966). The default value in the *Flight* programme is 560 kN m$^{-2}$ for the average stress between points A and C. This length range, which corresponds to an active strain of 0.26, is assumed to be the range allowed by the skeleton, because it is the range over which all cross-bridges are attached, as shown in the lower part of the diagram. Multiplying the stress and strain together gives the volume-specific work that would be done if the muscle were allowed to shorten infinitely slowly from point A to point C (146 kJ m$^{-3}$), and dividing this by the density of muscle (1060 kg m$^{-3}$) gives the mass-specific work (137 J kg$^{-1}$). When the muscle is shortening during locomotion, the stress is lower and so therefore is the specific work. According to Hill's equation (Box 7.1), if the muscle is allowed to shorten at a finite strain rate such that the efficiency of converting ATP energy into work is maximised, the stress falls to 0.56 $\sigma_{max}$ and the specific work (grey rectangle) is 76.7 J kg$^{-1}$. At a higher strain rate that maximises the power output, the stress is 0.30 $\sigma_{max}$ and the specific work (hatched rectangle) is 41.1 J kg$^{-1}$.

falls to zero (left), but the normal range of length permitted by the skeleton in the intact animal is from point A to point C.

## 7.2.3 REPETITIVE CONTRACTION AND THE WORK LOOP

The raw experimental data for Figure 7.2 were force on the *Y*-axis and length on the *X*-axis, rather than stress and strain as shown. In that form, the diagram looks the same, but the dimensions are different. If the muscle shortens from length A to length C, exerting a constant amount of tension as it does so, then the amount of work done is the force times the distance shortened. If the muscle now relaxes, and the apparatus stretches it back to length A, then a lesser amount of work is done by the apparatus on the muscle, equal to the same distance times a lower tension. The difference between the two amounts of work is the *cycle work*, meaning the net work done in one contraction. Any area on the diagram has the dimensions of distance times force (i.e. work), and the cycle work is the area of the loop that is drawn when the muscle first shortens at a constant tension, and is then lengthened at a lower tension. A muscle that is producing positive work in locomotion goes anti-clockwise around a *work loop* at each contraction, shortening at a higher tension than that at which it is lengthened. Its average mechanical power output over a number of cycles is the cycle work (area of the work loop) times the contraction frequency.

If we plot stress (force/area) against strain (distance shortened/extended length) as shown in Figure 7.2, then the dimensions of the work loop are

$$\frac{\text{Force}}{\text{Area}} \times \frac{\text{Distance}}{\text{Length}} = \frac{\text{Work}}{\text{Volume}}$$

The isometric stress times the maximum strain is the maximum possible *volume-specific work*, meaning the amount of work that can be done by unit volume of muscle in one contraction. When the term "specific work" is used without qualification, it usually refers to the *mass-specific work*, found by dividing the volume-specific work by the density of muscle. This should be clear from the units, joules per cubic metre for volume-specific work and joules per kilogram for mass-specific work.

Of course, the work loop does not have to be rectangular. The concept was introduced by Boettiger (1957) to describe the dynamics of the asynchronous flight muscles of dipteran flies, which he did by connecting a force transducer to one end of an isolated muscle and a length transducer to the other end. Then, with the length displayed on the X-axis of his oscilloscope and force on the Y-axis, the work loop was drawn directly on the screen. The work loops of these asynchronous muscles, which oscillate while being continuously stimulated, were oval in shape, and quite narrow, as the force while the muscle was shortening was not much more than when it was lengthening. The specific work of muscles of this type is much lower than that of vertebrate muscles, but despite that they can still produce high values of specific *power*, because they operate at higher frequencies. There is a substantial literature on work loops allegedly measured in the flight muscles of living birds, but at these larger sizes it is difficult or impossible to disentangle the properties of the muscle from the dynamic characteristics of the measuring apparatus. The discussion below refers to a rectangular work loop with zero stress during lengthening. Obviously, this is a simplified scenario, but when combined with the dynamic principles discovered by A.V. Hill in the 1930s (Box 7.1), it provides a readily intelligible basis for outlining the "engine" characteristics of a flying vertebrate's flight muscles, in a way that replaces the discussion of engines in aeronautical textbooks.

## 7.2.4 UNIFORMITY OF THE SLIDING-FILAMENT MECHANISM

Gordon et al. (1966) used frog muscle for the experiments on which Figure 7.2 is based, so one might wonder whether the results are representative of other vertebrate muscles, such as the flight muscles of birds and bats, or the swimming muscles of whales. It seems that they are. According to H.E. Huxley (1985), who surveyed this, the thickness and length of both the actin and myosin filaments are the same in the locomotor muscles of all vertebrates, and so is the scale of sarcomere lengths along the top of Figure 7.2. The cross-sectional geometry and the density of the filaments are also constant in different vertebrates, as is the force that each filament can exert. The most probable reason is that, given the basic hexagonal geometry, which is the same in all vertebrates but not in all other groups of animals, there is very little scope for reshaping the filaments in a way that would enhance their performance. It might be possible to make each myosin filament exert more force, by lengthening it and adding more cross-bridges at each end, but that would require the middle section to be stronger, and therefore thicker, and then the surrounding hexagon of actin filaments would no longer fit. The conclusion is that such quantities as isometric stress and active strain applied to the myofibrils (not to the whole muscle) can be regarded as constant in all flying vertebrates.

# 7.3 ● MUSCLE PERFORMANCE IN LOCOMOTION

## 7.3.1 ISOMETRIC STRESS AND ACTIVE STRAIN
## AS PERFORMANCE CONSTANTS

An *isometric* contraction is one in which the muscle is maximally stimulated by a stream of electrical impulses, but not allowed to shorten, so that its length remains constant. The isometric *stress* is the force exerted by each unit of cross-sectional area, and this is one of two quantities that determine the amount of work that a muscle can do in one contraction. The isometric stress may be considered constant in vertebrates, reflecting the force exerted by each myosin filament, since these are packed at a constant density of about $5.7 \times 10^{14}$ filaments per square metre (White and Thorson, 1975). However, it would be something of a challenge to measure this force at the level of the individual myosin filament, and measuring the stress over a whole muscle tends to underestimate the stress in the myofibrils. This is because the measured cross-sectional area includes components that do not contribute to the force, and also because it is difficult to be sure that an isolated muscle, set up as a physiological preparation, can exert as much force as it would under normal conditions in the intact animal. Estimates of isometric stress in the literature (Alexander and Bennet-Clark, 1977; Alexander, 1985) mostly range from 300 to 450 kNm$^{-2}$ for vertebrate muscles, and generally refer either to whole muscles or isolated muscle fibres. However, indirect estimates based on the known flight performance of very large birds (swans) indicate that the myofibrils must be capable of exerting an isometric stress of at least 560 kNm$^{-2}$ (Section 7.3.7). This figure, which is used as the default in *Flight*, is above the experimental range for vertebrates, but well below figures of 800 kNm$^{-2}$ reported for crayfish skeletal muscles, and 1400 kNm$^{-2}$ for bivalve shell closer muscles.

The second constant that determines muscle performance is the *active strain*, which is the distance through which the muscle shortens, divided by the initial (extended) length. The measurements of Gordon et al. (1966), on which Figure 7.2 is based, supply a good estimate of the upper limit of the active strain. We can take the length of a sarcomere at point A in Figure 7.2 (2.25 $\mu$m) to represent the extended length, as the skeleton will not normally allow the muscle to be extended beyond the greatest length where all cross-bridges are still connected. If the muscle shortens from point A to point C, the length of each sarcomere decreases by 0.58 to 1.67 $\mu$m. The maximum strain is therefore 0.58/2.25 = 0.26, when the muscle is working between points A and C, and that goes for sparrow and whale muscles alike. Multiplying 560 kNm$^{-2}$ by 0.26 gives 146 kJm$^{-3}$ for the maximum volume-specific work. If we divide this by the density of muscle (1060 kgm$^{-3}$), we get 137 Jkg$^{-1}$ for the maximum mass-specific work. This is represented by the area under the curve between points A and C in Figure 7.2.

The $Y$-scale in Figure 7.2 is "relative stress", meaning the ratio of the actual stress to the isometric stress. Of course, if the relative stress is 1, the muscle cannot shorten (by definition), and the rate at which it does work (the power) is zero. If the relative stress is zero, the muscle shortens freely along the $X$-axis in Figure 7.2, but produces no power in this case either, because no work is done. To produce positive power, the relative stress has to be between zero and 1, and it follows from Hill's equation (Box 7.1) that the maximum instantaneous power is produced at a relative stress of 0.30, while adenosine triphosphate (ATP) energy is converted into work with maximum *efficiency* at a higher relative stress of 0.56. In these two special cases, the specific work is proportional to the area of the box in Figure 7.2 between points A and C and between the X-axis and the horizontal line for maximum power or maximum efficiency. The specific work for maximum power (hatched area at the bottom) is 41.1 Jkg$^{-1}$, and that for maximum efficiency (grey area) is 76.7 Jkg$^{-1}$. Maximum power is attained by lowering the stress so that less work is done, but doing the work faster (or more often in the case of repetitive contraction).

**BOX 7.1 The force–velocity relationship.**

The study of muscle mechanics was put on a quantitative basis in a famous paper by A.V. Hill (1938), in which he analysed the activity of isolated frog muscles, at a time long before the molecular basis of muscular contraction was known. The method was to remove the muscle from the animal, and install it in an apparatus that applied a measured, constant tension force for the muscle to pull against, and measured the speed at which it shortened when stimulated.

**Variable definitions for this box**

| | |
|---|---|
| $\alpha$ | Stress constant |
| $\beta$ | Strain rate constant |
| $\lambda$ | Active strain |
| $\sigma_{max}$ | Isometric stress |
| $\sigma$ | Active stress |
| $\psi$ | Strain rate |
| $\psi_{max}$ | Maximum strain rate |

**The force–velocity relationship**

The original form of Hill's equation related the speed of shortening to the tension in an *isotonic* contraction, that is, one in which the tension against which the muscle pulled was held constant and measured by the apparatus. As the tension was increased, so the speed of shortening decreased, until at some value of the tension, the muscle was unable to shorten, and the speed was zero. This tension is the *isometric* tension, meaning that the muscle stays at the same length. In order to make the same equation apply to muscles of any size and shape, Hill (1950) later re-expressed it to relate *strain rate* (rather than speed) to *stress* (rather than force). The strain rate ($\psi$) is the speed divided by the extended length, and the stress ($\sigma$) is the tension divided by the cross-sectional area of the muscle (which is itself directly related to the force that each myosin filament exerts). In this form, Hill's equation is:

$$\psi = \beta(\sigma_{max} - \sigma)/(\sigma + \alpha) \qquad (1)$$

**BOX 7.1** *Continued.*

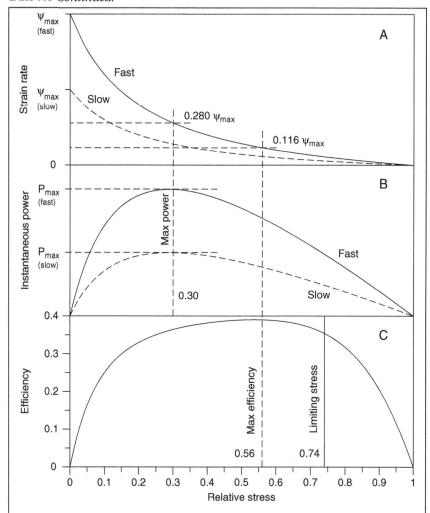

FIGURE 7.3 Muscle properties plotted against "relative stress", defined as the tensile stress exerted by the muscle while shortening, divided by the isometric stress. (A) Strain rate is the speed of shortening divided by the extended length. The maximum strain rate ($\psi_{max}$), obtained when the muscle shortens freely against zero stress, is a variable property of the muscle, which distinguishes "fast" from "slow" muscles. It also determines the rate of fuel energy consumption when the muscle is exerting steady tension but doing no external work (Box 7.2). (B) The instantaneous power output while the muscle is shortening shows a maximum when the relative stress is 0.30, and is higher for a fast muscle than for a slow one. (C) The efficiency of converting ATP energy into work peaks at nearly 0.39 at a relative stress of 0.56, whether the muscle is fast or slow. 0.74 is a limiting value of the relative stress, above which declining efficiency offsets any gain of power due to increasing the stress. After Pennycuick (1992).

BOX 7.1 *Continued.*

where $\sigma_{max}$ is the isometric stress, $\alpha$ is a constant with the dimensions of stress, and $\beta$ is another constant with the dimensions of strain rate. This curve is a hyperbola, and it is shown in Figure 7.3A. If the stress is zero, then the muscle shortens freely at its *maximum strain rate*, which is:

$$\psi_{max} = \beta\sigma_{max}/\alpha \qquad (2)$$

Multiplying $\psi_{max}$ by the extended length gives the speed at which the muscle shortens with zero tension. Hill called this the muscle's "intrinsic speed", indicating that its value is a property that characterises the muscle as slow or fast. $\psi_{max}$ differs from the intrinsic speed in having the dimensions of inverse time ($\mathbf{T}^{-1}$), rather than those of speed ($\mathbf{LT}^{-1}$). Its advantage is that it does not depend on the length of the muscle, whereas the intrinsic speed does. At the molecular level, the value of $\psi_{max}$ comes down to the rate constants that determine the frequency with which myosin cross-bridges detach and reattach from binding sites on adjacent actin filaments. These are not fixed, but are slowly adjustable (during growth, for example), so as to match the muscle to the characteristics of its load.

**Matching the muscles to the load**

An animal has only a limited amount of control over the strain rate at which a muscle shortens during locomotion, as this is determined mainly by the mechanical characteristics of the load to which it is attached. It depends strongly on the size of the animal (higher in smaller animals). "Matching" the muscle to its load consists in adjusting the maximum strain rate ($\psi_{max}$). For example, the flight muscles of a bird that requires a wingbeat frequency of 5 Hz (like a cormorant) must be able to complete one cycle of contraction and relaxation in one wingbeat period, lasting 0.2 s. In maximal exertion, the muscle has to shorten through a strain of about 0.25 in about half the wingbeat period (0.1 s), which amounts to a strain rate of $0.25/0.1 = 2.5\ \mathrm{s}^{-1}$. For the muscle to deliver maximum power, the strain rate at which the muscles shorten needs to be about 28% of $\psi_{max}$ (Figure 7.4), in other words, $\psi_{max}$ needs to be set to $2.5/0.28 = 8.9\ \mathrm{s}^{-1}$. The bird has only a limited capacity to alter its wingbeat frequency to suit the characteristics of its muscles, as the wings cannot be driven far from the natural wingbeat frequency, which is set by the physics of the beating wings (Box 7.3). The matching adjustment consists in setting the rate constants of the sliding filaments, to give the required value for the maximum strain rate. It is not known whether this adjustment is fully determined genetically, or whether (as seems likely) the setting can be modified on a short time scale, by exercise training or otherwise. During a long migratory flight, the reduction of body mass caused by consumption of fuel is expected to lead in turn to a reduction of wingbeat frequency, and this would require a corresponding in-flight adjustment of the maximum strain rate of the flight muscles, if maximum efficiency is to be maintained.

## 7.3.2 THE FORCE–VELOCITY RELATIONSHIP

To see where these numbers come from, we have to consider the dynamics of muscular contraction, a subject that was explored in a famous series of papers by A.V. Hill and colleagues in the 1930s, before the invention of the electron microscope led to the discovery of the sliding-filament mechanism. If the muscle is allowed to shorten instead of being held at a constant length, the stress that it develops falls below the isometric stress, as shown in Figure 7.3A. The two curves in the figure were calculated from Hill's equation, which relates the speed of shortening to the tension in an "isotonic" contraction, that is, one in which the apparatus holds the force constant while the muscle shortens. A variant of Hill's equation is given in Box 7.1, which expresses the same relationship in terms of the *strain rate* (rather than speed) at which the muscle shortens and the stress (rather than force) that it exerts. The strain rate is the speed, divided by the initial length of the muscle. It can be thought of as speed expressed in muscle lengths per second, but it is not necessary to measure strain rate in units of m s$^{-1}$ per metre, as some physiologists have been known to do. The metres cancel, and the units are simply s$^{-1}$ or "per second", corresponding to the dimensions of inverse time ($\mathbf{T}^{-1}$).

To avoid being too long-winded, I shall refer to stress and strain rate by the Greek letters commonly used as variable names, that is, $\sigma$ (sigma) for stress and $\psi$ (psi) for strain rate. The X-axis of the three graphs of Figure 7.3 is "relative stress" ($\sigma_{rel}$) which means the actual stress ($\sigma$), divided by the isometric stress ($\sigma_{max}$). $\sigma_{max}$ can be considered constant for all vertebrate skeletal muscles, as its value is set by the properties of the myosin filaments (above). The stress may exceed $\sigma_{max}$ when an active muscle is forcibly lengthened, but this is outside the scope of this account. With that restriction, $\sigma_{rel}$ is a dimensionless measure of stress, whose value goes from 0 to 1. At the right-hand side of Figure 7.3A, $\sigma_{rel} = 1$, and the strain rate is zero because the muscle cannot shorten (by definition). At the left-hand end ($\sigma_{rel} = 0$), the muscle shortens freely, at some *maximum strain rate* ($\psi_{max}$).

$\psi_{max}$ is *not* the same for different muscles. On the contrary, $\psi_{max}$ is a very important characteristic that distinguishes one muscle from another. Two curves are shown in Figure 7.3A for a "fast" and a "slow" muscle, which differ in having different values of $\psi_{max}$. The value of $\psi_{max}$ is adjustable in the individual muscle over a wide range, and this adjustment is essential to match the muscle's properties to those of the load against which it operates (below).

### 7.3.3 POWER AND EFFICIENCY

The *rate* at which a muscle does work, as it simultaneously exerts tension and shortens, is the instantaneous power output, while shortening is in progress. Multiplying the stress by the strain rate during shortening gives the instantaneous volume-specific power, and this is plotted as a function of the relative stress in Figure 7.3B. It is zero when $\sigma_{rel} = 0$, because the muscle is shortening against zero force, and also at $\sigma_{rel} = 1$, because then the strain rate is zero. The relative stress for maximum power can be found from Hill's equation, and marks the peak of the curve in Figure 7.3B at $\sigma_{rel} = 0.30$. The work done over a given amount of strain is the same for a fast muscle (high $\psi_{max}$) as for a slow one, but the slow muscle takes longer to do it, and so its instantaneous power output at any given value of $\sigma$ is less.

In the same paper in which he introduced the force–velocity relationship now known as Hill's equation, Hill (1938) also studied the time course of the tiny amounts of heat produced by the muscle. An active muscle produces heat whether it is doing work or not, showing that ATP energy, stored inside the isolated fibre, is being consumed. If the relative stress is above zero but less than 1, the muscle produces both heat and work. The "efficiency" with which the muscle converts fuel energy into work is the ratio of the amount of work produced to the sum of the work and the heat (Box 7.2). This is zero at $\sigma_{rel} = 0$, and also at $\sigma_{rel} = 1$, and Hill's results showed that it passes through a rather broad peak in between. Figure 7.3C is calculated from a later quantitative sliding-filament theory by A.F. Huxley (1957), which successfully reproduced Hill's results, and also other results that were discovered later, a *tour de force* which has been described in a complete but highly readable form by McMahon (1984). The efficiency in Huxley's theory refers to the consumption of ATP energy inside the muscle fibre, not to the consumption of fat or carbohydrate fuel. According to this theory, the efficiency peaks at 0.39 when the relative stress is 0.56. Although a fast muscle produces work faster than a slow one, it also produces heat faster, and the efficiency curve of Figure 7.3C is the same for either fast or slow muscles.

### 7.3.4 MUSCLE POWER AND WINGBEAT FREQUENCY

Because of the intermittent action of muscle, doing work only while it is shortening, the instantaneous power during shortening is not the appropriate variable to match the power available from the muscles to that required by the flapping wings. For that, we need to estimate the *average* power available from the muscle, which comes down to estimating the work that it can do in one contraction, and the frequency with which it can contract. Muscles of the same type, whether large or small, develop the same stress and strain in the myofibrils, and therefore the same specific work. Mass-specific *power* is the average mechanical power output per unit mass of muscle, and is equal to the specific work times the contraction frequency. Since the flight muscles drive the wings directly, their contraction frequency is the same as the wingbeat frequency. The animal has only a limited amount of control over the wingbeat frequency, as it cannot deviate too far from a natural frequency which is determined by the morphology of the wings, the strength of gravity, and the air density (Box 7.3). Large birds flap their wings at lower frequencies than small ones, and it follows that they have less specific power available from their flight muscles, even though the specific work may be the same. *Flight* calculates both the specific work and the specific power required at a given speed, and monitors changes in both during long migratory flights (Chapter 8).

BOX 7.2 **Efficiency of muscle.**

In the same paper in which he studied the force-velocity relationship, Hill (1938) also measured the amounts of heat produced when muscles contract, and its time course. Noting that all of the fuel energy consumed when a muscle contracts is converted either into work or into heat, Hill defined the efficiency in terms of the work and heat generated, as:

$$\text{Efficiency} = \text{Work}/(\text{Work} + \text{Heat}) \tag{1}$$

According to Gnaiger (1989), all of the input energy is not necessarily converted into work and heat, if entropy changes are also taken into account. However, Hill's definition of efficiency holds for the conversion of ATP energy into work in the myofibrils (although he was unaware of their existence at the time), and it is also a good approximation for the aerobic (but not anaerobic) conversion of fuel energy into ATP energy. Hill's results were later accounted for quantitatively by Huxley (1957), in terms of the molecular dynamics of the sliding filaments, and this theory was reviewed and assessed in the light of a large amount of later evidence by McMahon (1984). McMahon's Equation (4.32) is a molecular version of Equation (1) in Box 7.1, relating stress to strain rate, and his Equation (4.37) relates the rate of consumption of ATP energy by the sliding filaments to strain rate. These equations are somewhat complicated, in effect relating the external stress produced by the muscle to the strain rate, the sarcomere length, the density and spacing of crossbridges, the maximum deflection of a crossbridge, the amount of ATP energy consumed by a crossbridge in one cycle of attachment and detachment, the maximum work done (three quarters of the ATP energy), and three rate constants (adjustable from one muscle to another) that determine the rates of attachment and detachment of crossbridges. These equations were later combined and simplified to give the efficiency (Pennycuick, 1991), but it should be noted that this published version contains errors in the equations (rectified below), owing to an editorial accident.

BOX 7.2 *Continued.*

## Variable definitions for this box

$p_{ATP}$   Volume-specific rate of consumption of ATP energy
$\sigma$   Active stress
$\sigma_{max}$   Isometric stress
$\sigma_{rel}$   Relative stress
$\psi$   Strain rate
$\psi_{max}$   Maximum strain rate
$\psi_{rel}$   Relative strain rate ($\psi / \psi_{max}$)
$\eta$   Efficiency

## Efficiency of the sliding filaments

McMahon's (1984) Equations (4.32) and (4.37) can be combined to give the efficiency ($\eta$) in terms of a single variable, the "relative strain rate" ($\psi_{rel}$). This is the actual strain rate ($\psi$) divided by the maximum strain rate ($\psi_{max}$), which is itself the adjustable property that distinguishes a fast muscle from a slow one (Box 7.1). The efficiency refers to the conversion of ATP energy (not fuel energy) into work, and is defined as the ratio of the volume-specific mechanical power during shortening (the stress times the strain rate) to the volume-specific "ATP power" ($p_{ATP}$), which is the rate of consuming ATP energy per unit volume of muscle:

$$\eta = \sigma\psi / p_{ATP}$$
$$= (48/13)\psi_{rel}[(1 - 4\psi_{rel}))(1 - \exp(-1/4\psi_{rel})) \tag{2}$$
$$(1 + 0.13\psi_{rel})]/[(3/13) + 4\psi_{rel}(1 - \exp(-1/4\psi_{rel}))]$$

The efficiency peaks (Figure 7.4) just below 0.4 at a low value of $\psi_{rel}$ (about 0.12). To look at it another way, if the muscle is to convert ATP energy into work with maximal efficiency, then $\psi_{max}$ must be adjusted to about 8 times the strain rate at which the muscle normally shortens during locomotion. The strain rate in flapping flight is determined by the wingbeat frequency (Box 7.3). The flight muscles have to be matched to their load, by adjusting $\psi_{max}$ to suit the strain rate imposed by the wings and the air (Box 7.1).

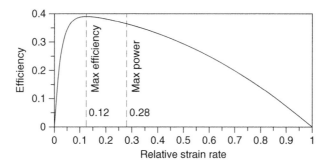

FIGURE 7.4 If efficiency is plotted against "relative strain rate" (ratio of strain rate to $\psi_{max}$ for the particular muscle), the peak efficiency is the same as Figure 7.3C, at a relative strain rate of about 0.12. After Pennycuick (1992).

BOX 7.2 *Continued.*

Since $\psi$ in Equation (2) is related by Hill's equation (Box 7.1) to the stress ($\sigma$), the efficiency in Equation (2) can also be expressed as a function of the dimensionless *relative stress* ($\sigma_{rel}$), which is the ratio of the actual stress to the isometric stress ($\sigma_{max}$). This curve (Figure 7.3C) is more symmetrical than the curve of efficiency versus relative strain rate, with an ill-defined maximum at $\sigma_{rel} = 0.56$. It shows that a muscle can operate without much loss of efficiency at any stress between about 20% and 80% of $\sigma_{max}$, but that the efficiency falls steeply to zero at either end of the curve, where the stress is either so high that the muscle can hardly shorten or so low that it has insufficient resistance to shorten against.

**Efficiency of mitochondria**

The conversion efficiency of mitochondria was included in classical studies of human exercise physiology (Wilkie, 1968; Margaria, 1976), with an approximate value of 60% for the conversion of the energy liberated by oxidising a substrate (fuel energy) to ATP energy. Combining this with a peak efficiency near 40% for converting ATP energy to work (above) gives an estimate of 24% for the overall efficiency of converting fuel energy to work, under optimal conditions. In bird migration, where conversion efficiency is most important, the fuel substrate is nearly all fat, with a small percentage (probably 5%) of the energy coming from protein. It is not known whether the different substrates used by human athletes would lead to a large difference in the conversion efficiency of the mitochondria (one way or the other), but there is no reason at present to think so.

**Overall conversion efficiency**

*Flight* calculates the mechanical work that the flight muscles are required to produce, and then asks what mass of fat is required to account for that amount of work. The *energy density* of fat from bomb calorimetry is used to find the fuel energy available from a given mass of fat, and the *conversion efficiency* is the fraction of this energy that is assumed to appear as work. This is difficult to measure, but two experimental results have been published (Tucker, 1972; Bernstein et al, 1973), using birds flying in a tilting wind tunnel, and wearing oxygen masks. The efficiency was determined by an "incremental" method, tilting the tunnel by a small amount, and comparing the change in the rate of oxygen consumption with the change in mechanical power due to the gradient. Only the increments of mechanical and chemical powers had to be measured, not the absolute value of either. There was a good deal of scatter, but the mean values of both experiments were near to 0.23, and this is the default value used in *Flight*.

There have been claims that birds are spectacularly inefficient, based on measurements of chemical power of birds flying horizontally in wind tunnels, compared with calculated (not measured) mechanical power. This is wholly invalid. A low estimate of efficiency from this method actually means an unexpectedly high measurement of chemical power, which can be due to all kinds of reasons, not least poor training, with birds not flying steadily during the experiment.

BOX 7.2 *Continued.*

## Maintenance of tension

As noted at the beginning of this chapter, animals use muscles for two distinct functions, converting fuel energy into work and maintaining steady tension in an element of variable length. A muscle that is maintaining a steady force does no work, but consumes fuel energy because the crossbridges continually attach and detach whenever the muscle is active, whether or not it is allowed to shorten. The efficiency, as defined above in terms of producing work, is zero, but this is not a fair reflection of the muscle's success in performing its function, which is to maintain tension in this case, not to do work. The "cost" of maintaining tension can be represented as the ratio of the chemical power expended to the steady force maintained. This ratio has the dimensions of power/force, the same as speed ($\mathbf{LT^{-1}}$). It is not difficult to guess that the cost of maintaining tension is connected in some way with the speed at which the muscle could shorten, if it were allowed to, and indeed it has long been known that slow muscles are more economical for maintaining tension than fast ones. This characteristic is measured by the maximum strain rate ($\psi_{max}$), which has dimensions of inverse time ($\mathbf{T^{-1}}$), as does the ratio of volume-specific power to stress. The ratio of the volume-specific ATP power ($p_{ATP}$) to the isometric stress ($\sigma_{max}$), when the strain rate is set to zero, can be determined from McMahon's equations (Pennycuick, 1991) and turns out to be:

$$p_{ATP}/\sigma_{max} = \psi_{max}/16 \qquad (3)$$

There is no need to make measurements of force and metabolic rate to determine the metabolic cost of maintaining tension. All you have to do is measure $\psi_{max}$, the maximum strain rate. This is a measurement that a determined observer can make on his own biceps muscle. Divide it by 16, and multiply by the volume of the muscle to get the absolute ATP power (because $p_{ATP}$ is volume-specific). To get the metabolic power, you have to allow for the efficiency of the mitochondria (try 60%).

The flexor muscles of a concert pianist's fingers require a high $\psi_{max}$ to perform their function, and this makes them very inefficient at maintaining steady force, in the event that the pianist is required to hang from a window ledge by his finger tips. Conversely, orangutans require slow finger flexor muscles as they hang by their fingers from branches for hours at a time, and this makes them incapable of rapid and intricate finger movements. The pectoralis muscles of gliding birds have both of these conflicting requirements at the same time, as they have to hold the wing steadily in the horizontal position when the bird is gliding, but shorten at a strain rate that is set by the flapping frequency in powered flight. Birds like vultures and albatrosses, that spend a lot of time gliding, but also have to be able to flap their wings when necessary, have a pectoralis muscle that is divided into two parts, a dark red superficial part and a deeper part which is lighter in colour. It is assumed that the darker part is a fast muscle used for flapping the wing, and the lighter part is a slow, tonic muscle used to hold the wing level while gliding. If both parts can operate aerobically, the deeper colour would indicate a higher specific chemical power in the fast part.

BOX 7.2 *Continued.*

> Equation (3) indicates that a muscle that cannot shorten at all ($\psi_{max} = 0$) would not require any power to maintain a steady tension, and indeed a passive wire or tendon will do this without consuming any energy, if the length does not need to be adjustable. Albatrosses, apparently alone among birds, have a tendon sheet in parallel with the pectoralis muscle, which restrains the wing from rising above the horizontal position when the humerus is pulled fully forwards, but unlocks to allow the wing to flap, when the humerus is pulled back by a small amount. In other soaring birds, gliding flight is not as "effortless" is it may appear. The bird has to support its weight on its elbows in a manner somewhat similar to a man between two filing cabinets.

## 7.3.5 MATCHING THE MUSCLE TO THE LOAD

If a bird beats its wings through a fixed angle, but varies the wingbeat frequency, then the strain rate during the downstroke is directly proportional to the wingbeat frequency. The Huxley sliding-filament theory can be used to produce a curve of efficiency versus strain rate (Figure 7.4). The efficiency is zero when the strain rate is zero (isometric contraction), and also when the muscle shortens at its maximum strain rate (because the stress is zero). Unlike the curve of efficiency versus relative stress, this curve is strongly skewed. The efficiency peaks at a low strain rate, about 0.12 $\psi_{max}$, while maximum power calls for a higher strain rate, about 0.28 $\psi_{max}$, where the efficiency is still high, about 93% of the maximum. Unlike the isometric stress, which can be considered constant for all muscles of the same type, the maximum strain rate ($\psi_{max}$) varies from one muscle to another (above). The actual strain rate in flight is constrained by the size of the animal and its wing morphology, and the rate constants have to be set (by adjusting the biochemistry of the sliding-filament mechanism) so that the muscle's maximum strain rate is matched to the actual strain rate, imposed upon it by the load. Hummingbird flight muscles would effectively be in permanent isometric contraction if installed in a swan, because their maximum strain rate is too high. Conversely, swan muscles in a hummingbird would not be able to shorten fast enough to keep up with the wings, and would not generate any stress.

A particular bird requires a lower wingbeat frequency for maximum efficiency in cruising flight than for maximum power in takeoff and climb. Pilots who are familiar with variable-pitch propellers will see a direct analogy. For take-off, the pilot sets the propeller blades in fine pitch, so that they offer a low resistance to the air, and allow the engine to run at its maximum revolutions. For economical cruising, the blades are set to a coarser pitch, which forces the engine to slow down and increase its torque. Likewise a bird sets its wings at takeoff to allow the muscles to shorten with a low stress and high strain rate, maximising the wingbeat frequency and the power, but when cruising it sets up higher stress and lower frequency for maximum efficiency.

There is not enough information to allow the *Flight* programme to simulate changes of wingbeat frequency, as the wingbeat amplitude (the angle through which the wings beat) also affects the stress, strain and power, and is under the bird's control to some extent. *Flight* calculates a single value for the wingbeat frequency (Box 7.3), which seems

---

BOX 7.3 **Wingbeat frequency.**

A bird's wingbeat frequency in flapping flight (wingbeats per second) determines the amount of power available from each gram of flight muscle, and also has aerodynamical implications which are mentioned in Chapter 4. To some degree, the wingbeat frequency is under the bird's control, but obviously small birds beat their wings at higher frequencies than large ones with wings of similar shape. Among birds of similar mass, those with small wings (ducks) beat them at higher frequencies than those with larger wings (gulls). One may surmise that there is a "natural" frequency, at which a particular bird can beat its wings most easily, and that this is determined by the physics of the beating wings. Actually calculating such a natural frequency, if it exists, is a much harder proposition than calculating power. However, there is another approach. Instead of attempting to calculate the wingbeat frequency directly, one can consider its *dimensions*, and also the dimensions of any variables believed to be involved in determining its value.

**Variable definitions for this box**
$B$    Wing span
$f$    Wingbeat frequency
$g$    Acceleration due to gravity
$m$    All-up mass
$S$    Wing area
$\rho$    Air density

The dimensions of all quantities encountered in mechanics can be expressed in terms of only three primary quantities, the choice of which is to some extent arbitrary. The convention in physics is to select mass, length and time as the three primary quantities (written **M**, **L** and **T**), and to represent the dimensions of all other quantities in terms of combinations of those three. For example, velocity has the dimensions of length divided by time (written $\mathbf{LT}^{-1}$), while acceleration has the dimensions of length divided by time-squared ($\mathbf{LT}^{-2}$). Any equation says that the expression on the left-hand side of the "equals" sign is numerically identical to the expression on the right, and if the equation represents physical quantities, then the

**BOX 7.3** *Continued.*

dimensions must also be the same on both sides. Having noted that the physical quantity "frequency" has the dimensions of inverse time ($T^{-1}$), the problem is to list those physical variables that are likely to determine the wingbeat frequency and find a way to combine them, so that the result also has dimensions $T^{-1}$. I will skip ahead to the solution:

$$f \propto m^{3/8} g^{1/2} B^{-23/24} S^{-1/3} \rho^{-3/8} \tag{1}$$

where $m$ is the all-up mass, $g$ is the acceleration due to gravity, $B$ is the wing span, $S$ is the wing area and $\rho$ is the air density. The symbol "$\propto$" means "is proportional to"; in other words, the frequency is equal to the expression on the right, multiplied by an unknown constant. The dimensions of the five variables on the right-hand side are as follows:

$m$   $\mathbf{M}$
$g$   $\mathbf{LT}^{-2}$
$B$   $\mathbf{L}$
$S$   $\mathbf{L}^2$
$\rho$   $\mathbf{ML}^{-3}$

If Proportionality 1 is correct, and the unknown constant is dimensionless, then the dimensions must be the same on both sides of the proportionality sign:

$$\mathbf{T}^{-1} = (\mathbf{M})^{3/8}(\mathbf{LT}^{-2})^{1/2}(\mathbf{L})^{-23/24}(\mathbf{L}^2)^{-1/3}(\mathbf{ML}^{-3})^{-3/8} \tag{2}$$

We can separately collect the exponents of mass, length and time on the right-hand side:

Mass:   $3/8 - 3/8 = 0$
Length:   $1/2 - 23/24 - 2/3 + 9/8 = (12 - 23 - 16 + 27)/24 = 0$
Time:   $-1$

The exponents of mass cancel, and so do those of length. This leaves inverse time, confirming that Proportionality 1 is dimensionally correct, and also that any unknown constant is dimensionless.

Unlikely as it may seem at first sight, the expression on the right-hand side of Proportionality 1 is indeed a frequency. Two data sets consisting of observed speeds and wingbeat frequencies of a heterogeneous set of wild bird species, whose masses, wing spans and wing areas were approximately known for each species (but not for individual birds), were used in the original derivation (Pennycuick 1990, 1996) to adjust the exponents to give a best fit to the field data. It was first noted that $g$ is the only variable whose dimensions ($\mathbf{LT}^{-2}$) contain time, so that the required dimensions for the result ($\mathbf{T}^{-1}$) can only be obtained if the exponent of $g$ is $1/2$. That left four other exponents to be determined, for $m$, $B$, $S$ and $\rho$, and only two conditions, that the dimensions of length and mass must both be zero. However, it turned out that the scope for adjusting the four remaining exponents was quite limited because of various constraints, within which a "best fit" solution was found by examining partial regression coefficients of the observed

**BOX 7.3** *Continued.*

frequency on mass, wing span and wing area. It turned out that the constant of proportionality is indistinguishable from 1. In other words, for predictive purposes, Proportionality 1 may be replaced by an equation:

$$f = m^{3/8} g^{1/2} B^{-23/24} S^{-1/3} \rho^{-3/8} \tag{3}$$

Figure 15.13 in Chapter 15 is from a later data set, which also included measurements of speed. It showed that Equation (3) predicted the actual wingbeat frequency quite well, except in passerine species that use "bounding" flight, where the actual frequency was higher than predicted. This intermittent flight style requires the bird to pull up during the flapping phase, which results in an increase of gravity, and when this is taken into account, the observed frequencies match the predictions for these species also (Chapter 15). The speeds in all cases appeared to be close to the minimum power speed for each species. Thus, the frequency calculated from Equation (3) is used in *Flight* as an estimate of the wingbeat frequency of a bird flying at its minimum power speed.

### Allometry of wingbeat frequency

It is an error to conclude from Equation (3) (as some authors have done) that the wingbeat frequencies of a set of birds of different body mass should vary with the 3/8 power of the body mass. This is because two other variables in Equation (3), $B$ and $S$, themselves vary allometrically with the mass. From Equation (3):

$$\begin{aligned} f &\propto m^{3/8} \times m^{-(1/3)(23/24)} \times m^{-(2/3)(1/3)}, \\ &= m^{(27-23-16)/72} = m^{-1/6} \end{aligned} \tag{4}$$

Bigger birds have lower wingbeat frequencies, as everybody knows. When comparing species, and not taking account of systematic trends in wing span and wing area, we expect that the wingbeat frequency will vary with the $-1/6$ power of the mass. However, the situation in an individual bird is different again. Here, the mass may increase as a result of feeding and fat deposition, or decrease during a long flight because of the consumption of fat, but the wing measurements stay the same. In this case, the wingbeat frequency varies with the square root of the mass: For an individual:

$$f \propto m^{1/2} \tag{5}$$

Why not $m^{3/8}$? Because the extra 1/8 in the exponent of mass comes from the wing's moment of inertia, an additional variable which was "hidden" in the original formulation because it cannot be measured without killing the bird (Pennycuick, 1990). When the bird gains or loses mass, the wing moment of inertia is assumed to remain constant, like the span and the area. Proportionality 5 is used to adjust the wingbeat frequency of a long-distance migrant in the "Migration" section of *Flight*, as the bird consumes fuel and loses mass.

### Measuring wingbeat frequency from video

Standard frame rates for analogue video are 25 Hz in Europe and 30 Hz in America, but in equipment that records interlaced frames, it is sometimes

BOX 7.3 *Continued.*

possible to separate the two fields that make up each frame, so obtaining pictures with reduced vertical resolution, at rates of 50 and 60 Hz, respectively. These rates are sufficient for large and medium-sized birds, but marginal for small passerines. High-speed video cameras exist, but these are expensive, specialised devices that are mostly not ideal for field use.

An ideal sequence for measuring wingbeat frequency, either in the field or in the wind tunnel, consists of 20 or so wingbeats of *steady* flapping, and the measurement consists of counting the number of frames for a given number of wingbeats. A video editing system that keeps count of the frames is essential for this. Counting wingbeats in a consistent way depends on selecting an identifiable event that can be used to mark the beginning of each wingbeat cycle. The best marker is the beginning of the downstroke, at the moment when the wing rotates at the shoulder to develop a positive angle of attack, and bends upwards as the lift develops and pulls against the pectoralis muscle. This happens quickly, and the appearance of the wing alters abruptly, when seen from a variety of different directions. Single-step the video to determine the frame number at the first such transition (number zero) then run it in slow motion, counting transitions, and single-step again to find the frame number of the last one in the sequence.

If the bird is bounding, or flap-gliding with clear-cut transitions between flapping and gliding, then the wingbeat frequency should be measured within a period of steady flapping, not averaged over periods when the bird is flapping and periods when it is not.

to agree with values observed in cruising flight (Pennycuick, 2001). This "natural" wingbeat frequency is determined by the mass and wing measurements of the bird, and strength of gravity and the air density. The properties of the flight muscles have to be matched to it, and one might suppose that the maximum strain rate ($\dot{\psi}_{max}$) of the flight muscles would be set to give maximum efficiency at the cruising wingbeat frequency. However, if this were so, the bird would have to increase its wingbeat frequency by a factor of 2.4 when maximum power is required for takeoff and initial climb. Birds do not appear to do that, probably because the stresses at the wing root would become excessive if the frequency were increased so far above the cruising level. It is more likely that the maximum strain rate is normally set so that the cruising wingbeat frequency is only slightly below that for maximum power. That would make maximum power readily available when needed, and still allow the bird to approach the maximum-efficiency point in cruising flight, by reducing the wingbeat frequency and/or amplitude, and increasing the stress. In considering climbing performance (below), it is assumed that the muscles are already contracting near the frequency for maximum power when the bird is in level flight. When reduced power is required, as in a gentle descent, some birds (especially gulls and terns) flap steadily at a reduced frequency, while others flap at the cruising frequency, but do so intermittently, gliding between periods of flapping.

## 7.3.6 SCALING OF MUSCLE POWER OUTPUT AND POWER MARGIN

The notion of "scaling" (Box 7.4) refers to the changes that occur in different variables if you "scale" an animal up or down, meaning enlarge or diminish it, keeping the shape unchanged. The most casual observation of finches, pigeons and herons is enough to reveal the

BOX 7.4 **Scaling of power available and required.**

**Variable definitions for this box**

| | |
|---|---|
| $A$ | Area |
| $f$ | Contraction frequency |
| $l$ | Length |
| $m$ | Mass |
| $P$ | Power |
| $Q$ | Cycle work |
| $V_{mp}$ | Minimum-power speed |
| $v$ | Volume |

The notion of "scaling" refers to a set of objects that are "geometrically similar" to one another. This is straightforward if the objects are simple geometrical shapes like cubes and spheres. For a set of spheres of different size, we say that the "surface area varies with the square of the radius" meaning that if you double the radius, the surface area goes up by a factor of 4 ($2^2$), and if you triple the radius, the surface area goes up by a factor of 9 ($2^3$), and so on. In shorthand, this is written:

$$A \propto l^2 \qquad (1)$$

where $A$ stands for "any area", $l$ stands for "any length", and the symbol "$\propto$" means "is proportional to". Proportionality 1 works equally well if $A$ is the area of one face of a cube and $l$ is the length of a side, in fact it works with objects of any shape, so long as we compare corresponding areas and corresponding lengths, in different-sized object of the same shape. Likewise, volumes ($v$) vary with the cube of the length:

$$v \propto l^3 \qquad (2)$$

If the density of all the objects is the same, then the mass ($m$) varies directly with the volume, and with the cube of the length:

$$m \propto v \propto l^3 \qquad (3)$$

BOX 7.4 *Continued.*

Scaling relationships in biology are most commonly expressed in terms of the mass, rather than the length. Inverting the above relationships, one can say that the length varies with the one-third power of the mass, and the area with the two-thirds power of the mass:

$$l \propto m^{1/3} \text{ and } A \propto m^{2/3} \tag{4}$$

A.V. Hill (1950) applied this type of reasoning to hypothetical sets of geometrically similar animals, and also extended it to dynamical quantities, such as speed, power, rate of heat production and so on. The *cycle work* ($Q$), meaning the amount of work that a muscle can do in one contraction, is directly proportional to the mass of the muscle. The power that the muscle can produce is the cycle work times the contraction frequency, which is the same as the flapping frequency in flying animals. If birds were all geometrically similar, then the flapping frequency ($f$) would vary with the $-1/6$ power of the mass (Box 7.3):

$$f \propto m^{-1/6} \tag{5}$$

The power that the flight muscles can produce would vary with the five-sixths power of the mass:

$$P = Qf \propto m \times m^{-1/6} = m^{5/6} \tag{6}$$

Thus, if you scale up a small duck to a larger duck by doubling all linear measurements, the mass increases by a factor of 8 ($2^3$), but the power available from the flight muscles increases by a factor of $8^{5/6} = 5.7$. The larger duck has less power available, per unit of its body mass. On the other hand, the power *required* to fly at $V_{mp}$ increases by a larger factor than the body mass. The reader may like to try scaling up the teal in the "Preset birds" in *Flight*. With the measurements given, and sea level air density, its minimum mechanical power is 2.43 W, but if you scale it up to a super-teal, with eight times the mass (1.88 kg), twice the wing span (1.16 m), and the same aspect ratio as before (7.40), the minimum power is now 27.6 W, up by a factor of 11.4, which is $8^{7/6}$. The super-teal requires more power to fly at $V_{mp}$, relative to its body mass, than the original, because $V_{mp}$ itself has gone up from 11.4 to 16.2 m s$^{-1}$. On the other hand, it has less power available from its flight muscles, relative to its body mass, because its wingbeat frequency has gone down by a factor of 0.71, from 7.92 to 5.62 Hz, i.e. by a factor of $8^{-1/6}$.

To some degree, bigger birds can defeat these trends by not being geometrically similar to their smaller relations. The wing spans of birds within a family usually vary with about the 0.37 power of the mass, rather than the one-third power as expected, whereas the wing areas vary with about the 0.63 power of the mass instead of the two-thirds power as expected; in other words, large birds usually have longer and narrower wings than smaller members of the same group. This reduces the slope of power required, but does little if anything to increase that of power available.

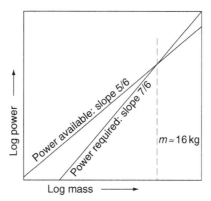

FIGURE 7.5 Logarithmic plot of power required to fly at $V_{mp}$, and power available from the flight muscles, versus body mass in geometrically similar birds. As real birds are not geometrically similar, the slope of power required may be slightly less steep than shown, and that of power available slightly steeper. There seems to be a practical upper limit around 16 kg for birds that can fly horizontally under present environmental conditions. This is a purely mechanical argument that does not involve aerobic capacity. After Pennycuick (1969).

general trend, that a larger bird beats its wings at a lower frequency than a smaller one. In a series of geometrically similar birds, the power available from the flight muscles would increase less steeply than the body mass, because of the decreasing wingbeat frequency (Figure 7.5), whereas the power required to fly at the minimum power speed would increase more steeply than the body mass. These differing trends imply that there is an upper limit to the mass of an animal that flies by flapping its wings. Above some maximum mass, more power is required than is available. Still bigger birds may be able to glide, but cannot fly horizontally. In practice, the maximum all-up mass of birds that are able to fly horizontally would appear to be around 16 kg, because still larger birds would have a "negative power margin". This means that the power available from the flight muscles is less than the power required to fly at $V_{mp}$, but it does not necessarily mean that such a bird would be unable to fly. One can imagine a huge bird that could walk up a hill like a hang-glider pilot, and launch itself off the top, and then carry on by soaring (Chapter 10). This same limit applies to ornithopters of the type proposed by Leonardo da Vinci, in which the flapping wings are driven directly by the pilot's legs, but not to pedal-powered aircraft like MacCready's Gossamer Condor and Gossamer Albatross. In these fixed-wing designs, a propeller is driven through a gear train, so that it can rotate at its own optimum angular velocity, without constraining the contraction frequency of the pilot's muscles.

Small birds have sufficient power to fly at speeds below and above $V_{mp}$, and still smaller birds have enough to fly at air speeds all the way down to zero (hovering). Some large kingfisher species seem to be the largest birds that can do this, and the still smaller humming-birds are specialised for prolonged, aerobic hovering. In the context of the evolution of flight, a non-flying animal has the best chance to develop powered flight if its body mass is in the range 10–100 g, since flying animals of that size have sufficient power margin to tolerate wings that do not yet work very well, without requiring wingbeat frequencies that are only possible for insects.

BOX 7.5 **Rate of climb.**

If the maximum mechanical power that can be produced by the flight mus-cles exceeds that required to fly horizontally, then the excess power can be used to increase the bird's potential energy, i.e. to climb. The maximum rate of climb at any speed is given by dividing the excess power available at that speed by the bird's all-up weight. The maximum excess power (and hence rate of climb) is obtained by flying at $V_{mp}$, because that is by definition the speed at which the least power is required to fly horizontally. It seems than many or most migrants begin a long flight by climbing to a cruising altitude of a few thousand metres, when fully loaded with fat, before level-ling off for the cruise phase of flight, and as this climb is likely to take a mat-ter of hours, the flight muscles must presumably be operating aerobically. This means that the mitochondria fraction in the flight muscles must be high enough to sustain the power required in the climb, and considerably higher than that needed in the cruise phase of level flight. No doubt future field observers will measure changes in the mitochondria fraction of the flight muscles of migrants, but until someone does that, the following account should be regarded as provisional.

**Variable definitions for this box**

| | |
|---|---|
| $F_{musc}$ | Flight muscle fraction |
| $f$ | Wingbeat frequency |
| $g$ | Acceleration due to gravity |
| $k_{mito}$ | Inverse power density of mitochondria |
| $m$ | All-up mass |
| $m_{musc}$ | Flight muscle mass (including mitochondria) |
| $P_{ex}$ | Excess mechanical power available for climbing |
| $P_{max}$ | Maximum power available from the flight muscles |
| $P_{mech}$ | Mechanical power required for horizontal flight |
| $p_m$ | Mass-specific power in the whole muscle including mitochondria |
| $q_m$ | Specific work required from the myofibrils for horizontal flight |
| $q_{mlim}$ | Limiting specific work available from the flight muscles |
| $V_z$ | Rate of climb (vertical component of air speed) |
| $\zeta$ | Volume fraction of mitochondria in flight muscles (level flight) |
| $\zeta_{climb}$ | Volume fraction of mitochondria in flight muscles (aerobic climb) |

BOX 7.5 *Continued.*

| | |
|---|---|
| $\lambda$ | Active strain |
| $\rho_{musc}$ | Density of muscle |
| $\sigma$ | Stress |
| $\sigma_{max}$ | Isometric stress |

It is argued in Box 7.1 that a muscle that is producing the maximum power of which it is capable must be matched to the wings (by setting its maximum strain rate) so that when it is operating at the wingbeat frequency used in climb, the stress is ~30% of the isometric stress ($\sigma_{max}$). For an active strain $\lambda$, the upper limit to the specific work ($q_{mlim}$) available from the myofibrils is

$$q_{mlim} = 0.30\sigma_{max}\lambda/\rho_{musc}, \tag{1}$$

where $\rho_{musc}$ is the density of muscle. The maximum power available $P_{max}$ is then

$$P_{max} = q_{mlim}m_{musc}(1 - \zeta)f, \tag{2}$$

where $m_{musc}$ is the mass of the flight muscles, including mitochondria and $f$ is the wingbeat frequency. $\zeta$ (Greek zeta) is the volume fraction of mitochondria, and as other muscle components are neglected $(1 - \zeta)$ is the volume fraction of myofibrils. It is implicitly assumed in *Flight* that mitochondria and myofibrils have the same density ($\rho_{musc}$), and that the mass fractions of the two components are the same as their volume fractions. At any given speed, the excess power available for climbing ($P_{ex}$) is found by subtracting the power required to fly horizontally ($P_{mech}$) from the maximum power:

$$P_{ex} = P_{max} - P_{mech}, \tag{3}$$

and the maximum rate of climb $V_z$ at that speed is

$$V_z = P_{ex}/mg \tag{4}$$

where $m$ is the all-up mass and $g$ is the acceleration due to gravity.

When it comes to calibrating predicted performance against observations, it is better to expand Equation (4) so as to express the rate of climb ($V_z$) in terms of the difference between the limiting specific work *available* [$q_{mlim}$ from Equation (1)] and the specific work *required* for horizontal flight ($q_m$), which is

$$q_m = P_{mech}/(m_{musc}(1 - \zeta)f). \tag{5}$$

$q_m$ and $P_{mech}$ are calculated by *Flight*'s Power Curve calculation (Chapter 3) and tabulated in the Excel output. The excess power for climbing is:

$$P_{ex} = (q_{mlim} - q_m)m_{musc}(1 - \zeta)f, \tag{6}$$

This has to be divided by the all-up weight ($mg$) to get the rate of climb, which means that the flight muscle mass is replaced by the flight muscle *fraction* ($F_{musc}$), where

$$F_{musc} = m_{musc}/m, \tag{7}$$

BOX 7.5 *Continued.*

so that the rate of climb is:

$$V_z = (q_{mlim} - q_m)F_{musc}(1 - \zeta)f/g. \tag{8}$$

Equation (8) shows that calculating the rate of climb does not require a value for the flight muscle mass as such, only for the flight muscle fraction. The default assumption in *Flight*'s Migration calculation is that enough flight muscle mass is consumed in flight to hold the specific work in the myofibrils constant. Unlike the alternative assumptions offered, this assumption results in the flight muscle fraction remaining nearly constant during a long period of level flight, although a large amount of fat is consumed, as is some protein from other organs. This is in general agreement with field observations of the consumption of flight muscle tissue in flight, and also of its replacement during the refuelling process (Lindström and Piersma, 1993). It means that, even though the flight muscle mass may be very different in different specimens of the same species, the flight muscle fraction usually is not, and that this measurement may be obtained from fat or thin birds without any need for large samples or full carcase analysis.

**Definition of Flight Muscle Fraction**
Readers who encounter unexpectedly high values of the flight muscle fraction in the literature should pay close attention to the Methods section of the paper, as there are some authors who use the term "flight muscle fraction" to refer to a different ratio, that of the flight muscle mass to the "lean mass", which means the all-up mass minus the fat mass. This ratio does not remain even approximately constant as the bird slims down or fattens up, and it has no significance for calculating the rate of climb or any other aspect of performance. Performance calculations require the ratio of the flight muscle mass to the *all-up* mass, meaning everything that the bird has to lift, including fat, crop contents, mud on the feet, and any hardware with which it may be burdened, such as rings and radio transmitters. Likewise, the fat fraction is the ratio of the fat mass to the all-up mass, *not* to the lean mass.

**Changes of mitochondria fraction**
In *Flight*'s Migration calculation, the mitochondria fraction ($\zeta$) in cruising is found for each 6-minute interval of flight by rearranging Equation (6) of Box 7.6:

$$\zeta = p_m k_{mito} \rho_{musc}, \tag{9}$$

where $p_m$ is the mechanical power per unit mass of the whole muscle (myofibrils and mitochondria). $p_m$ is calculated for level flight, at the starting configuration of mass, fat and flight muscle, and it is assumed that the bird consumes mitochondria in the course of the flight, as necessary to hold the power density in the mitochondria constant. If the bird starts its migration by climbing aerobically to a cruising height of a few thousand metres, as many or most migrants apparently do, then a higher mitochondria fraction ($\zeta_{climb}$) will be needed for the initial climb. This in turn reduces the power available from the muscles, and needs to be taken into account when calculating the maximum rate of climb.

BOX 7.5 *Continued.*

The value of $\zeta$ given by Equation (9) comes from the aerodynamical power *required* to fly level at the selected cruising speed, whereas the value needed to calculate the maximal aerobic rate of climb from Equation (8) has to come from the power *available* when the muscles are operating at maximum power. This depends on the isometric stress ($\sigma_{max}$), the active strain ($\lambda$), the wingbeat frequency ($f$) and the inverse power density of the mitochondria ($k_{mito}$), as in Equation (4) of Box 7.6. At the required stress for maximum power (0.30 $\sigma_{max}$), the mitochondria fraction needed is:

$$\zeta_{climb} = 0.30k_{mito}\sigma_{max}\lambda f / (1 + 0.30k_{mito}\sigma_{max}\lambda f). \tag{10}$$

The difference between the mitochondria fraction required for aerobic climb at maximum power [Equation (10)] and that required for level flight [Equation (9)] depends on the power margin that the bird has in level flight. For example, the Whooper Swan considered in the main text of this chapter requires nearly full power to fly level at $V_{mp}$, and has almost no power margin. It is capable of a marginal rate of climb of 0.039 m s$^{-1}$, and to maintain that aerobically, it would have to increase the mitochondria fraction in its flight muscles to 0.152, from the cruising value of 0.149, which is not a measurable difference. On the other hand, the Great Knot considered in Chapter 8, a much smaller bird with about the same flight muscle fraction, should be capable of climbing at 2.50 m s$^{-1}$ when fully loaded with fat and ready for departure. To maintain this rate of climb aerobically, it would require a mitochondria fraction of 0.30, as compared to 0.089 for level flight. This difference should be readily measurable. In the absence of field observations, we may surmise that ultra long-distance migrants like knots do boost their mitochondria fraction before departure, and consume the excess mitochondria after levelling off at the cruising height.

### Rate of climb in *Flight*

From Version 1.17 of *Flight*, the Excel output from the Power Curve calculation tabulates the maximum rate of climb as a function of speed, and the Summary screen displays the highest possible rate of climb, obtained by flying at $V_{mp}$ with maximum power output from the muscles. It also shows the mitochondria fraction required for level flight at $V_{mp}$ from Equation (9) and that for maximum aerobic rate of climb from Equation (10). The maximum rate of climb must be zero or above for the bird to be capable of sustained horizontal flight. If it is not, then the most that the bird can achieve by maximum exertion is to delay its descent. This can be used to calibrate the assumed value of the isometric stress, from which $q_{mlim}$ is calculated [Equation (1)]. The default value of the isometric stress (560 kN m$^{-2}$) was obtained in this way from the observation that an exceptionally large whooper swan (*Cygnus cygnus*), whose wing span and area were known, was able to take off and fly normally when his all-up mass was 13.5 kg. His flight muscle fraction was of course not measured, but was estimated to be 0.131 from a sample of 5 whoopers that were dissected after fatal collisions with power wires. *Flight*'s Migration programme does not cover climb and descent, but simply launches the bird at the chosen cruising height. More information about how real migrants manage climb and

BOX 7.5 *Continued.*

descent would be needed before this could be realistically simulated. Observations of the mitochondria fraction in the flight muscles would provide some clues.

The wingbeat frequency in a bird that is climbing to cruising altitude at the beginning of a long flight may be a little more than the value estimated by *Flight* for cruising flight, but probably not very much more, as the flight muscles are likely to be matched to a frequency only slightly below the maximum that the wing structure will tolerate. Using the wingbeat frequency provided by *Flight* may slightly underestimate climbing performance. For example, Hedenström and Alerstam (1992) reported that the wingbeat frequencies of knots departing from West Africa at the start of a long migratory flight averaged about 9% higher than the value given by the same formula that is used by *Flight* to estimate the cruising wingbeat frequency, but there are uncertainties in this estimate.

### 7.3.7 MAXIMUM RATE OF CLIMB AND ISOMETRIC STRESS

If a bird has more power available from its flight muscles than is required to fly horizontally, then it can convert some of the work that the muscles do into potential energy, that is, it can climb. A bird that can only just fly must have enough power available to maintain at least zero rate of climb at its minimum power speed, while a margin of power available over power required will allow it to maintain or increase its height over a range of speeds that extends below and above $V_{mp}$. *Flight's* Power Curve calculation tabulates the maximum rate of climb against air speed (introduced in Version 1.17). This calculation (Box 7.5) requires estimates of (1) the isometric stress, (2) the active strain, (3) the bird's muscle fraction (not muscle mass) and (4) the wingbeat frequency that it uses when climbing.

Three of these four variables can be estimated with confidence as follows. (1) The experiment of Gordon et al. (1966), on which Figure 7.2 is based, established 0.26 as a good estimate of the strain in maximal exertion (above). (2) The flight muscle fraction can be estimated by dissecting dead specimens of a given species, regardless of whether the birds are light or heavy. The reason that this is possible is explained in Chapter 8, and indicated in Box 7.5. Caution is needed in taking data from the literature, as some authors have used the term "flight muscle fraction" to mean the ratio of flight muscle mass to lean mass, which is the wrong ratio for performance calculations. These require the ratio of flight muscle mass to *all-up* mass, that is the total mass of everything that the bird has to lift. (3) The wingbeat frequency at $V_{mp}$ is provided by *Flight's* Power Curve calculation (Box 7.3). *Flight's* estimates of rate of climb, being based on the natural frequency, should be good for migrating birds that are climbing to their cruising altitude, but may underestimate the ultimate performance that can be sustained by briefly increasing the wingbeat frequency in moments of emergency.

That leaves the isometric stress, for which published estimates based on dividing isometric tension by the measured cross-sectional area of a muscle have given a rather wide range of figures from 300 to 450 kN m$^{-2}$ (Alexander and Bennet-Clark, 1977; Alexander, 1985). The term "stress" is used in this book and in *Flight* to refer to the force exerted by the myofibrils only, not to the force averaged over the cross-section of both myofibrils and other muscle components that do not contribute to the force. This is difficult to measure directly, but it can be estimated from the measurements of a very large bird that is known to be marginally capable of level flight. Figure 7.6 shows maximum rate of climb calculated by *Flight* for a very large whooper swan (*Cygnus cygnus*), for values of the isometric stress from 500 to 580 kNm$^{-2}$. This particular swan is in *Flight*'s Preset Birds database, but his mass was set to 13.5 kg for this calculation, as this was the highest mass at which he was known to be able to fly normally (Table 7.2). The line intersects zero rate of climb at an isometric stress between 540 and 550 kN m$^{-2}$. Figure 7.7 shows curves of rate of climb versus air speed for three values of the isometric stress, of which the highest one (for 560 kN m$^{-2}$) indicates that this swan would be able to maintain level flight up to 24.6 m s$^{-1}$ but no faster. This is 1.17 times his estimated minimum power speed. This is consistent with the behaviour of wintering whoopers at Caerlaverock, which clearly require maximal effort to take off and clear obstructions, but are able

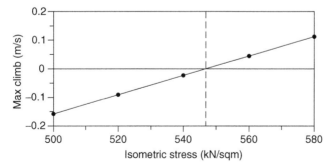

FIGURE 7.6 Maximum rate of climb (when flying at $V_{mp}$), calculated by *Flight* for a very large Whooper Swan who was known to be able to fly when his all-up mass was 13.5 kg (Table 7.2). Each point represents a different value for the isometric stress, which can be set in the Power Curve Setup screen. The line intersects zero rate of climb when the isometric stress is between 540 and 550 kN m$^{-2}$.

TABLE 7.2 Whooper Swan JAP (Figures 7.6 and 7.7).

| | |
|---|---|
| Measurements and assumptions | |
| All-up mass | 13.5 kg |
| Wing span | 2.56 m |
| Wing area | 0.756 m$^2$ |
| Aspect ratio | 8.67 |
| Flight muscle fraction | 0.131 |
| Fat fraction | 0.200 |
| Active strain | 0.26 |
| | |
| Calculated performance at sea level (1.23 kg m$^{-3}$) | |
| Minimum power speed ($V_{mp}$) | 21.1 m s$^{-1}$ |
| Wingbeat frequency | 3.44 Hz |
| Specific work in myofibrils at $V_{mp}$ | 40.2 J kg$^{-1}$ |
| Chemical power at $V_{mp}$ | 1.02 kW |
| Effective lift:drag ratio at $V_{mp}$ | 11.9 |
| Ratio of chemical power to BMR at $V_{mp}$ | 48.1 |
| Maximum rate of climb at $V_{mp}$ | 0.045 m s$^{-1}$ |
| Maximum air speed in level flight | 24.6 m s$^{-1}$ |
| Maximum range speed ($V_{mr}$) | 33.3 m s$^{-1}$ |
| Maximum rate of climb at $V_{mr}$ | −0.486 m s$^{-1}$ |

The letters JAP refer to this swan's telescope-readable leg ring. I am indebted to Jenny Earle of the Wildfowl and Wetlands Trust for the following biographical information. JAP was a highly successful swan, who migrated between the breeding grounds in Iceland and the Wildfowl and Wetlands Trust's reserve at Caerlaverock, Scotland each year between 1983 and 2001, and brought 36 cygnets to Caerlaverock during that time. He was identifiable by eye as the largest swan in the flock. He was caught and weighed at Caerlaverock in most winters. His highest recorded mass was 13.5 kg on 14 January 1993 and his wing measurements were taken on the same occasion. He was tracked by the Argos satellite system in 1995 on his spring migration to Iceland (Pennycuick et al., 1996b).

to keep going over level terrain, once they are airborne and up to speed. This is the reason why 560 kN m$^{-2}$ was selected as *Flight*'s default isometric stress. Both this value and the default active strain (0.26) can be adjusted from the Power Curve Setup screen.

## 7.4 ⬝ ADAPTATIONS FOR AEROBIC FLIGHT

The characteristics of flight muscles that determine their performance as engines have been discussed so far with hardly a mention of physiology or the support systems. The physiology of flight, as usually understood, is about the consumption of fuel and oxygen, and the disposal of heat. For continuous aerobic operation, the support systems have to be capable of supplying fuel and oxygen to the flight muscles, and removing waste products and heat, at rates that are set by the required mechanical power, which itself depends on the bird's mass and wing morphology, the density of the air in which it flies and the

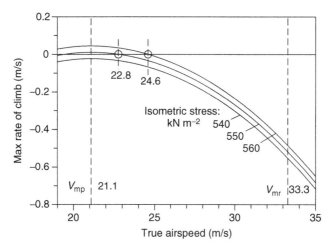

FIGURE 7.7 The same calculation as Figure 7.6, but plotting maximum rate of climb against air speed, for three different values of the isometric stress. At the middle value (550 kN m$^{-2}$), the swan is barely able to maintain level flight at speeds very near $V_{mp}$. At 560 kN m$^{-2}$, the swan can maintain height from below the minimum power speed (21.1 m s$^{-1}$) up to 24.6 m s$^{-1}$, which is 1.17 times the estimated $V_{mp}$. The value of 560 kN m$^{-2}$ is selected as the default for *Flight*. This figure refers to the stress across the myofibrils (not the whole cross-section of the flight muscles), and it should be valid for any vertebrate, not just for whooper swans.

strength of gravity. Physiological limitations may restrict performance variables such as speed, rate of climb, or the maximum height at which level flight is possible, to values that are below those that are mechanically possible under given conditions.

## 7.4.1 AEROBIC MUSCLES

Muscles themselves can be classified as "aerobic" or "anaerobic", the main difference being that an aerobic muscle contains mitochondria in among the contractile filaments (myofibrils), whose function is to replenish ATP on a continuous basis, so as to maintain a constant level in the cytoplasm. The mitochondria increase the cross-sectional area and volume of the cell, so reducing the stress that the muscle as a whole can exert, and the specific work and power that it can produce. The fraction of the cell's volume that has to consist of mitochondria, and the resulting dilution of the specific power, depend on the contraction frequency. The ultimate limit of specific power depends only on the properties of the mitochondria, and not on those of the sliding filaments (Box 7.6 and Figure 7.8). Muscle fibres containing approximately equal volumes of mitochondria and myofibrils are known in both

BOX 7.6 **Aerobic muscles.**

Exercise is *aerobic* if it can be sustained for a prolonged period, without incurring an oxygen debt. Some birds can apparently sustain level, aerobic flight continuously, for as long as a week (Chapter 8). Two requirements have to be met, to sustain continuous flight beyond a few minutes. First, within each muscle fibre, there have to be sufficient mitochondria to regenerate ATP from ADP, at the same rate that the contractile filaments consume it. Second, support systems external to the muscle have to supply fuel and oxygen for the mitochondria to consume, and remove the carbon dioxide and heat that the muscle generates. The support systems are basically the blood system and the lungs.

The addition of mitochondria to the contents of a muscle cell, alongside the contractile filaments, results in an increase in the cross-sectional area of the cell, without any increase in the tension that the muscle exerts. The mitochondria "dilute" the acto-myosin array, resulting in a decrease of isometric stress, and also of the active stress while the muscle is contracting. This in turn decreases the specific work and specific power, if these are measured relative to the volume or mass of the whole muscle, rather than to that of the contractile filaments alone.

**Variable definitions for this box**

| | |
|---|---|
| $f_{op}$ | Operating frequency |
| $k_{mito}$ | Inverse power density of mitochondria |
| $p_m$ | Mass-specific power output |
| $p_v$ | Volume-specific power output |
| $P_{mech}$ | Mechanical power output |
| $v$ | Volume of muscle |
| $v_c$ | Volume of contractile filaments |
| $v_t$ | Volume of mitochondria |
| $\zeta$ | Volume fraction of mitochondria |
| $\lambda$ | Active strain |
| $\rho_{musc}$ | Density of muscle |
| $\sigma$ | Stress in the contractile filaments |

**Volume of myofibrils and operating frequency**

In an aerobic muscle, a volume ($v_c$) of contractile filaments is required to produce the mechanical power ($P_{mech}$), and in addition a volume ($v_t$) of mitochondria is needed to keep the contractile filaments supplied with ATP. Other muscle components are assumed to be small enough in volume to be neglected in this account. The volume of myofibrils needed for a given level of power can be found by dividing the power by the volume-specific power of the myofibrils, which is itself the product of their active stress ($\sigma$) and strain ($\lambda$) while shortening, and the *operating frequency* ($f_{op}$):

$$v_c = P_{mech}/(\sigma \lambda f_{op}) \tag{1}$$

The operating frequency is defined as the frequency at which the muscle is *adapted* to contract aerobically. It is a property of the muscle which can be estimated by measuring the volume fraction of mitochondria from

BOX 7.6 *Continued.*

electron micrographs (below). $f_{op}$ is *not* some arbitrary frequency at which a muscle is forced to contract by an experimental apparatus, as some experimenters have assumed.

### Volume of mitochondria required

We next assume that a fixed volume of mitochondria is required to regenerate ATP at a rate sufficient to support each watt of mechanical power. This number is called the *inverse power density* of mitochondria, and is denoted by $k_{mito}$. It determines the volume ($v_t$) of mitochondria that is needed to sustain the mechanical power output $P_{mech}$:

$$v_t = P_{mech} k_{mito} \qquad (2)$$

The default value for $k_{mito}$ in *Flight* is $1.2 \times 10^{-6}$ m$^3$ W$^{-1}$, meaning that 1.2 ml of mitochondria is assumed to be required for each watt of mechanical power produced by the muscle. This estimate was obtained by measuring the percentage volume of mitochondria in various bird flight muscles, whose volume-specific power output could be estimated (Pennycuick and Rezende 1984). Presumably $k_{mito}$ would decrease at higher temperatures. This is not an issue when comparing homeothermic animals with similar body temperatures, but it has implications for the evolution of homeothermy in the first place (below).

If we neglect other cell components, then the total volume ($v$) of the muscle is the sum of the volumes of contractile filaments and mitochondria:

$$v = v_c + v_t = P_{mech}[(1 + k_{mito}\sigma\lambda f_{op})/\sigma\lambda f_{op}]. \qquad (3)$$

and the volume fraction of mitochondria ($\zeta$) needed is:

$$\zeta = v_t/v = k_{mito}\sigma\lambda f_{op}/(1 + k_{mito}\sigma\lambda f_{op}). \qquad (4)$$

If $\zeta$ is measured from electron micrographs, and estimates are available for $\sigma$, $\lambda$ and $k_{mito}$, then the operating frequency ($f_{op}$) can be estimated as:

$$f_{op} = \zeta/[k_{mito}\sigma\lambda(1 - \zeta)]. \qquad (5)$$

The operating frequency corresponding to a given value of $\zeta$ is much higher in insects with asynchronous flight muscles than in vertebrates, because the active strain of such muscles is less than 10% of that in their vertebrate counterparts.

### Volume-specific power of the whole muscle

The volume-specific power output of an aerobic muscle is less than that of the contractile filaments by the factor $1/(1 + k_{mito}\ \sigma\lambda f_{op})$:

$$p_v = \sigma\lambda f_{op}/(1 + k_{mito}\sigma\lambda f_{op}). \qquad (6)$$

At low operating frequencies, the denominator in Equation (6) is only just over 1, and the volume-specific power is nearly proportional to the operating frequency. As the operating frequency increases, diminishing returns set in, and the volume-specific power levels off at an asymptotic

BOX 7.6 *Continued.*

value of $1/k_{mito}$, representing an infinitesimally small contractile component operating at infinite frequency, embedded in a muscle consisting otherwise entirely of mitochondria. In practice, the highest volume fraction of mitochondria found in both vertebrates (hummingbirds) and flying insects is about 0.5, that is, equal volumes of contractile filaments and mitochondria.

**Aerobic performance and temperature**

The mass-specific power ($p_m$) that an aerobic muscle can sustain depends on the mitochondria fraction ($\zeta$), the inverse power density ($k_{mito}$) of the mitochondria and the density of muscle ($\rho_{musc}$), but not on the mechanical properties of the contractile filaments:

$$p_m = \zeta/(k_{mito}\rho_{musc}). \tag{7}$$

For a given volume fraction of mitochondria, the mass-specific power is inversely proportional to $k_{mito}$. As noted above, one would expect $k_{mito}$ to be a negative function of temperature, that is, less mitochondria would be needed to sustain a given level of mass-specific power at a higher temperature. To put this another way, the higher the temperature at which the muscles operate, the less total muscle mass (contractile filaments and mitochondria) is required for a given level of power output. This is actually the only reason why it is advantageous for active animals to be homeothermic. A high power output from anaerobic muscles depends on matching the maximum strain rate to suit the load, and that can be done at any temperature within a wide range, as in fishes that live in cold water. While any muscle that has to be accurately matched to its load needs to run at a *constant* temperature, a *high* temperature is only advantageous for aerobic muscles, to keep the volume of mitochondria down.

**Vertebrate versus insect muscles**

Vertebrate muscles are said to be *synchronous*, which refers to their response to stimulation. Each contraction of a vertebrate muscle (i.e. each wingbeat in a flying vertebrate) is initiated by a short burst of nerve impulses, each of which triggers a wave of electrical activity in the membrane enclosing the muscle fibre, which in turn activates the myofibrils. At the end of the downstroke, the stimulation stops, allowing the myofibrils to relax, so that they can be passively extended during the upstroke. The minimum time needed for this cycle of activation and relaxation is around 20 ms, so limiting the maximum contraction frequency of hummingbird muscles to around 50 Hz. Honeybees, however, beat their wings at much higher frequencies, around 250 Hz, and some small midges can exceed 2 kHz. Their flight muscles are *asynchronous*, meaning that the muscle is held in the active state by a sustained sequence of nerve impulses at a comparatively low frequency. While the muscle is "on", the wings beat continuously at a higher frequency, which is determined by the mechanical resonance resulting from the inertia of the wings, and the elasticity of the thorax to which they are attached. Clipping the tips off a bee's wings retunes its wingbeat to a higher frequency, whereas no such resonance is involved in determining a bird's wingbeat frequency.

BOX 7.6 *Continued.*

As asynchronous muscles work at higher frequencies than vertebrate muscles, they can develop similar levels of specific power with less strain. The active strain is only about 0.02 in the flight muscles of bees and flies. The length of the sarcomere hardly changes at all in operation, and the myosin filaments extend almost from Z-line to Z-line. The cross-section has a different arrangement from that shown in vertebrates, with three actin fibres per myosin filament rather than two (Figure 7.1). It is not clear whether there is a difference in the isometric stress, but it is known that the stress does not drop to zero while the muscle is lengthening. The active stress for calculating specific work and power in insect muscles is the difference between the stress while shortening and that while lengthening. The low strain means that the flight muscles of insects cannot drive the wings directly as those of vertebrates do. They work by distorting the exoskeleton of the thorax by a small amount, and this small movement is then geared up by the wing articulation, to rotate the base of the wing by a larger amount.

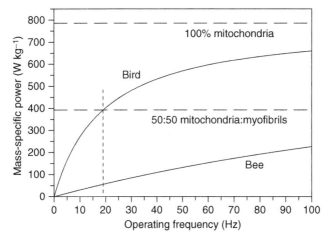

FIGURE 7.8 Mass-specific power, averaged over many cycles of contraction and relaxation, for a muscle that contains a sufficient volume of mitochondria to support continuous aerobic activity. The higher the operating frequency for which the muscle is adapted, the larger the ratio of mitochondria to myofibrils required. If this ratio is 50:50, the specific power is just under 400 W kg$^{-1}$ regardless of the mechanical characteristics of the muscle, but the operating frequency at which this occurs is about 19 Hz for bird muscles, and about 250 Hz for insect muscles, assuming that the isometric stress is the same for both, and the active strain is much smaller in insect asynchronous flight muscles. The unattainable asymptote towards which both curves tend (nearly 800 W kg$^{-1}$), is independent of the properties of the myofibrils. It represents a muscle that consists entirely of mitochondria, except for an infinitesimally small contractile component, contracting at an infinite frequency. After Pennycuick and Rezende (1984).

hummingbirds and insects, and this seems to be the practical upper limit for the relative volume of mitochondria.

Aerobic muscles are invariably red, for three reasons. First, the mitochondria themselves apparently have a red colour in bulk. Second, the cytoplasm of aerobic muscle cells contains myoglobin, a red respiratory pigment related to haemoglobin, whose function is to buffer the pressure of dissolved oxygen within the cell. Third, an aerobic muscle requires a blood supply that can provide fuel and oxygen, and remove carbon dioxide and heat, at a sufficient rate to match the muscle's needs in continuous operation, and the copious blood capillaries contribute to the red colour. Muscles that are specialised for anaerobic operation in short bursts of "sprint" activity do not have these requirements. They are paler in colour, and are often characterised as "white".

BOX 7.7 **Bird and bat respiratory systems.**

Birds have some capabilities that are not shared by bats, especially in high-altitude performance, and these depend on the unique arrangement of the avian respiratory system, which is very different from that of mammals. This account is summarised from reviews of the anatomy by Duncker (1985) and of the physiology by Scheid (1982). The lungs of birds are small, dense organs that are firmly fixed to the ribs on the dorsal side of the body cavity, and (unlike mammal lungs) they do not expand when the bird breathes in. The incoming air passes through the lungs, and into a system of air sacs beyond them (Figure 7.9). The air sacs expand and contract as the bird breathes in and out, but gas exchange takes place only in the lungs, not in the air sacs. The walls of the air sacs are thin and transparent, with no blood supply, and this makes them difficult to see in dissection. They can be made visible by sucking the air out of a dead bird's respiratory system through the trachea, and then injecting latex under atmospheric pressure. They are then found to ramify everywhere, around and between the organs in the body cavity. There are seven air sacs in all, comprising three pairs (thoracic, anterior abdominal and posterior abdominal), and one medial air sac, the inter-clavicular (Figure 7.9A). This last is connected to the lungs of both sides, and sends branches into the bones, not only the cavities of the hollow humeri but also fine channels inside the pectoral and pelvic girdles, the vertebrae and the skull. Branches of the inter-clavicular air sac also penetrate into the interior of the pectoralis muscles (See Chapter 5, Figure 5.13).

**Bronchi, lungs and air sacs**
Figure 7.9B shows first of all that air enters and leaves via the same route, the primary bronchus, as the bird breathes in and out, so that the system as a whole is tidal, as is a mammal's lung. In the primary bronchus, the flow reverses direction at each breath, but once inside the system, the air moves around inside it in one direction only. The primary bronchus leads to a system of secondary bronchi, which have been drastically simplified in

BOX 7.7 *Continued.*

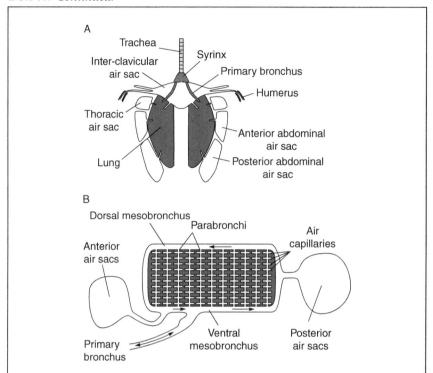

FIGURE 7.9 (A) Air enters and leaves a bird's respiratory system through the trachea, which divides at the syrinx (the voice organ of birds) into two primary bronchi, leading to the lungs (pink), which are compact organs attached to the ribs along the dorsal wall of the body cavity. Their volume does not change when the bird breathes. Instead, air is drawn through the lungs by expansion and contraction of the air sacs, which are thin-walled, non-respiratory structures with no significant blood supply of their own. The inter-clavicular air sac sends branches into the interior of the hollow humeri, and also into fine cavities in the bones of the limb girdles, vertebral column and skull, and the interior of the pectoralis muscles. (B) Primary bronchi deliver air to the lungs, mesobronchi are channels interconnecting different parts of the system, and parabronchi are an array of thin, parallel channels, with blind air capillaries branching off them. Air flows downwards along the parabronchi in the diagram, both when the bird breathes in and when it breathes out (see Figure 7.10A). Gas exchange takes place in the air capillaries, which are embedded in a mesh of blood capillaries.

Figure 7.9 into dorsal and ventral mesobronchi. When the bird breathes in, the air flows along the ventral mesobronchus to the posterior group of air sacs, comprising the anterior and posterior abdominal air sacs, but when the bird breathes out, the expired air comes from the anterior air group of air sacs (thoracic and inter-clavicular). In between is the lung, where the actual gas exchange takes place in the parabronchi. These are an array of fine, parallel air channels that connect the dorsal and ventral mesobronchi. The air flows through the parabronchi in the same direction, both when the

BOX 7.7 *Continued.*

bird is breathing in and when it is breathing out. Oxygen diffuses from the lumen of each parabronchus into a series of blind "air capillaries", which are embedded in a mesh of blood capillaries. Carbon dioxide likewise diffuses from the blood capillaries into the air capillaries, whence it diffuses into the parabronchus and is carried away.

**Heat disposal from the air sacs**

There are no valves directing the flow in the bronchial system, and no obvious structures that would force the air to flow through the parabronchi in one direction, or indeed to prevent it from flowing in and out of the air sacs without passing through the parabronchi at all. It seems that the parabronchi and the air sacs are effectively in parallel, and that the proportion of the inspired air that passes through the parabronchi is determined by the resistance of the parabronchi to air flow. This can be adjusted by smooth muscles in the walls of the parabronchi, which can vary their diameter by a factor of two. Since the resistance of a tube to the flow of fluid along it is inversely proportional to the fourth power of the diameter, this would change the resistance of the parabronchi by a factor of 16. The respiratory system actually combines two functions, gas exchange in the parabronchi, and evaporative heat loss from the thin, moist non-respiratory walls of the air sacs. By constricting the parabronchi, and increasing the depth and/or frequency of its breathing, a bird can increase the rate of heat disposal without excessively increasing the rate of gas exchange. Birds do this when heat-stressed. Unlike bats, they can fly without overheating in air temperatures that are too high for convective heat disposal, provided that the relative humidity is not too high, and water is available. Heat is not carried away from the air sacs by the blood, as they have no significant blood supply. Hence the air sacs have to penetrate everywhere that cooling is needed, and the water vapour must escape by diffusion from the finer passages.

**Gas exchange in the parabronchi**

Although the respiratory system as a whole is tidal, the air flow through the parabronchi is uni-directional, and this raises the possibility that the parabronchi might act as counter-current gas exchangers, with air flowing in one direction and blood in the other. Many experiments have been done on this, and Scheid (1982) considered that the system should properly be described as a "cross-current" arrangement, as the actual blood capillaries do not flow parallel to the parabronchi. Functionally, however, the arrangement that he describes is a true counter-current gas exchanger. The arterioles bring deoxygenated blood in at the end of the parabronchi where air is flowing out, and run parallel to the parabronchi, dwindling to nothing by the time they reach the other end (Figure 7.10A). Capillaries connect each arteriole to a parallel venule, which increases in diameter as it collects blood from successive capillaries. The gas exchange occurs where the blood capillaries form a meshwork in close contact with the air capillaries. According to this interpretation, oxygen is progressively withdrawn from the air as it passes along the parabronchus, so that the partial pressure of oxygen decreases from one air capillary to the next, along the array. The blood flows in the opposite direction to the air, and the partial pressure of oxygen that it

BOX 7.7 *Continued.*

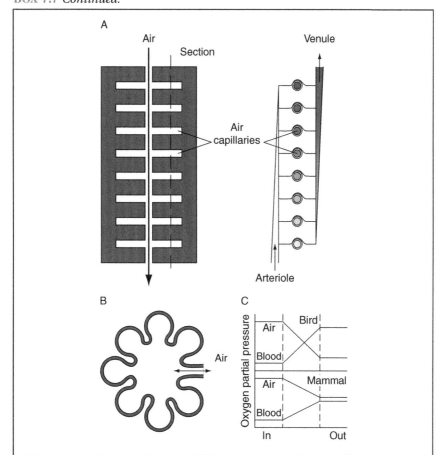

FIGURE 7.10 (A) A parabronchus (left) is a fine channel with air flowing along it, always in the same direction. Gas exchange occurs in the blind air capillaries that branch off it on both sides. The dashed line marks the plane of the section at right, through the air capillaries. Each air capillary is supplied by blood capillaries (thin lines) that carry blood from an arteriole to a venule. The oxygen gets depleted in the parabronchus, and the air capillaries, as the air moves from top to bottom of the diagram. The arteriole supplies deoxygenated blood to all the blood capillaries, and the blood entering the venule is progressively more oxygenated (red tint) as it flows from bottom to top. (B) An alveolus of a mammal lung is a blind sac, with an elastic wall containing a plexus of blood capillaries. When it expands and fills with fresh air, the oxygen partial pressure in the cavity drops while that in the blood rises, until there is insufficient pressure difference to maintain further movement of oxygen. (C) In any lung, the partial pressure of oxygen in the air entering the lung is higher than that in the blood. The counter-current arrangement of the bird lung allows the blood leaving the lung to reach a higher oxygen partial pressure than the expired air, whereas this is not possible in a mammal lung.

contains increases along the venule. The partial pressure of oxygen in each air capillary exceeds that in its associated blood capillary by a roughly

BOX 7.7 *Continued.*

constant amount from one end of the parabronchus to the other, although the air in the parabronchus is progressively depleted of oxygen as it goes.

The advantage of this arrangement is that it allows the partial pressure of oxygen in the blood to rise above that in the air as both air and blood leave the lung. This is not the case in the alveoli of mammal lungs, which are expandable blind cavities, with a blood plexus in the elastic walls (Figure 7.10B). When the alveolus expands to suck in fresh air, the oxygen partial pressure in the lumen drops as that in the surrounding blood capillaries rises, but the transfer of oxygen requires a gradient of partial pressure from the air into the blood, and stops when the two partial pressures meet (Figure 7.10C). This means that a bird can maintain a higher level of oxygen concentration in its blood than can a mammal, when the partial pressure of oxygen in the inspired air is low, as it is at high altitudes. Tucker (1968a) put sparrows and mice in a hypobaric chamber, and lowered the air pressure to simulate an altitude of 6100 m ASL, where the pressure is less than half that at sea level. In these conditions the sparrows continued to hop about and make short flights, whereas the mice were comatose.

### Fractal properties of the parabronchial system

Another conjectural property of the parabronchial arrangement is that the fine structure of the air and blood capillaries may have fractal properties, meaning that its geometry is "self-similar" over a certain range of scale. The consequences of this were explored by Mandelbrot (1982). In this case, it would not be meaningful to talk of the "surface area" of the air capillaries, because their surface does not necessarily have any such property as "area", meaning a measurable quantity with dimensions $L^2$. In Euclidean geometry, this is distinct from a "volume" having dimensions $L^3$, but the dimensions of a fractal surface are $L^d$, where $d$ does not have to be an integer. In Mandelbrot's terminology, the exponent $d$ is called the "dimension" of a line or surface. $d$ is between 1 and 2 for fractal lines such as coast lines, and between 2 and 3 for fractal surfaces such as mountainsides and (perhaps) bird lungs. This would have consequences for the rate at which a lung of given mass can absorb oxygen, when the whole structure is scaled to different sizes (Chapter 13).

### Breathing

Air is forced in and out of the respiratory systems of both birds and mammals by changes of volume. A mammal's lungs are surrounded by the rib cage, which is divided from the abdominal cavity by a curved, muscular diaphragm. The mammal breathes in by contracting and flattening the diaphragm, so increasing the volume of the rib cage, and forcing air to flow into the lungs under atmospheric pressure. A bird has no diaphragm, but the volume of the whole body cavity can be changed by intercostal muscles that flex or extend the joints between the ribs and the ventral ribs. This moves the sternum and the synsacrum towards or away from each other, so that these two curved surfaces, with their concave sides facing inwards, act as the two halves of a bellows (Chapter 5, Figure 5.12). Those air sacs

BOX 7.7 *Continued.*

that are enclosed in the body cavity, and not surrounded by bone, expand and contract, forcing air to flow through the lungs, although the lung volume is constant.

## 7.4.2 BIRD AND BAT RESPIRATORY SYSTEMS

The lungs of bats work in the same way as those of other mammals including ourselves, and are subject to the same limitations, but those of birds are somewhat different (Figure 7.9). They are often described as "through-draught" lungs, as opposed to the "dead-end" or "tidal" lungs of mammals. There is an element of truth in this, although a bird's respiratory system as a whole is tidal like ours, insofar as air alternately enters and leaves through the trachea. Bird lungs are not distensible sacs like those of a mammal, but small, dense organs whose volume is constant, with fine air channels (the parabronchi) through which air flows in one direction only, whether the bird is breathing in or out. It appears that this arrangement is responsible for the high-altitude capabilities of birds, which depend on maintaining a high enough partial pressure of oxygen in the blood for strenuous activity, at air pressures and densities that are too low for mammals to function normally (Box 7.7).

The tidal bellows action that ventilates a bird's lungs depends on a system of distensible, thin-walled *air sacs*, which lie beyond the lungs, but are not themselves respiratory. The air sacs also serve an entirely different function, as the sites of a highly effective and controllable system of evaporative cooling. The air sacs ramify everywhere in the body cavity where they are in intimate contact with all of the internal organs, and also penetrate into many of the bones, and into the interior of the pectoral muscles. It would appear that evaporation of water from the moist inside walls of the air sacs removes heat directly from these interior sites, without the intervention of the blood system, an arrangement that is in some respects reminiscent of the insect tracheal system, although it involves only water and heat, not respiratory gas exchange. A heat-stressed bird can increase the ventilation of its air sacs without also increasing the ventilation of the lungs. Birds may often be seen flying around with their beaks open when the air is hot, or fluttering the throat pouch when on the ground. This ventilates the inner surface of the gular sac, and also the upper part of the trachea, where heat is lost by evaporation. Most birds do not have any method of external evaporative cooling, corresponding to the sweat glands of mammals, although a few (storks and New World vultures) urinate down their legs when overheated, apparently as a cooling procedure.

The primary method of heat disposal in bats is convection, which requires a plentiful supply of blood to be passed through a surface that is exposed to a flow of air that is cooler than the blood. The wing membranes of bats, with their copious and controllable blood supply, and huge surface area exposed to the air flow, are ideal convective cooling surfaces. Heat passes from the blood to the air, and is convected away. In birds, the thinly feathered skin surfaces on the underside of the inner part of the wing, and on the sides of the body below the wings, are used for convective cooling. When the bird is not flying, the wings are folded, so covering up both of these surfaces and preventing loss of heat, but in flight both are exposed to the air flow. Although the cooling area is much less than that of bat wing membranes, it could be argued that convection is the primary method of cooling for both birds and bats, since evaporative cooling requires expenditure of water, which may be in short supply in arid environments. Birds rely on convective cooling so long as the air is cool enough, but differ from bats in having the option to resort to controlled evaporative cooling when necessary. Migrating birds have some control over the air temperature, as a bird can reach cooler air by climbing, at the rate of about 1 °C per 100 m of height, and thus increase the effectiveness of convective cooling. Also, the oxidation of fat in flight generates water as a by-product, and this seems to be sufficient to replace water losses in a number of migratory bird species that fly continuously for several days and nights without stopping, with no source of water.

## 7.4.3 THE CIRCULATORY SYSTEM

Fuel, respiratory gases and heat are generated in one part of the bird's body (the source), and conveyed by the blood to another part (the sink). The flight muscles are the sink for a large and steady flow of fuel (from the fat storage organs) and oxygen (from the lungs), and the source for carbon dioxide (going to the lungs), and heat (going to the skin and/or the respiratory system). In round numbers, it appears that the heart, which supplies the mechanical power needed to circulate the blood, is about 1% of the body mass in birds and bats, as compared to 0.5% in non-flying mammals. The power needed by the heart can be found by multiplying the pressure difference in the aorta above that of the venous return by the volume rate of blood flow. The required mass of the heart could be estimated, if the heartbeat frequency could be estimated. The mechanical power required from the flight muscles can be estimated (Chapter 3), but there is no theory that connects the power required from the heart with that required from the flight muscles. It seems intuitively reasonable to expect that birds would require big hearts, and unsurprising that the heart mass should decline during long migratory flights, in a similar way to that of the flight muscles, but there is no quantitative basis to account for this. These changes are known to occur (Piersma, 1998) but they cannot be accounted for in *Flight*.

## 7.4.4 MASS AND POWER REQUIREMENTS OF SUPPORT SYSTEMS

Because of scale relationships, it is more difficult for large birds than for small ones to get sufficient power from their flight muscles to fly. This is for purely mechanical reasons (Box 7.4). One might expect to see disproportionately large flight muscles in large birds, but this does not seem to be the case (Chapter 13, Figure 13.8). Swans are the largest birds capable of migrating by continuous flapping flight, and the mass of their flight muscles is typically about 13% of the all-up mass, a lower flight-muscle fraction than in many smaller birds. Mechanical limitations due to their large size limit migrating swans to air speeds only slightly above their minimum power speeds (above), and they land from time to time on the ground or water, but it is not clear whether the primary limitation for continuous flight is in the muscles themselves or in the support systems. Increasing the capacity of the lungs would increase the total mass, so requiring more flight muscle to maintain the minimum power speed, and it is conceivable that the increased power requirements of the lungs and heart themselves would be greater than those of the additional muscle. This remains conjectural, in the absence of any quantitative theory that allows one to calculate what mass of the lungs and heart are needed, to support a given level of mechanical power output in the flight muscles.

# 8

# SIMULATING LONG-DISTANCE MIGRATION

The *Flight* programme uses a time-marching computation to simulate a long migratory flight, revising the bird's mass and body composition at 6-minute intervals as
fuel is consumed. Sufficient protein is taken from the flight muscles to hold the
mechanical conditions in the muscles constant as the mass declines. Predictions
are in satisfactory agreement with field data on long-distance migrants. Range calculations are greatly simplified by expressing energy reserves in the form of "energy
height", which applies to potential and kinetic energy, as well as fuel energy.

Tiny birds fly airline distances without refuelling, although they take
longer about it than airliners, flapping steadily along over oceans and
deserts as day follows night, and one climatic zone replaces another.
Should we be surprised that such small creatures can fly so far? Not
really. A migrating bird (or aircraft) does work against drag, and the
amount of work it has to do is the distance times the drag. The fuel
consumed has to account for the work done. If a bird (or aircraft) starts
with a given proportion of its all-up mass consisting of consumable
fuel, and the drag force is proportional to the weight, then birds of all
sizes will go the same distance before they run out of fuel. This line
of thinking is enshrined in Breguet's range equation, which dates from

the early days of aviation, and is named after the French aviation pioneer Louis-Charles Breguet (Anderson 1997).

# 8.1 ⇒ ESTIMATING RANGE

## 8.1.1 FUEL FRACTION AND LIFT:DRAG RATIO

Although Breguet's analysis was somewhat simplified (Box 8.1), it focused attention on two fundamental generalisations which apply to anything that flies. These remain as valid now as they were in 1922, when Breguet presented them to a London audience, in a famous lecture that pointed the way for the development of commercial aviation (Breguet 1922). Breguet's first fundamental point was that an aircraft's range is directly proportional to the *ratio* of the lift developed by the wings (which supports the weight) to the total drag, and his main message to aircraft designers in 1922 was that much work needed to be done at that time on reducing drag. His second point was that his range equation did not mention the mass of either the aircraft or the fuel, but only the fuel *fraction*, which is the ratio of the fuel mass to the all-up mass. The equation says that a willow warbler would go the same distance (relative to the air) as a jumbo jet, provided that both start with the same fraction (say 25%) of the all-up mass as consumable fuel, and both use fuel of the same energy density, convert fuel energy to work with the same efficiency, and have the same lift: drag ratio. The warbler would take longer to get there, but its eventual range (relative to the air) would be the same. Actually, warblers do not go as far as jumbo jets, but Breguet's equation predicts this, if the two main reasons are known. The first is that there are several effects that combine to make the lift:drag ratio less by a factor of 2–4 for small birds than for large aircraft, and the second is that the energy density of bird fuel (fat) is only about 83% that of jet fuel. Breguet's two messages for the modern ornithologist are that the key to estimating a bird's range is to calculate its *lift:drag ratio*, and that the fuel reserve is measured by the fat *fraction*, not the fat mass.

## 8.1.2 THE TIME-MARCHING MIGRATION COMPUTATION IN *FLIGHT*

It is not practical to adapt Breguet's equation to give an instant estimate of a migrating bird's range, because of some of the things that migrating birds are known to do. They slim down their bodies as fat is consumed, so progressively increasing the effective lift:drag ratio,

BOX 8.1 **The Breguet range equation.**

The classical range calculation is for a fixed-wing aircraft, flying horizontally, in which the wing provides a lift force that supports the weight, while a separate engine-driven propeller provides a horizontal thrust force that balances the drag. As fuel is consumed, the weight decreases, and so do both the lift and the drag, but the ratio of lift to drag (or of thrust to weight) is assumed to remain constant. The work done by the engine is found by integrating the product of the thrust and the distance flown, and this is assumed to be proportional to the mass of fuel consumed. The range is found by integrating the distance flown until the fuel is finished. Birds differ from fixed-wing aircraft in that they get both lift and thrust from their flapping wings, they consume other body components in flight besides the fuel (notably part of the engine), and the ratio of lift to drag tends to increase during a long flight, but the principle is the same if the flight is broken into short segments. The *Flight* program does this, and deals with the complications by integrating distance flown and fuel consumed numerically rather than analytically.

**Variable definitions for this box**

| | |
|---|---|
| $D$ | Drag |
| $e$ | Energy density of fuel |
| $F$ | Fuel fraction |
| $g$ | Acceleration due to gravity |
| $L$ | Lift |
| $m$ | All-up mass |
| $m_1, m_2$ | Mass at beginning and end of flight |
| $N$ | Lift:drag ratio |
| $P_{mech}$ | Mechanical power |
| $t$ | Time |
| $V$ | True airspeed |
| $Y$ | Distance flown |
| $\eta$ | Conversion efficiency |

The function of the wing in a fixed-wing aircraft in level flight is to develop a lift force ($L$) perpendicular to the flight path (i.e. vertical in this special case), which balances the weight ($mg$), with the smallest possible amount of drag ($D$) acting backwards along the flight path. The fuselage and other non-lifting parts add to the drag but do not contribute to the lift. The overall ratio of lift to drag ($L/D$) is a measure of the aircraft's "aerodynamic efficiency". We name this $N$, where

$$N = \frac{L}{D} \qquad (1)$$

For the case of a bird, $N$ has to be defined in a different way (Chapter 3, Box 3.4), because of the flapping wings, and metabolic complications, but its significance for calculating the range is the same. In level flight, the rate

BOX 8.1 *Continued.*

($P_{mech}$) at which the engine has to do mechanical work to overcome the drag is simply the drag times the speed:

$$P_{mech} = DV, \tag{2}$$

and the lift:drag ratio is $mg/D$, where $mg$ is the weight, so that

$$P_{mech} = \frac{mgV}{N}. \tag{3}$$

The aircraft starts its journey with an all-up mass of $m_1$. Part of this initial mass consists of consumable fuel, with an energy density $e$. The energy density is the chemical energy liberated by oxidising unit mass of the fuel. At any particular point in the flight, the mass is declining at a rate which is proportional to the mechanical power:

$$\frac{dm}{dt} = \frac{-P_{mech}}{e\eta} = \frac{-mgV}{e\eta N}, \tag{4}$$

where $\eta$ is the efficiency with which the engine converts fuel energy into work. If $Y$ is the distance travelled, then the speed $V = dY/dt$. Dividing this by $dm/dt$ from Equation (4), the distance flown per unit mass of fuel consumed at any point during the flight is:

$$\frac{dY}{dm} = \frac{-e\eta N}{mg}. \tag{5}$$

The range is the total distance flown while the mass declines from $m_1$ to a smaller value $m_2$, assuming that this is entirely due to the consumption of fuel. The range can be found by integrating Equation 5:

$$Y = \left(\frac{e\eta N}{g}\right) \ln\left(\frac{m_1}{m_2}\right). \tag{6}$$

It is more convenient to express the ratio $m_1/m_2$ in terms of the "fuel fraction" ($F$), which is the mass of fuel to be consumed, divided by the initial all-up mass. In that case:

$$Y = \left(\frac{e\eta N}{g}\right) \ln\left[\frac{1}{(1-F)}\right]. \tag{7}$$

This formula is known as "Breguet's equation" after its discoverer, the French aviation pioneer Louis-Charles Breguet (Anderson 1997).

With due deference to its historical importance in aeronautics, we have to recognise that Equation (7) does not give a satisfactory prediction of the range attainable by migrating birds. One reason is that the assumption that $N$ remains constant throughout a long flight may be satisfactory for an aircraft whose external shape is unaffected by the consumption of fuel, but in birds the body slims down as fat is consumed, and that causes a progressive increase in $N$. Also a bird can consume part of its engine (the flight muscles) as it flies, and get additional energy by doing so, and this dramatically increases the range. Rather than trying to incorporate these complications in an analytical equation, the approach used in *Flight* is the time-marching computation, revising the bird's body composition after every 6 minutes of flight.

and they consume part of the engine (the flight muscles) as the power requirements dwindle, using the protein as supplementary fuel. *Flight* takes account of these complications by using a "time-marching computation" rather than an analytical range equation. The programme calculates a snapshot of the bird's body composition and performance characteristics at the start of the flight, including the power and the *effective* lift:drag ratio, which replaces the fixed-wing concept of the lift:drag ratio for rotary or flapping wings (Chapter 3, Box 3.4). Then it estimates the distance flown and the amount of fuel used in a short period of flight (6 minutes). Then it revises the mass, body frontal area and so on to take account of the fuel used, and recalculates the power and the effective lift:drag ratio for the next 6 minutes of flight, and repeats this until all the fat has been consumed. No range equation is needed, and complicated rules can easily be incorporated in the process of revising the speed and body composition (Figure 8.1).

## 8.2 — ULTRA LONG-DISTANCE MIGRANTS

*Flight's* migration simulation predicts the time course of many variables which can, in principle, be measured in real long-distance migrants, and these predictions follow from assumptions which can be varied by the user. By comparing the predictions with field data, alternative hypotheses can be tested about (for example) the way in which the bird manages its fuel reserves in flight. Getting suitable field data is easier said than done, but the examples in this chapter illustrate the potential of this approach. Table 8.1 gives the necessary input details for simulating known flights of two ultra long-distance migrants, whose routes are shown on the map of Figure 8.1. They are a migration stage of 5,420 km from Australia to China that is believed to be flown non-stop by Great Knots (*Calidris tenuirostris*) on their way to their Siberian breeding grounds in the northern spring, and on the same map, the route of the Alaskan Bar-tailed Godwit (*Limosa lapponica*) which leaves the Alaskan peninsula in the northern autumn with over half of its body mass as fat, and flies non-stop 10,300 km to the northern tip of New Zealand (Piersma and Gill 1998). Using default assumptions for input variables other than those in Table 8.1, *Flight* predicts that both birds start with enough fat to get well beyond their destinations, and that the Godwit should just about be able to reach the South Pole.

The Excel output from *Flight's* simulation of the Great Knot's migration from Pennycuick and Battley (2003) is reproduced in full in Table 8.2. The first section of the table shows the input data from

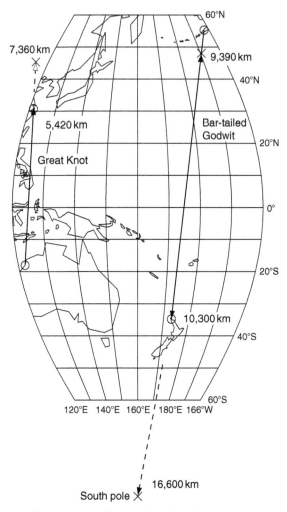

FIGURE 8.1 Routes flown non-stop by two ultra long-distance migrants, the Great Knot (*Calidris tenuirostris*) from Pennycuick and Battley (2003) and the Alaskan Bar-tailed Godwit (*Limosa lapponica*) from Piersma and Gill (1998). The Godwit had insufficient fuel to fly back non-stop from New Zealand to Alaska, but is known to stage in Australia in this direction. Note that the projection distorts directions. The lines of longitude are great circles, but the straight lines showing the birds' routes are not. Circles: Points where samples of birds were collected. Crosses: Calculated distance from start, where bird runs out of fuel.

Table 8.1, with the air density set for 3000 m in the International Standard Atmosphere, and other variables set to their default values. The rest of the table is the output, showing 30 variables whose changing values have been calculated every 6 minutes throughout the flight, and tabulated at intervals of 6 hours. The reader who wishes to repeat

TABLE 8.1 Data for Great Knot and Alaskan Bar-tailed Godwit.

| Observed before flight: | Great Knot | Bar-tailed Godwit |
|---|---|---|
| Wing span (m) | 0.586 | 0.748 |
| Wing area (m$^2$) | 0.0397 | 0.0568 |
| Aspect ratio | 8.65 | 9.85 |
| All-up mass (kg) | **0.233** | 0.367 |
| Fat mass (kg) | **0.0898** | 0.201 |
| Flight muscle mass (kg) | **0.0336** | 0.055 |
| Airframe mass (kg) | **0.110** | 0.111 |
| Fat fraction | 0.385 | 0.548 |
| Flight muscle fraction | 0.144 | 0.150 |
| Airframe fraction | 0.472 | 0.302 |
| **After flight (Great Knot):** | **Observed** | **Predicted** |
| All-up mass (kg) | **0.125** | **0.134** |
| Fat mass (kg) | **0.0107** | **0.0166** |
| Flight muscle mass (kg) | **0.0255** | **0.0215** |
| Airframe mass (kg) | **0.0888** | **0.0955** |

Mass components observed in the Great Knot before flight are in bold type, as are the corresponding figures, both observed and predicted, after the 5420-km migration (from Pennycuick and Battley 2003). The Bar-tailed Godwit was observed before departure from Alaska, but not on arrival in New Zealand (from Piersma and Gill 1998). The mass subdivisions are defined in Chapter 1, Box 1.2. Note that the mass fractions for each bird add up to 1.

this exercise will find the data for the Great Knot in *Flight's* Preset Birds database, from which they can be entered directly into the programme. To get an Excel spreadsheet the computer must have Microsoft Excel installed, otherwise the programme will output the same information in the form of a text file.

## 8.2.1 TESTING HYPOTHESES

Some of the features that are represented in *Flight's* migration simulation are described in the following paragraphs, and illustrated by the graphs accompanying the commentary on the Great Knot's flight in Box 8.2. The programme incorporates current hypotheses or guesses about how a migrating bird works, which are explained elsewhere in this book. Discrepancies between its predictions and field observations can be used to test, and if necessary correct, the hypotheses and assumptions underlying the model. For example, it predicts the total amounts of mass consumed in the course of the flight from the stored fat, the flight muscles and other parts of the body (the "airframe"), and Table 8.1 lists the observed and predicted masses of these three components at the end of the flight. The birds' all-up mass declined by

TABLE 8.2 Great Knot Migration.

| | A | B | C | D | E | F | G | H | I | J | K |
|---|---|---|---|---|---|---|---|---|---|---|---|
| 1 | Migration summary for | | | | | Calidris tenuirostris | | | | (Non-Passerine) | |
| 2 | Sex: | Male and female | | Age: | Adult | | From Flight Version 1.16 (2004) by C.J. Pennycuick | | | | |
| 3 | Variable values for this run: | | | | | Saved as: | C:\FlightOutput\ExcelFiles\Migration\GK_0002.xls | | | | |
| 4 | Empty body mass (kg) | | | | 0.233 | | | | | | |
| 5 | Flight muscle mass (kg) | | | | 0.03122 | | Flight muscle fraction | | | 0.134 | |
| 6 | Fat mass (kg) | | | | 0.08971 | | Fat fraction | | | 0.385 | |
| 7 | Airframe mass (kg) | | | | 0.1121 | | Airframe fraction | | | 0.481 | |
| 8 | Crop contents mass (kg) | | | | 0 | | Gravity (m/sec-squared) | | | 9.81 | |
| 9 | All-up mass (kg) | | | | 0.233 | | Induced power factor | | | 1.2 | |
| 10 | Wing span (m) | | | | 0.587 | | Profile power constant | | | 8.4 | |
| 11 | Wing area (sq m) | | | | 0.0396 | | Profile power ratio | | | 0.9654 | |
| 12 | Aspect ratio | | | | 8.701 | | Respiration factor | | | 1.1 | |
| 13 | Air density (kg/cubic m) | | | | 0.909 | | Conversion efficiency | | | 0.23 | |
| 14 | Altitude (m ASL in Standard Atmosphere) | | | | 3000 | | | | | | |
| 15 | Body frontal area | | | | 0.003081 | | Flight style | | Continuous flapping | | |
| 16 | Body drag coefficient | | | | 0.1 | | Power fraction | | | 1 | |
| 17 | Mitochon lnv power density (cubic m/W) | | | | 1.20E-06 | | BMR factor | | | 1 | |
| 18 | Fat energy density (J/kg) | | | | 3.90E+07 | | Air speed control | | Standard | | |
| 19 | Dry protein energy density (J/kg) | | | | 1.83E+07 | | Starting ratio V:Vmp | | | 1.2 | |
| 20 | Protein hydration ratio | | | | 2.2 | | Calculation interval (min) | | | 6 | |
| 21 | Minimum energy from protein (%) | | | | 5 | | Output interval (hours) | | | 6 | |
| 22 | Protein burn criterion | | Constant specific work | | | | Distance to destination (km) | | | 0 | |
| 23 | Mitochondria control | | Constant mito power density | | | | Wingbeat frequency factor | | | 1 | |
| 24 | | | | | | | | | | | |
| 25 | As in Pennycuick & Battley. Data in Preset Birds database | | | | | | | | | | |
| 26 | | | | | | | | | | | |
| 27 | Time | Distance | True air speed (V) | Vmp | V:Vmp ratio | Wing Reynolds number | Body Reynolds number | Total Body mass | | Fat mass | Flt muscle mass | Airframe mass |
| 28 | | | | | | | | | | | | |
| 29 | Hrs | km | m/s | m/s | number | number | kg | kg | kg | kg | kg |
| 30 | 0 | 0 | 15.84 | 13.2 | 1.2 | 58640 | 54560 | 0.233 | 0.08971 | 0.03122 | 0.1121 |
| 31 | 6 | 361.4 | 17.47 | 13.01 | 1.342 | 64560 | 59240 | 0.2235 | 0.08268 | 0.03122 | 0.1096 |
| 32 | 12 | 751.4 | 18.58 | 12.83 | 1.448 | 68750 | 62190 | 0.2141 | 0.07578 | 0.03122 | 0.1071 |
| 33 | 18 | 1163 | 19.48 | 12.64 | 1.54 | 72040 | 64220 | 0.2049 | 0.069 | 0.03122 | 0.1046 |
| 34 | 24 | 1592 | 20.11 | 12.46 | 1.615 | 74480 | 65400 | 0.1959 | 0.06242 | 0.03104 | 0.1025 |
| 35 | 30 | 2023 | 19.85 | 12.27 | 1.618 | 73510 | 63610 | 0.1874 | 0.05614 | 0.02946 | 0.1018 |
| 36 | 36 | 2449 | 19.6 | 12.1 | 1.621 | 72590 | 61900 | 0.1794 | 0.05026 | 0.028 | 0.1011 |
| 37 | 42 | 2870 | 19.37 | 11.92 | 1.624 | 71710 | 60290 | 0.1719 | 0.04475 | 0.02665 | 0.1005 |
| 38 | 48 | 3286 | 19.14 | 11.76 | 1.628 | 70870 | 58760 | 0.1649 | 0.03957 | 0.02539 | 0.09991 |
| 39 | 54 | 3697 | 18.92 | 11.6 | 1.631 | 70060 | 57300 | 0.1583 | 0.0347 | 0.02422 | 0.09933 |
| 40 | 60 | 4103 | 18.71 | 11.45 | 1.635 | 69280 | 55910 | 0.152 | 0.03011 | 0.02313 | 0.09878 |
| 41 | 66 | 4505 | 18.51 | 11.3 | 1.639 | 68530 | 54580 | 0.1461 | 0.02578 | 0.02211 | 0.09825 |
| 42 | 72 | 4903 | 18.31 | 11.15 | 1.643 | 67820 | 53320 | 0.1406 | 0.02169 | 0.02115 | 0.09773 |
| 43 | 78 | 5296 | 18.13 | 11.01 | 1.647 | 67130 | 52110 | 0.1353 | 0.01781 | 0.02026 | 0.09724 |
| 44 | 84 | 5686 | 17.95 | 10.87 | 1.651 | 66470 | 50950 | 0.1303 | 0.01414 | 0.01941 | 0.09676 |
| 45 | 90 | 6072 | 17.78 | 10.74 | 1.655 | 65830 | 49840 | 0.1256 | 0.01066 | 0.01862 | 0.0963 |
| 46 | 96 | 6454 | 17.61 | 10.61 | 1.66 | 65220 | 48780 | 0.1211 | 0.007346 | 0.01788 | 0.09585 |
| 47 | 102 | 6833 | 17.45 | 10.48 | 1.665 | 64620 | 47760 | 0.1168 | 0.004197 | 0.01718 | 0.09542 |
| 48 | 108 | 7208 | 17.3 | 10.36 | 1.67 | 64060 | 46790 | 0.1127 | 0.001198 | 0.01652 | 0.09501 |
| 49 | 110.5 | 7363 | 17.24 | 10.31 | 1.672 | 63830 | 46390 | 0.1111 | 0 | 0.01625 | 0.09484 |
| 50 | | | | | | | | | | | |

108 g in the course of the flight, which was 9 g more than predicted. The predicted consumption of protein from the flight muscles was 4 g too high, that from the airframe was 7 g too low, and that of fat was 6 g too low. This is not bad agreement considering that the winds along the route were not known, and that there was a degree of uncertainty about the interval between the time the samples were collected and the actual

TABLE 8.2 Great Knot Migration.

| | A | B | C | D | E | F | G | H | I | J | K |
|---|---|---|---|---|---|---|---|---|---|---|---|
| 51 | Time | Mechanical | Chemical | Effective | Wingbeat | Reduced | Specific | Specific | Pbmr | Pchem: | Mitochon |
| 52 | | power | power | lift:drag | frequency | frequency | power | work | | Pbmr | mass frac |
| 53 | Hrs | W | W | ratio | Hz | | W/kg | J/kg | W | ratio | % |
| 54 | 0 | 2.604 | 13.48 | 11.68 | 9.189 | 0.123 | 93.29 | 10.15 | 0.9302 | 14.49 | 10.61 |
| 55 | 6 | 2.556 | 13.24 | 12.58 | 9 | 0.1094 | 91.4 | 10.16 | 0.9186 | 14.41 | 10.42 |
| 56 | 12 | 2.507 | 12.99 | 13.06 | 8.809 | 0.1006 | 89.44 | 10.15 | 0.9069 | 14.32 | 10.22 |
| 57 | 18 | 2.459 | 12.75 | 13.35 | 8.618 | 0.09387 | 87.53 | 10.16 | 0.8953 | 14.24 | 10.02 |
| 58 | 24 | 2.394 | 12.42 | 13.53 | 8.427 | 0.0888 | 85.54 | 10.15 | 0.8839 | 14.06 | 9.813 |
| 59 | 30 | 2.228 | 11.61 | 13.66 | 8.241 | 0.08798 | 83.66 | 10.15 | 0.8731 | 13.3 | 9.618 |
| 60 | 36 | 2.076 | 10.88 | 13.79 | 8.064 | 0.08717 | 81.86 | 10.15 | 0.8629 | 12.61 | 9.43 |
| 61 | 42 | 1.938 | 10.21 | 13.91 | 7.893 | 0.08638 | 80.13 | 10.15 | 0.8533 | 11.96 | 9.25 |
| 62 | 48 | 1.812 | 9.593 | 14.03 | 7.73 | 0.08561 | 78.47 | 10.15 | 0.8443 | 11.36 | 9.076 |
| 63 | 54 | 1.696 | 9.032 | 14.14 | 7.574 | 0.08484 | 76.88 | 10.15 | 0.8357 | 10.81 | 8.908 |
| 64 | 60 | 1.59 | 8.516 | 14.25 | 7.423 | 0.08408 | 75.35 | 10.15 | 0.8277 | 10.29 | 8.747 |
| 65 | 66 | 1.493 | 8.042 | 14.35 | 7.278 | 0.08334 | 73.88 | 10.15 | 0.82 | 9.807 | 8.59 |
| 66 | 72 | 1.403 | 7.605 | 14.44 | 7.138 | 0.0826 | 72.46 | 10.15 | 0.8128 | 9.357 | 8.439 |
| 67 | 78 | 1.321 | 7.202 | 14.53 | 7.003 | 0.08187 | 71.09 | 10.15 | 0.8059 | 8.937 | 8.292 |
| 68 | 84 | 1.244 | 6.829 | 14.61 | 6.873 | 0.08115 | 69.76 | 10.15 | 0.7993 | 8.544 | 8.151 |
| 69 | 90 | 1.173 | 6.484 | 14.69 | 6.747 | 0.08043 | 68.49 | 10.15 | 0.7931 | 8.175 | 8.013 |
| 70 | 96 | 1.108 | 6.164 | 14.76 | 6.625 | 0.07972 | 67.25 | 10.15 | 0.7872 | 7.83 | 7.88 |
| 71 | 102 | 1.047 | 5.866 | 14.82 | 6.506 | 0.07901 | 66.05 | 10.15 | 0.7815 | 7.506 | 7.75 |
| 72 | 108 | 0.9901 | 5.589 | 14.88 | 6.392 | 0.07831 | 64.89 | 10.15 | 0.7761 | 7.201 | 7.624 |
| 73 | 110.5 | 0.9677 | 5.479 | 14.91 | 6.345 | 0.07802 | 64.41 | 10.15 | 0.7739 | 7.08 | 7.573 |
| 74 | | | | | | | | | | | |
| 75 | Time | Fat | Protein | Total | Protein | BMR | Fat | Flight | Airframe | Fat energy | Fat + Prot |
| 76 | | burn | burn | fuel burn | burn frac | burn frac | fraction | muscle | fraction | height | energy ht |
| 77 | Hrs | kJ | kJ | kJ | % | % | | fraction | | km | km |
| 78 | 0 | 0 | 0 | 0 | 0 | 0 | 0.385 | 0.134 | 0.481 | 444.5 | 514.4 |
| 79 | 6 | 274.4 | 14.19 | 288.5 | 4.917 | 6.92 | 0.3699 | 0.1397 | 0.4902 | 423.3 | 485.4 |
| 80 | 12 | 543.5 | 28.36 | 571.8 | 4.959 | 6.939 | 0.3539 | 0.1458 | 0.5001 | 399.4 | 455.8 |
| 81 | 18 | 807.5 | 42.26 | 849.8 | 4.973 | 6.959 | 0.3367 | 0.1524 | 0.5107 | 375.4 | 425.4 |
| 82 | 24 | 1064 | 55.94 | 1120 | 4.994 | 6.995 | 0.3186 | 0.1584 | 0.523 | 350.7 | 394.7 |
| 83 | 30 | 1309 | 68.84 | 1378 | 4.995 | 7.063 | 0.2996 | 0.1572 | 0.5432 | 325.6 | 363.8 |
| 84 | 36 | 1538 | 80.9 | 1619 | 4.996 | 7.168 | 0.2801 | 0.1561 | 0.5637 | 300.6 | 333.6 |
| 85 | 42 | 1753 | 92.22 | 1846 | 4.997 | 7.293 | 0.2603 | 0.155 | 0.5846 | 275.7 | 304.1 |
| 86 | 48 | 1955 | 102.8 | 2058 | 4.997 | 7.431 | 0.24 | 0.154 | 0.6059 | 250.9 | 275.1 |
| 87 | 54 | 2145 | 112.8 | 2258 | 4.998 | 7.576 | 0.2193 | 0.153 | 0.6276 | 226.3 | 246.6 |
| 88 | 60 | 2324 | 122.3 | 2446 | 4.998 | 7.727 | 0.1981 | 0.1521 | 0.6498 | 201.8 | 218.7 |
| 89 | 66 | 2493 | 131.2 | 2624 | 4.998 | 7.882 | 0.1764 | 0.1513 | 0.6723 | 177.5 | 191.2 |
| 90 | 72 | 2653 | 139.6 | 2792 | 4.998 | 8.039 | 0.1543 | 0.1505 | 0.6952 | 153.2 | 164.2 |
| 91 | 78 | 2804 | 147.5 | 2951 | 4.998 | 8.198 | 0.1316 | 0.1497 | 0.7186 | 129.1 | 137.6 |
| 92 | 84 | 2947 | 155.1 | 3102 | 4.999 | 8.358 | 0.1085 | 0.149 | 0.7425 | 105 | 111.4 |
| 93 | 90 | 3083 | 162.2 | 3245 | 4.999 | 8.52 | 0.08485 | 0.1483 | 0.7668 | 81.07 | 85.55 |
| 94 | 96 | 3212 | 169 | 3381 | 4.999 | 8.682 | 0.06066 | 0.1477 | 0.7916 | 57.22 | 60.09 |
| 95 | 102 | 3335 | 175.5 | 3510 | 4.999 | 8.845 | 0.03592 | 0.1471 | 0.8169 | 33.45 | 34.97 |
| 96 | 108 | 3452 | 181.6 | 3633 | 4.999 | 9.008 | 0.01061 | 0.1465 | 0.8428 | 9.757 | 10.15 |
| 97 | 110.5 | 3499 | 184.1 | 3683 | 4.999 | 9.076 | 0 | 0.1463 | 0.8537 | 0 | 0 |
| 98 | | | | | | | | | | | |
| 99 | The current version of Flight for Windows is available as a free download from: | | | | | | | | | | |
| 100 | http://www.bio.bristol.ac.uk/people/pennycuick.htm | | | | | | | | | | |
| 101 | Please send comments and bug reports to c.pennycuick@bristol.ac.uk | | | | | | | | | | |

times of departure and arrival. If larger discrepancies had been seen, there might have been a case for revising some of the default values for input variables, such as the body drag coefficient, and the assumptions that determine the relative amounts of protein taken from different sources. Some variations were tried in the original paper, but no case was made for changing any of the defaults (Pennycuick and Battley 2003).

BOX 8.2 **A simulated migratory flight by a Great Knot.**

This box is a commentary on a single run of the Migration calculation in *Flight*, described by Pennycuick and Battley (2003), whose output is presented in full in Table 8.2. Two sets of data were collected by Phil Battley from the same population of Great Knots (*Calidris tenuirostris*), before and after migrating from Broome on the north-west coast of Australia to the coast of China near Shanghai. The birds are believed to fly this route non-stop in a single flight taking between 4 and 5 days. The great-circle distance between the two sample stations, marked by circles in Figure 8.1, is 5420 km. The reader who wishes to repeat the calculation will find the starting data for the Great Knot in the "Preset Birds" database that comes with the *Flight* programme. The effect of varying the values of different input variables can be tried simply by returning to the Migration Setup screen, making any changes required, and running the programme again. Any input values that were changed since the previous run will appear in red in the Excel output.

Extracts from the output are presented here in the form of a dozen graphs selected from the 30 columns of data available, all plotted against distance

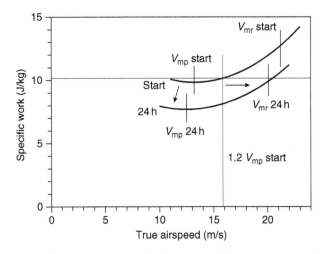

FIGURE 8.2 Data from two points in the Great Knot's Migration output (Table 8.2) have been used to run *Flight*'s power curve calculation for the start of the flight (upper curve), and after flying for 24 hours (lower curve). In this simulation the altitude was constant at 3000 m ASL throughout the flight. The curves represent specific work in the flight muscles (not power), but they are almost the same shape as power curves, with a minimum at $V_{mp}$. The selected muscle burn criterion was "Constant Specific Work" (the default), and the starting curve was used to select the value of the specific work ($10.1\,J\,kg^{-1}$) which was sufficient to fly at 1.2 $V_{mp}$ at the starting mass. Under the "Standard" speed control rules, the specific work was held constant initially by allowing the speed to increase, and later by consuming material from the flight muscles. After 24 hours of flight, using up fat, the same value of the specific work (thin horizontal line) was sufficient to fly about $4\,m\,s^{-1}$ faster that at the start. This speed is beyond the (decreasing) value of the maximum range speed $V_{mr}$ on the lower curve. After this the speed tracks $V_{mr}$ (which continues to decrease), and the specific work is held constant by consuming protein from the flight muscles.

BOX 8.2 *Continued.*

flown. Variables in the output table are referred to by the cell in Table 8.2 in which the column of output starts, for instance B30 is the distance flown, starting at 0 and ending at 7363 km when the bird consumes the last of its fat. This is, of course, *air* distance. The air mass through which the bird flies is the frame of reference, relative to which *Flight* computes its progress. If the earth's surface also moves relative to the air mass while the flight is in progress (wind) then the ground distance is different, exceeding the air distance with a following wind, and less with a headwind. The flight time from the start can be read in the left-hand column of each of the three blocks of output (A30, A54 and A78). The calculated points from the Excel output are shown in the graphs as small solid circles, joined together by "smooth links", which are purely cosmetic. Four-digit precision has been used only because graphs do not look smooth when plotted from the default three-digit output.

### Airspeed control

Before launching the migration simulation, the programme requires a set of rules to determine the speed at which the bird will fly at different stages of its flight. If the standard option for Air Speed Control is selected in the Migration Setup screen, the programme will begin by calculating the minimum power speed ($V_{mp}$) at the start of the flight, and set the initial speed to a multiple of $V_{mp}$, for which the default value is 1.2. It will then find the specific work in the flight muscles at this speed (Chapter 7). If "Constant Specific Work" is chosen as the Muscle Burn Criterion, it will hold this initial value constant for the rest of the flight. Figure 8.2 shows how the specific work at 1.2 $V_{mp}$ is located initially at 10.1 J kg$^{-1}$. The curve of specific work against speed drops as the bird gets lighter, and after about 24 hours of flight, the original value of the specific work is enough to maintain the maximum range speed ($V_{mr}$). After this, the speed tracks the decreasing value of $V_{mr}$ for the rest of the flight, as there is no range advantage in flying any faster. The specific work is held constant by consuming excess flight muscle.

### Climb and descent

The initial climb is not represented in this version of the Migration calculation. The bird is simply launched at its assumed cruising altitude, which is 3000 m in this case. The assumption that the fully loaded bird is only just capable of level flight at 1.2 $V_{mp}$ is justified in the case of swans (Chapter 7), but not so easily reconciled with Table 8.1. The Great Knot requires a specific work of 10.1 J kg$^{-1}$ from the myofibrils of its flight muscles, whereas they should be able to produce 41.1 J kg$^{-1}$ when operating at maximum power (Chapter 7). Other medium-sized birds also seem to have a generous margin of muscle power available over that required for level flight, even when burdened with large fat loads, and it now seems likely that long-distance migrants use this power margin for the initial climb to cruising altitude. The rate of climb calculation described in Box 7.5 was added to *Flight* in Version 1.17. It takes due account of the additional mitochondria that would be needed in the flight muscles to sustain an aerobic climb, and gives an initial maximum rate of climb at sea level of 2.5 m s$^{-1}$ for the Great Knot, when fully loaded and flying at $V_{mp}$. This is at the high end of tracking radar observations by Hedenström and Alerstam (1992) in southern Sweden, and Piersma et al. (1997) in West Africa of waders and ducks departing on what

**BOX 8.2** *Continued.*

were presumed to be long migratory stages. Rates of climb were mostly between 1 and 2 m s$^{-1}$, with a few above 2 m s$^{-1}$. This suggests that most birds prefer to climb at a somewhat faster airspeed than that for maximum rate of climb, as do most pilots. The real bird has a range of options, which would be difficult to encapsulate in *Flight*'s Migration calculation, although field observations of mitochondria fraction might allow the possibilities to be narrowed down somewhat.

In terms of energy height (below), the effect is that the bird converts some of its energy height into actual height at the beginning of the flight, and converts its actual height into distance at the end. Although it is true that more energy is expended on the initial climb, when the bird is heavy, than is retrieved on the final descent when it is light, there is little effect on the distance flown, because this depends on height (or energy height) not on energy as such (Box 8.3).

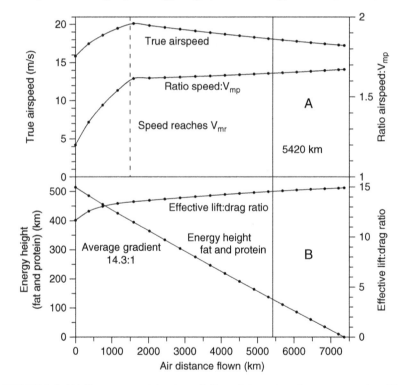

FIGURE 8.3 (A) True airspeed (scale at left) and the ratio of true airspeed to $V_{mp}$ (scale at right) plotted against distance flown from the Migration output for the Great Knot. The dashed vertical line is the point where the airspeed converges on $V_{mr}$, after which the specific work is held constant by consuming flight muscle, instead of increasing speed. The solid vertical line through both graphs is the distance to the destination. (B) Likewise for energy height for combined fat and protein fuel (left scale) and effective lift:drag ratio (right scale). The effective lift:drag ratio is the gradient of the energy height graph. Data from the Great Knot simulation in Table 8.2.

BOX 8.2 *Continued.*

### Speed, energy height and effective lift:drag ratio

The time course of several variables is shown in Figures 8.3–8.5. The airspeed (Figure 8.3 upper graph, and C30) builds up for the first 24 hours and 1600 km of the flight, until it converges with the maximum range speed ($V_{mr}$), which is coming down. The ratio of $V_{mr}$ to $V_{mp}$ (E30) is around 1.6 in the early part of the flight, and the ratio nearly levels off at that point, continuing to increase very slightly. After this, the bird progressively reduces speed so as to continue flying at $V_{mr}$, and also starts to consume protein from the flight muscles (J30). In each 6-minute period, it consumes sufficient flight muscle to hold the specific work constant (H54), at the value that it had at the beginning of the flight (10.1 J kg$^{-1}$).

The energy height for both fat and protein together (Figure 8.3 lower graph, and K78) starts at 514 km at departure, and declines to zero after the bird has flown 7363 km. This gives an average gradient of 14.3:1, which is also the average effective lift:drag ratio for the whole flight. The virtual descent is not so perfectly linear as it at first appears. Its gradient at any point is the effective lift:drag ratio (D54), which starts at 11.7:1 and increases in the early hours of the flight, as the airspeed converges on to $V_{mr}$. It continues to increase gradually while the bird maintains $V_{mr}$, because the consumption of fat and muscle slims down the body's frontal area, so reducing the drag of the body. After increasing to about 13.5:1 after 24 hours, it eventually reaches 14.9:1 by the end of the flight.

### Mass, flight muscle fraction and specific power

In the upper graph of Figure 8.4, the all-up mass (H30) declines smoothly from its starting value of 0.233 kg to less than half of that when the programme terminates because all the fat has been consumed. The fat fraction (G78) was 0.385 at the start of the flight (a high value but not extreme), whereas a bigger fraction than that (52%) of the starting mass disappears by the end of the flight. The remaining mass loss is made up of protein taken partly from the flight muscles and partly from other unspecified organs, which no doubt include the heart, digestive system leg muscles and so on.

The flight muscle *fraction* (H78) increases for the first 24 hours, because fat is being consumed but flight muscle tissue is not (above). The bird starts consuming flight muscle after flying for about 24 h, when the speed converges on $V_{mr}$. By then the flight muscle fraction is 0.158, and thereafter it declines very gradually to 0.146, very near its initial value of 0.144. In other words, the flight muscle fraction does not change very much in the course of a long migratory flight. Conversely, migrants accumulate both fat and protein during stopovers, also resulting in only minor variation of the flight muscle fraction. A practical consequence is that although the flight muscle *mass* may vary wildly in healthy birds of the same species at different levels of condition, the flight muscle *fraction* only varies a little. Consequently, this measurement can be obtained by dissecting birds that have been shot, or that have met with accidents, without worrying overmuch about variations in condition.

**BOX 8.2** *Continued.*

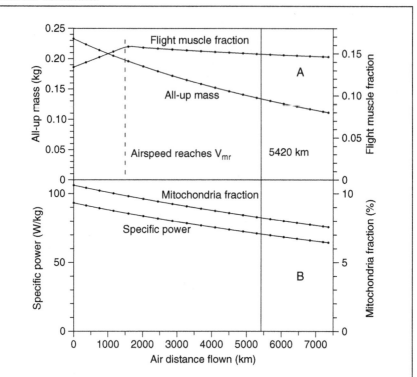

FIGURE 8.4 (A) The flight muscle fraction increases during the initial acceleration phase, because fat is consumed but flight muscle protein is not. Once the consumption of flight muscle begins, the flight muscle fraction declines, but only slightly. The "Constant specific work" assumption agrees quite well with the empirical observation that the flight muscle fraction of migrants does not vary much, despite large changes of all-up mass. (B) The mitochondria fraction in the flight muscles is closely related to the specific power, and both decline progressively throughout the flight. Mitochondria fraction is a quantity that could be observed. Data from the Great Knot simulation in Table 8.2.

In the lower graph of Figure 8.4, the specific power in the flight muscles (G54) declines steadily throughout the flight, if the specific work (H54) is held constant, as it was in this programme run. As an alternative option, the programme allows the specific power to be held constant. If that option were selected, the flight muscle mass would decrease more steeply than seen here, while the specific work would progressively increase throughout the flight, neither of which results accord with observations. The mitochondria fraction in the flight muscles depends directly on the specific power and is expected to decline throughout the flight if the specific work is held constant. This is a prediction that could be tested, if flight muscle samples suitable for electron microscopy were collected in future field studies.

BOX 8.2 *Continued.*

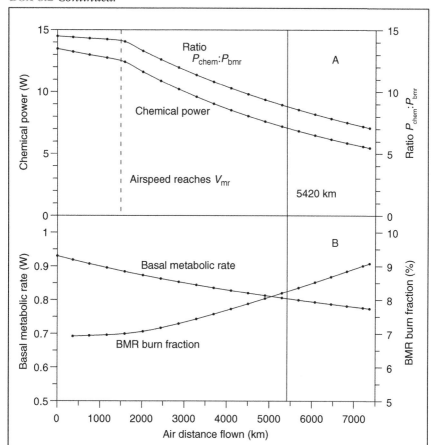

FIGURE 8.5  (A) The chemical power declines throughout the flight, and so does the ratio of the chemical power to the basal metabolic rate (BMR). If the chemical power is expressed as a multiple of BMR, it declines from about 14 to 7 times BMR. This ratio is also higher in large birds than in small ones. This traditional method of expressing chemical power conveys no useful information, and obscures the original power measurements. (B) The calculated BMR is based on the mass excluding fat, and declines throughout the flight. The cumulative fraction of the total energy that is expended on basal metabolism increased from about 7 to 9%. These percentages would be higher in a smaller bird and lower in a larger one. Data from the Great Knot simulation in Table 8.2.

## Chemical power and basal metabolism

The chemical power (Figure 8.5 upper graph and C54) is the rate of consumption of fuel energy. This is what physiologists measure by such methods as the rate of oxygen consumption, and doubly labelled water. In this simulation the chemical power declines from the beginning of the flight because the bird is getting lighter, and also because the speed is initially increasing towards $V_{mp}$, so increasing the effective lift:drag ratio. However,

BOX 8.2 *Continued.*

as the speed increases, the power (mechanical and chemical) moves up the ever-steepening power curve, which offsets these gains until the speed reaches $V_{mr}$. The small decrease in chemical power seen in the first day or so of the flight is actually due to a decrease in wingbeat frequency (E54) which itself results from the decrease in all-up mass. So long as both the flight muscle mass (J30) and the specific work (H54) are being held constant under the options selected, the mechanical power is proportional to the wingbeat frequency, and the chemical power nearly so. Once the airspeed is tracking the (decreasing) value of $V_{mr}$, and protein is being removed from the flight muscles, the chemical power drops more steeply, and ends the flight at only 41% of its starting value.

The BMR (Figure 8.5 lower graph and I54) is calculated from the body mass excluding fat, and is therefore assumed to decline as protein is consumed from the flight muscles and other organs. It is actually unknown what happens to the BMR in flight, but under these assumptions, the basal metabolism burn fraction (F78) is the fraction of the total fuel energy consumed so far in the flight that is due to basal metabolism. This is undefined at the beginning of the flight, before any fuel energy has been consumed, so the curve only starts 6 hours into the flight. The curve climbs from about 7% to 9% as the flight proceeds (note the suppressed zero on this graph). The BMR in *Flight* is a somewhat crude statistical estimate, which is added to the computed mechanical power as an "overhead", so it is not to be expected that this fraction would have any particular significance. Estimates depend strongly on the size of the bird, from around 2.5% for a whooper swan to about 25% for a chaffinch.

## 8.2.2 BURNING THE ENGINE

Migrating birds typically set out with a substantial fat fraction, that is with a proportion of the body mass consisting of fat which is available for consumption in flight. As the flight proceeds, the bird gets lighter as the fat is progressively used up, and the power required from the flight muscles, which depends strongly on the all-up mass, declines even more steeply than the mass. If the bird has just enough flight muscle to cover its power requirements at departure, then it soon has excess power available, as the mass declines. In aircraft, the simplest option is to run the engine(s) at reduced power, but this is inefficient as aircraft engines are designed to operate most efficiently at near-maximum power. Alternatively a multi-engined aircraft can shut down one or more engines, which allows the aircraft to run the remaining engine(s) at a higher power level, but leaves it burdened by the weight and drag of the inactive engines. A migrating bird has a third option that no aircraft can emulate. Rather than shutting down part of its flight muscles, it can oxidise the excess muscle tissue, and use it as supplementary fuel. It was shown by Lindström and Piersma (1993) that migrating birds do indeed consume both fat and protein in flight, and replace both during stopovers. "Burning the engine" not only provides some extra fuel energy, but also gets rid of both weight and drag, so further reducing the power required, and stretching the distance that the bird can fly on the remaining fuel.

### 8.2.3 ALTERNATIVE MUSCLE BURN CRITERIA

How much muscle tissue, exactly, is enough to continue flying? *Flight* provides two alternative *muscle burn criteria* (selectable by the user) to determine how much muscle is still needed at each 6-minute interval. Excess muscle is consumed, and the energy released is treated as a "credit", by putting back an equivalent amount of fat. At the beginning of the flight, the mechanical power is calculated and divided by the mass of the myofibril component of the flight muscles to get the mass-specific power, that is, the average mechanical power produced by unit mass of contractile filaments. This in turn is divided by the wingbeat frequency to get the mass-specific work which is the amount of work done by unit mass of myofibrils in each wingbeat. Both of these variables decline as the mass declines if no muscle tissue is consumed, but the specific work declines less steeply than the specific power, because the wingbeat frequency also declines (Chapter 7, Box 7.3).

The default muscle burn criterion in *Flight* is "Constant Specific Work", meaning that just enough muscle is consumed to hold the specific work constant, at the value it had at the beginning of the flight. The alternative is "Constant Specific Power", in which enough muscle is consumed to hold the specific *power* constant at its initial value. It turns out that if Constant Specific Work is selected, the muscle mass (including mitochondria) declines at a rate that keeps the flight muscle *fraction* approximately constant, which is in general agreement with field observations, whereas Constant Specific Power results in a steeper decline of the flight muscle fraction than is observed (Pennycuick 1998a). Also, maintaining Constant Specific Work results in a progressive decline in the specific power, whereas maintaining Constant Specific Power causes the specific work to increase in the course of the flight. Since the specific work is directly related to the mechanical properties of the myofibrils (Chapter 7), a procedure that forces it to increase above its initial value is questionable, whereas the alternative procedure, which causes the specific power to decline, raises no such questions. The programme also provides a third option, in which no muscle tissue is consumed at all, but this produces results at variance with field observations on long-distance migrants, with possible exceptions in very large birds such as swans.

## 8.2.4 MITOCHONDRIA IN THE FLIGHT MUSCLES

The Migration calculation in *Flight* represents the flight muscles as a two-component system, consisting of the contractile filaments (myofibrils) which do the work, and the mitochondria which oxidise fat in aerobic flight, and supply ATP as the immediate source of energy for the myofibrils (Chapter 7, Box 7.6). It is assumed that both myofibrils and mitochondria are consumed as muscle mass is reduced, but the mitochondria have their own default criterion, which is to hold the specific power (or "power density") constant, referred to the mass of mitochondria. There are no field data to go on here, as no field observer has preserved muscle samples for electron microscopy, to observe the mitochondria fraction. If and when someone does this, *Flight* will predict the expected mitochondria fraction, according to various combinations of options, and any discrepancies will shed some light on the way in which migrating birds manage the components of their flight muscles.

## 8.2.5 AIRSPEED CONTROL

To travel the maximum air distance per unit of fuel energy consumed, a bird has to fly at its maximum range speed ($V_{mr}$), which is typically between 1.6 and 1.8 times the minimum power speed ($V_{mp}$) (Chapter 3). As fuel is used up, both $V_{mp}$ and $V_{mr}$ decrease, and the ratio of $V_{mr}$ to $V_{mp}$ increases (but only a little). One might suppose that the appropriate strategy for a bird that requires to cover the maximum distance for its starting fuel load is to accelerate to $V_{mr}$ immediately after take-off, and then reduce speed progressively so as to track the (declining) maximum range speed throughout the flight. However, extreme long-distance migrants commonly accumulate fat until their body mass is 40% fat or even more (55% for the Alaskan Bar-tailed Godwit), meaning that the bird takes on fat until it is barely able to fly. A bird that can only just fly is limited to speeds near the minimum power speed (Chapter 3), although as the power curve is almost horizontal near $V_{mp}$, increasing the speed to 1.2 $V_{mp}$ increases the power by only 1 or 2%. The standard Air Speed Control option in *Flight* assumes that the bird is capable of flying at 1.2 $V_{mp}$ at the start of the flight, but no faster. It begins by estimating $V_{mp}$ for the bird in its initial configuration, before it has used up any fat, and sets the initial airspeed to 1.2 $V_{mp}$. The flight then proceeds in two phases, as follows:

In Phase 1, no flight muscle tissue is consumed. Instead, as excess muscle power becomes available because of the declining mass, it is used to increase the speed, while holding the specific work or the specific power constant, depending on the Muscle Burn Criterion selected. Meanwhile, $V_{mr}$ is coming down, and Phase 1 ends when the increasing airspeed converges with $V_{mr}$. There is no range advantage in speeding up beyond $V_{mr}$, so in Phase 2 the bird maintains $V_{mr}$ (which continues to decrease), and also holds the specific work (or specific power) constant at the starting value, by consuming muscle tissue as necessary, as above. A small bird, starting with a fat fraction that is not too high, may be able to maintain a speed faster than $1.2\ V_{mp}$ from the start, and in that case a higher value can be entered in the Migration Setup Screen for the starting ratio of $V_{mr}$ to $V_{mp}$. One extreme assumption is to get the starting ratio of $V_{mr}$ to $V_{mp}$ by clicking the box so labelled, and set it as the starting ratio of the airspeed to $V_{mp}$. That will cause the bird to fly at $V_{mr}$ for the whole flight, so getting the theoretical maximum range, but it should be understood that this is unlikely to be a practical option for any serious long-distance migrant. At the other extreme, no swan gets anywhere near being able to fly at $V_{mr}$, even when all of its fat has been consumed.

## 8.3 ⬤ THE CONCEPT OF ENERGY HEIGHT

A bird's *energy height* is a virtual height, to which its stored energy would lift it under certain assumptions (Pennycuick 2003 and Box 8.3). The stored energy can be of three kinds, fuel energy, potential energy and kinetic energy. Potential energy height is the easiest kind to visualise, as it is the same as ordinary height, above some reference level such as a ground or water surface. The bulk of bird migration takes place at heights between sea level and 3000 m above sea level (ASL). A bird that is using thermals as its source of energy for migration gains height (and also potential energy) by allowing itself to be carried up in a thermal. Then it converts the height into distance, gliding down on a gradient that depends on its aerodynamic efficiency (Chapter 10). 2000 m in a single climb would be typical in good weather conditions. An albatross rolling off the crest of the wave gains no height in the process, but it gains kinetic energy, which can optionally be converted into height. However, the height that it can gain by pulling up is two orders of magnitude less than the height gained in a typical thermal, say 20 m. A moderate store of fat, on the other hand, could supply enough fuel energy to raise the bird two orders of magnitude higher than a typical thermal, say 200 km. In the real world that is far above the level where the atmosphere is dense

BOX 8.3 **Energy height.**

A bird's *fuel energy height* is defined as the height to which the bird would be lifted, if all of its fuel energy were progressively converted into work, with the same efficiency that its muscles convert fuel energy into work, and used to lift the bird against gravity. The result is not the same as simply dividing the work by the weight (i.e. turning the converted fuel energy directly into potential energy), because of the requirement to do the imaginary lifting operation "progressively". The bird gets lighter as it is lifted, because of the fuel consumed, and therefore the height gained for each unit of work done increases.

**Variable definitions for this box**

| | |
|---|---|
| $e_{fat}$ | Energy density of fat |
| $e_{fp}$ | Energy density of combined fat and protein fuel |
| $e_{pwet}$ | Energy density of wet protein |
| $E$ | Chemical energy released by oxidation of fuel |
| $F_{fat}$ | Fat fraction |
| $F_{fp}$ | Fuel fraction for combined fat and protein fuel |
| $g$ | Acceleration due to gravity |
| $h$ | Height |
| $h_{fat}$ | Fat energy height |
| $k_p$ | Proportion of fuel energy from protein |
| $m$ | All-up mass |
| $m_{fat}$ | Fat mass |
| $m_{fp}$ | Combined fuel mass of fat and protein |
| $m_{pwet}$ | Mass of wet protein consumed |
| $N$ | Effective lift:drag ratio |
| $Y$ | Distance flown |
| $\eta$ | Conversion efficiency |

We begin with the restriction (later to be relaxed) that the bird's only fuel is fat with an energy density $e_{fat}$, meaning that when a mass $\Delta m_{fat}$ of fat is oxidised, an amount $\Delta E$ of chemical energy is released, where:

$$\Delta E = e_{fat} \Delta m_{fat}, \tag{1}$$

and also with the assumption that the reaction products are lost from the body, so that the all-up mass declines by $\Delta m_{fat}$. We may imagine the bird consuming its stored fat bit by bit, converting a proportion $\eta$ of the energy released into work, where the conversion efficiency ($\eta$) is a number between 0 and 1. At some point in the imagined ascent, the bird is lifted a small vertical distance $\Delta h$, for the consumption of a small amount of fuel energy $\Delta E$, of which a proportion $\eta$ is converted into work, and thence into potential energy. Equating the potential energy gained to the work done:

$$mg\Delta h = \eta \Delta E, \tag{2}$$

where $m$ is the current value of the bird's all-up mass, and $g$ is the acceleration due to gravity. The fuel energy ($\Delta E$) comes from reducing the mass

BOX 8.3 *Continued.*

of stored fat (and also the all-up mass), by a small amount ($\Delta m$), which is converted into energy by multiplying the energy density ($e_{fat}$) of fat:

$$\Delta E = -e_{fat}\Delta m. \tag{3}$$

The minus sign indicates that the mass decreases when the energy is supplied. Combining Equations (2) and (3):

$$\frac{\Delta h}{\Delta m} = \frac{-(e_{fat}\eta)}{mg}. \tag{4}$$

The height to which the bird would be lifted by the consumption of fat corresponding to a given part of its initial mass can be found by integration. Suppose that the bird starts at ground level with all-up mass $m$, which includes a mass $m_{fat}$ of fat, then the energy height $h_{fat}$ to which it is lifted when all of the fat has been consumed is:

$$h_{fat} = \left(\frac{e_{fat}\eta}{g}\right) ln\left[\frac{m}{(m - m_{fat})}\right]. \tag{5}$$

This can be expressed more conveniently in terms of the initial *fat fraction* ($F_{fat}$), where

$$F_{fat} = \frac{m_{fat}}{m}. \tag{6}$$

Note that $m$ is the all-up mass, *not* the "lean mass". This because $m$, including fat, is the mass that the bird has to lift when it flies. In terms of $F_{fat}$, Equation (5) becomes:

$$h_{fat} = \left(\frac{e_{fat}\eta}{g}\right) ln\left[\frac{1}{(1 - F_{fat})}\right]. \tag{7}$$

If this is multiplied by $N$, the lift:drag ratio, it gives the migration range according to Breguet's equation (Box 8.1). Breguet's equation could be written:

$$Y = h_{fat}N, \tag{8}$$

where $Y$ is the distance flown. The first factor ($h_{fat}$) represents the fuel reserve, taking into account the efficiency with which it can be converted into work, and the strength of gravity. The second factor ($N$) represents the bird's aerodynamic efficiency. It is the slope on which the bird comes "down" from its initial energy height, like the glide ratio in gliding flight.

### Protein as supplementary fuel

It would appear that the oxidative machinery in the mitochondria is not capable of metabolising fat alone, but must get a certain fraction of the total energy released (perhaps 5%) from oxidising protein (Jenni and Jenni-Eiermann 1998). Protein in living tissue is always hydrated, and we assume that the water of hydration is lost from the body when the protein is oxidised, as well as the mass of the protein itself. The energy density of *wet* protein ($e_{pwet}$) is what determines the mass lost when the protein is oxidised, and is nearly seven times less than $e_{fat}$. If the fuel energy released is divided

BOX 8.3 *Continued.*

between a proportion $k_p$ from protein and $1 - k_p$ from fat, then the mass of wet protein ($m_{pwet}$) lost in the course of oxidising a mass $m_{fat}$ of fat is:

$$m_{pwet} = m_{fat} \left( \frac{e_{fat}}{e_{pwet}} \right) \left[ \frac{k_p}{(1 - k_p)} \right] \tag{9}$$

The combined mass ($m_{fp}$) of fat and protein consumed is:

$$m_{fp} = m_{fat} + m_{pwet} = m_{fat} \left[ 1 + \left( \frac{e_{fat}}{e_{pwet}} \right) \left( \frac{k_p}{(1 - k_p)} \right) \right]. \tag{10}$$

The energy density ($e_{fp}$) of the combined fuel is intermediate between that of fat and wet protein:

$$e_{fp} = \frac{e_{fat} e_{pwet}}{[k_p e_{fat} + (1 - k_p) e_{pwet}]}. \tag{11}$$

The fuel fraction ($F_{fp}$) for the combined fat and protein fuel can be found from the fat fraction by dividing Equation (10) by the all-up mass:

$$F_{fp} = F_{fat} \left[ 1 + \left( \frac{e_{fat}}{e_{pwet}} \right) \left( \frac{k_p}{(1 - k_p)} \right) \right]. \tag{12}$$

The energy height for the combined fuel can then be found from Equation (7), using $F_{fp}$ instead of $F_{fat}$ for the fuel fraction, and $e_{fp}$ instead of $e_{fat}$ for the fuel energy density.

Figure 8.6 shows the energy height as a function of the fat fraction on the alternative assumptions that fat is the only fuel (lower curve), or that 5% of the energy comes from protein and 95% from fat (upper curve). It might be supposed that the energy heights resulting from these two assumptions would differ by a constant percentage, but they do not. The percentage increase in energy height (and therefore also in range) resulting from the use of protein as supplementary fuel increases from a gain of 10% for a moderate starting fat fraction of 0.20, to 33% for an extreme starting fat fraction of 0.55. This is because the energy density of wet protein, considered as a fuel, is much lower than that of fat, because of the water of hydration. Although only a small fraction of the total energy comes from protein, its consumption gets rid of a lot of mass, so reducing the amount of energy required per metre of distance flown. "Burning the engine" is a highly effective method of stretching the range, more so than appears at first sight. The process can be followed step by step from the computer output and commentary in Box 8.2.

**Regaining energy height at stopovers**
The consumption of both protein and fat during long flights implies that muscle tissue has to be built up at the same time that fat is stored, as a bird prepares for a long flight, or refuels during a stopover during a multistage migration. It is clear from physiological experiments that birds do this (Lindström and Piersma 1993), and presumably they put on flight muscle tissue fast enough so that they remain able to fly at any stage during the recovery process. This is a topic that needs to be investigated with the energy height concept in mind.

enough to sustain flight, but if it were possible to fly at such a height, the bird would be able to glide for perhaps 3000 km on the way down, which is enough to migrate across the Sahara. Fat is the only form in which a bird can store energy in sufficient quantity to fly a distance of this order, without replenishment (Figure 8.6).

## 8.3.1 ENERGY HEIGHTS FOR FAT AND PROTEIN FUEL

A bird's *fat energy height* is closely related to its fat *fraction* (not fat mass). It can be seen as one component of Breguet's range equation (Box 8.1). The other component is the effective lift:drag ratio, which is the gradient on which the bird comes "down" from its initial energy height, expressing the bird's aerodynamic efficiency in flapping flight (Chapter 3, Box 3.4). Stored fuel can raise a bird to energy heights of hundreds of kilometres, and therefore requires infrequent replenishment in the course of migration. In some cases, a single period of refuelling is sufficient for the migration flight and for the initial nesting activities after arrival. On the other hand, stored potential energy corresponds to a much smaller height, and therefore has to be replenished more often (as in thermal soaring), while an albatross has to replenish its kinetic energy at intervals of a minute or two at most (Chapter 10).

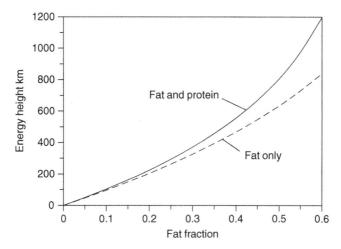

FIGURE 8.6 Fat energy height as a function of fat fraction, assuming that fat is the sole fuel (dashed line), or that 5% of the total energy comes from oxidising protein (solid line). This graph is the same for any bird, large or small. The proportional effect of consuming protein on the energy height becomes appreciable at fat fractions above about 0.2, and increases dramatically at higher fat fractions. After Pennycuick (2003).

The consumption of protein from the flight muscles (and elsewhere) increases the energy height corresponding to a given fat fraction (Box 8.3). The primary fuel remains fat, and the bird can fly until its fat (not its protein) is used up, but it seems that there are physiological constraints that require, in effect, that about 5% of the total fuel energy must come from oxidising protein (Jenni and Jenni-Eiermann 1998). One might imagine that this would increase the energy height for a given fat fraction by 5%, but this is not so. Besides providing supplementary fuel energy, the consumption of protein early in the flight also reduces the mass, so that the remaining fat corresponds to more energy height than it otherwise would have. The lower curve in Figure 8.6 shows the energy height corresponding to fat only, as a function of the fat fraction up to 0.60. It curves gently upwards, while the upper curve, showing the energy height if 5% of the total energy comes from protein, curves upwards more strongly, so widening the gap between the two curves. For a fat fraction of 0.2, the energy height for fat and protein is about 10% higher than that for fat alone, while for a fat fraction of 0.55, it is 33% higher. The range is extended by the same amount as the energy height, or actually somewhat more, as the reduction of body cross-sectional area due to consumption of fat and protein increases the lift:drag ratio.

## 8.3.2 ESTIMATING INITIAL ENERGY HEIGHT FROM BODY MASS ALONE

The practical question of how one estimates the energy height at the start of a migratory flight, without killing a lot of birds to determine their fat fractions, can also be addressed by *Flight's* migration calculation. The method is illustrated in Box 8.4 for that remarkable bird the African Lesser Flamingo (*Phoenicopterus minor*) studied by Tuite (1979). It can be applied to any bird in which a large number of individuals have been weighed, ranging from very fat pre-migratory birds to very thin ones. Box 8.4 shows how several runs of the programme were used to determine from Tuite's field data (which did not include carcase analyses) that Lesser Flamingos reach a maximum fat fraction of 0.46. Table 8.3 is the output for this starting value, and it looks exactly like Table 8.2 for the Great Knot. This is because both tables were output by the same programme. *Flight* is a general model that works for *any* bird (or bat or pterosaur), for which the input data are known from field observations, or can be set up with hypothetical values.

Some birds, such as geese, store fat in places that alter the external shape of the body, so that a field observer can assign a "fat index" score to an individual on the basis of its shape as seen through a telescope or on a photograph. Ecologists usually calibrate these scores in terms of the *mass* of stored fat which, besides being unnecessarily complicated, is not the appropriate variable to turn these scores into an estimate of energy height. The variable needed for that is the fat *fraction*, which is probably more directly related to the visible shape in the first place (Figures 8.7 and 8.8).

### 8.3.3 ENERGY HEIGHT AS AN INDICATOR OF CONDITION

Migrating birds, like aircraft, normally arrive at their destinations with some fuel in hand. The energy height on arrival can be seen as a measure of the bird's "condition", which can be directly compared across different species and different habitats. For example, an individual Alaskan Bar-tailed Godwit (Section 8.2 above) has recently been satellite-tracked around a complete annual circuit, consisting of three non-stop flights around the western Pacific, from New Zealand to north-eastern China (stopover), then to Alaska (breeding) and finally back to New Zealand (http://alaska.usgs.gov/science/biology/ shorebirds/barg_updates.html). This allows the *Flight* migration simulation from Pennycuick and Battley (2003) to be re-run with the actual distances. If the bird started on its southward migration from the Yukon-Kuskokwim delta in Alaska with a fat fraction of 0.548, corresponding to an energy height of 940 km, as previously assumed from Piersma and Gill (1998), then it would have arrived at the Firth of Thames in New Zealand (10,880 km) with 316 km of energy height in hand, as a reserve for dealing with headwinds or getting drifted off course. The first leg of the northward migration was almost as long, 9980 km to the Yalu Jiang Reserve in north-east China, where the godwit would have arrived at an energy height of 364 km, if it started from the same energy height as previously observed. It then flew 5550 km to the nesting grounds in Alaska, where it would have arrived at an energy height of 608 km, if it recovered its original 904 km during the stopover in Yalu Jiang. The godwit was clearly capable of flying the whole distance from New Zealand to Alaska non-stop (as it did that in the other direction), but the stopover in China allowed it to arrive on the breeding grounds with more energy height in hand, as a reserve for nesting. Those arctic-breeding species that arrive before the habitat is

TABLE 8.3 Model Flamingo Migration.

| | A | B | C | D | E | F | G | H | I | J | K |
|---|---|---|---|---|---|---|---|---|---|---|---|
| 1 | Migration summary for | | | | | Phoenicopterus minor | | | | (Non-Passerine) | |
| 2 | Sex: Male | | | Age: | Adult | | From Flight Version 1.16 (2004) by C.J. Pennycuick | | | | |
| 3 | Variable values for this run: | | | | | Saved as: | | | | | |
| 4 | Empty body mass (kg) | | | | 2.2 | | | | | | |
| 5 | Flight muscle mass (kg) | | | | 0.374 | | Flight muscle fraction | | | 0.17 | |
| 6 | Fat mass (kg) | | | | 1.012 | | Fat fraction | | | 0.46 | |
| 7 | Airframe mass (kg) | | | | 0.814 | | Airframe fraction | | | 0.37 | |
| 8 | Crop contents mass (kg) | | | | 0 | | Gravity (m/sec-squared) | | | 9.81 | |
| 9 | All-up mass (kg) | | | | 2.2 | | Induced power factor | | | 1.2 | |
| 10 | Wing span (m) | | | | 1.32 | | Profile power constant | | | 8.4 | |
| 11 | Wing area (sq m) | | | | 0.1936 | | Profile power ratio | | | 0.9333 | |
| 12 | Aspect ratio | | | | 9 | | Respiration factor | | | 1.1 | |
| 13 | Air density (kg/cubic m) | | | | 1.006 | | Conversion efficiency | | | 0.23 | |
| 14 | Altitude (m ASL in Standard Atmosphere) | | | | 2000 | | | | | | |
| 15 | Body frontal area | | | | 0.01374 | | Flight style | | Continuous flapping | | |
| 16 | Body drag coefficient | | | | 0.1 | | Power fraction | | | 1 | |
| 17 | Mitochon inv power density (cubic m/W) | | | | 1.20E-06 | | BMR factor | | | 1 | |
| 18 | Fat energy density (J/kg) | | | | 3.90E+07 | | Air speed control | | Standard | | |
| 19 | Dry protein energy density (J/kg) | | | | 1.83E+07 | | Starting ratio V:Vmp | | | 1.2 | |
| 20 | Protein hydration ratio | | | | 2.2 | | Calculation interval (min) | | | 6 | |
| 21 | Minimum energy from protein (%) | | | | 5 | | Output interval (hours) | | | 6 | |
| 22 | Protein burn criterion | | Constant specific work | | | | Distance to destination (km) | | | 0 | |
| 23 | Mitochondria control | | Constant mito power density | | | | Wingbeat frequency factor | | | 1 | |
| 24 | | | | | | | | | | | |
| 25 | Tuite's Model Flamingo | | | | | | | | | | |
| 26 | | | | | | | | | | | |
| 27 | Time | Distance | True air speed (V) | Vmp | V:Vmp ratio | Wing Reynolds number | Body Reynolds number | Total Body mass | Fat mass | Flt muscle mass | Airframe mass |
| 29 | Hrs | km | m/s | m/s | | number | number | kg | kg | kg | kg |
| 30 | 0 | 0 | 21.22 | 17.68 | 1.2 | 184700 | 167000 | 2.2 | 1.012 | 0.374 | 0.814 |
| 31 | 6 | 487.6 | 23.66 | 17.4 | 1.36 | 205700 | 183000 | 2.094 | 0.9337 | 0.374 | 0.7859 |
| 32 | 12 | 1017 | 25.29 | 17.1 | 1.479 | 219900 | 192400 | 1.99 | 0.8572 | 0.374 | 0.7584 |
| 33 | 18 | 1578 | 26.51 | 16.81 | 1.578 | 230700 | 198300 | 1.888 | 0.7829 | 0.373 | 0.7325 |
| 34 | 24 | 2146 | 26.08 | 16.52 | 1.579 | 227000 | 191800 | 1.793 | 0.7127 | 0.3507 | 0.7299 |
| 35 | 30 | 2705 | 25.67 | 16.24 | 1.58 | 223400 | 185600 | 1.705 | 0.6481 | 0.3302 | 0.7272 |
| 36 | 36 | 3255 | 25.28 | 15.98 | 1.582 | 220000 | 179800 | 1.624 | 0.5883 | 0.3115 | 0.7244 |
| 37 | 42 | 3797 | 24.91 | 15.73 | 1.583 | 216700 | 174400 | 1.549 | 0.5329 | 0.2944 | 0.7216 |
| 38 | 48 | 4331 | 24.55 | 15.49 | 1.585 | 213600 | 169300 | 1.479 | 0.4815 | 0.2787 | 0.7189 |
| 39 | 54 | 4858 | 24.21 | 15.26 | 1.586 | 210700 | 164500 | 1.414 | 0.4337 | 0.2643 | 0.7162 |
| 40 | 60 | 5377 | 23.88 | 15.04 | 1.588 | 207800 | 159900 | 1.353 | 0.389 | 0.2509 | 0.7135 |
| 41 | 66 | 5890 | 23.57 | 14.83 | 1.59 | 205100 | 155600 | 1.297 | 0.3474 | 0.2386 | 0.7109 |
| 42 | 72 | 6395 | 23.27 | 14.62 | 1.592 | 202500 | 151500 | 1.244 | 0.3083 | 0.2271 | 0.7083 |
| 43 | 78 | 6895 | 22.99 | 14.42 | 1.594 | 200000 | 147600 | 1.194 | 0.2717 | 0.2165 | 0.7058 |
| 44 | 84 | 7388 | 22.71 | 14.23 | 1.596 | 197600 | 143900 | 1.147 | 0.2374 | 0.2066 | 0.7034 |
| 45 | 90 | 7876 | 22.44 | 14.05 | 1.598 | 195300 | 140400 | 1.103 | 0.205 | 0.1974 | 0.701 |
| 46 | 96 | 8358 | 22.19 | 13.87 | 1.6 | 193100 | 137000 | 1.062 | 0.1745 | 0.1887 | 0.6987 |
| 47 | 102 | 8835 | 21.94 | 13.7 | 1.602 | 191000 | 133800 | 1.023 | 0.1458 | 0.1807 | 0.6964 |
| 48 | 108 | 9306 | 21.71 | 13.53 | 1.604 | 188900 | 130800 | 0.986 | 0.1186 | 0.1731 | 0.6942 |
| 49 | 114 | 9772 | 21.48 | 13.37 | 1.607 | 186900 | 127800 | 0.951 | 0.09288 | 0.1661 | 0.692 |
| 50 | 120 | 10230 | 21.26 | 13.21 | 1.609 | 185000 | 125000 | 0.9179 | 0.06851 | 0.1594 | 0.69 |
| 51 | 126 | 10690 | 21.05 | 13.06 | 1.612 | 183200 | 122400 | 0.8865 | 0.04539 | 0.1531 | 0.6879 |
| 52 | 132 | 11140 | 20.84 | 12.91 | 1.614 | 181400 | 119800 | 0.8566 | 0.02344 | 0.1472 | 0.686 |
| 53 | 138 | 11590 | 20.64 | 12.77 | 1.617 | 179700 | 117300 | 0.8283 | 0.002557 | 0.1417 | 0.684 |
| 54 | 138.8 | 11650 | 20.62 | 12.75 | 1.617 | 179400 | 117000 | 0.8246 | 0 | 0.1409 | 0.6838 |

producing enough food to support them (e.g. some species of geese) presumably require more energy height on arrival than those that nest in habitats that are already productive when they arrive.

**TABLE 8.3** Model Flamingo Migration.

| | A | B | C | D | E | F | G | H | I | J | K |
|---|---|---|---|---|---|---|---|---|---|---|---|
| 56 | Time | Mechanical | Chemical | Effective | Wingbeat | Reduced | Specific | Specific | Pbmr | Pchem: | Mitochon |
| 57 | | power | power | lift:drag | frequency | frequency | power | work | | Pbmr | mass frac |
| 58 | Hrs | W | W | ratio | Hz | | W/kg | J/kg | W | ratio | % |
| 59 | 0 | 30.45 | 150.4 | 13.24 | 5.564 | 0.1208 | 90.83 | 16.33 | 4.293 | 35.03 | 10.36 |
| 60 | 6 | 29.79 | 147.1 | 14.37 | 5.428 | 0.1058 | 88.62 | 16.33 | 4.22 | 34.86 | 10.13 |
| 61 | 12 | 29.11 | 143.8 | 14.93 | 5.291 | 0.09648 | 86.39 | 16.33 | 4.148 | 34.67 | 9.901 |
| 62 | 18 | 28.36 | 140.1 | 15.24 | 5.155 | 0.08958 | 84.15 | 16.32 | 4.075 | 34.38 | 9.668 |
| 63 | 24 | 26.04 | 128.9 | 15.47 | 5.023 | 0.08874 | 82 | 16.32 | 4.009 | 32.17 | 9.445 |
| 64 | 30 | 23.97 | 119 | 15.7 | 4.899 | 0.08792 | 79.96 | 16.32 | 3.946 | 30.15 | 9.232 |
| 65 | 36 | 22.12 | 110.1 | 15.91 | 4.78 | 0.08713 | 78.04 | 16.32 | 3.888 | 28.31 | 9.03 |
| 66 | 42 | 20.46 | 102 | 16.12 | 4.668 | 0.08637 | 76.21 | 16.32 | 3.834 | 26.62 | 8.837 |
| 67 | 48 | 18.96 | 94.85 | 16.33 | 4.562 | 0.08562 | 74.47 | 16.32 | 3.784 | 25.07 | 8.653 |
| 68 | 54 | 17.61 | 88.34 | 16.53 | 4.461 | 0.0849 | 72.82 | 16.32 | 3.736 | 23.64 | 8.477 |
| 69 | 60 | 16.39 | 82.45 | 16.72 | 4.364 | 0.08419 | 71.24 | 16.32 | 3.692 | 22.33 | 8.309 |
| 70 | 66 | 15.28 | 77.09 | 16.91 | 4.272 | 0.0835 | 69.73 | 16.32 | 3.651 | 21.12 | 8.147 |
| 71 | 72 | 14.27 | 72.22 | 17.09 | 4.183 | 0.08282 | 68.29 | 16.32 | 3.612 | 20 | 7.992 |
| 72 | 78 | 13.35 | 67.78 | 17.27 | 4.099 | 0.08217 | 66.91 | 16.32 | 3.575 | 18.96 | 7.844 |
| 73 | 84 | 12.51 | 63.71 | 17.44 | 4.018 | 0.08152 | 65.59 | 16.32 | 3.54 | 18 | 7.701 |
| 74 | 90 | 11.73 | 59.98 | 17.61 | 3.94 | 0.08089 | 64.32 | 16.32 | 3.507 | 17.1 | 7.563 |
| 75 | 96 | 11.03 | 56.55 | 17.77 | 3.865 | 0.08027 | 63.1 | 16.32 | 3.477 | 16.27 | 7.43 |
| 76 | 102 | 10.37 | 53.4 | 17.93 | 3.794 | 0.07966 | 61.93 | 16.32 | 3.447 | 15.49 | 7.302 |
| 77 | 108 | 9.771 | 50.49 | 18.08 | 3.725 | 0.07906 | 60.8 | 16.32 | 3.419 | 14.77 | 7.179 |
| 78 | 114 | 9.216 | 47.81 | 18.22 | 3.658 | 0.07847 | 59.71 | 16.32 | 3.393 | 14.09 | 7.059 |
| 79 | 120 | 8.701 | 45.32 | 18.36 | 3.594 | 0.07789 | 58.66 | 16.32 | 3.368 | 13.46 | 6.944 |
| 80 | 126 | 8.225 | 43.01 | 18.5 | 3.532 | 0.07732 | 57.65 | 16.32 | 3.344 | 12.86 | 6.832 |
| 81 | 132 | 7.783 | 40.87 | 18.63 | 3.472 | 0.07675 | 56.67 | 16.32 | 3.322 | 12.31 | 6.724 |
| 82 | 138 | 7.371 | 38.88 | 18.76 | 3.414 | 0.07619 | 55.73 | 16.32 | 3.3 | 11.78 | 6.619 |
| 83 | 138.8 | 7.319 | 38.63 | 18.77 | 3.406 | 0.07612 | 55.6 | 16.32 | 3.297 | 11.72 | 6.605 |
| 84 | | | | | | | | | | | |
| 85 | Time | Fat | Protein | Total | Protein | BMR | Fat | Flight | Airframe | Fat energy | Fat + Prot |
| 86 | | burn | burn | fuel burn | burn frac | burn frac | fraction | muscle | fraction | height | energy ht |
| 87 | Hrs | kJ | kJ | kJ | % | % | | fraction | | km | km |
| 88 | 0 | 0 | 0 | 0 | 0 | 0 | 0.46 | 0.17 | 0.37 | 563.4 | 679.5 |
| 89 | 6 | 3056 | 158 | 3214 | 4.918 | 2.861 | 0.4459 | 0.1786 | 0.3753 | 539.8 | 645.3 |
| 90 | 12 | 6041 | 315.2 | 6356 | 4.959 | 2.868 | 0.4307 | 0.1879 | 0.3811 | 515.1 | 610.4 |
| 91 | 18 | 8933 | 471.8 | 9405 | 5.016 | 2.883 | 0.4146 | 0.1975 | 0.3879 | 489.6 | 575 |
| 92 | 24 | 11670 | 614 | 12290 | 4.998 | 2.917 | 0.3974 | 0.1955 | 0.407 | 463.2 | 539.2 |
| 93 | 30 | 14190 | 746.8 | 14940 | 4.998 | 2.974 | 0.38 | 0.1936 | 0.4264 | 437.1 | 504.6 |
| 94 | 36 | 16520 | 869.5 | 17400 | 4.998 | 3.04 | 0.3622 | 0.1918 | 0.446 | 411.2 | 470.9 |
| 95 | 42 | 18680 | 983.1 | 19670 | 4.999 | 3.113 | 0.344 | 0.1901 | 0.4659 | 385.5 | 438.2 |
| 96 | 48 | 20690 | 1089 | 21780 | 4.999 | 3.189 | 0.3255 | 0.1884 | 0.486 | 360.1 | 406.3 |
| 97 | 54 | 22560 | 1187 | 23740 | 4.999 | 3.267 | 0.3067 | 0.1869 | 0.5064 | 334.9 | 375.2 |
| 98 | 60 | 24300 | 1278 | 25570 | 4.999 | 3.347 | 0.2874 | 0.1854 | 0.5271 | 309.9 | 344.8 |
| 99 | 66 | 25920 | 1364 | 27280 | 4.999 | 3.428 | 0.2678 | 0.184 | 0.5482 | 285.1 | 315.2 |
| 100 | 72 | 27440 | 1444 | 28890 | 4.999 | 3.509 | 0.2479 | 0.1826 | 0.5695 | 260.5 | 286.2 |
| 101 | 78 | 28870 | 1519 | 30390 | 4.999 | 3.591 | 0.2276 | 0.1813 | 0.5911 | 236.1 | 257.9 |
| 102 | 84 | 30210 | 1590 | 31800 | 4.999 | 3.673 | 0.2069 | 0.1801 | 0.613 | 211.9 | 230.1 |
| 103 | 90 | 31470 | 1656 | 33130 | 4.999 | 3.756 | 0.1858 | 0.1789 | 0.6353 | 188 | 203 |
| 104 | 96 | 32660 | 1719 | 34380 | 5 | 3.838 | 0.1644 | 0.1777 | 0.6579 | 164.2 | 176.3 |
| 105 | 102 | 33780 | 1778 | 35560 | 5 | 3.921 | 0.1425 | 0.1766 | 0.6808 | 140.6 | 150.2 |
| 106 | 108 | 34840 | 1834 | 36680 | 5 | 4.004 | 0.1203 | 0.1756 | 0.7041 | 117.2 | 124.6 |
| 107 | 114 | 35850 | 1886 | 37730 | 4.999 | 4.087 | 0.09766 | 0.1746 | 0.7277 | 93.97 | 99.42 |
| 108 | 120 | 36800 | 1936 | 38730 | 4.999 | 4.17 | 0.07464 | 0.1737 | 0.7517 | 70.93 | 74.69 |
| 109 | 126 | 37700 | 1984 | 39680 | 5 | 4.253 | 0.0512 | 0.1727 | 0.776 | 48.06 | 50.38 |
| 110 | 132 | 38550 | 2029 | 40580 | 5 | 4.336 | 0.02735 | 0.1719 | 0.8007 | 25.36 | 26.47 |
| 111 | 138 | 39370 | 2072 | 41440 | 5 | 4.419 | 0.003081 | 0.171 | 0.8259 | 2.822 | 2.933 |
| 112 | 138.8 | 39470 | 2077 | 41550 | 5 | 4.43 | 0 | 0.1709 | 0.8292 | 0 | 0 |
| 113 | | | | | | | | | | | |
| 114 | The current version of Flight for Windows is available as a free download from: |
| 115 | http://www.bio.bristol.ac.uk/people/pennycuick.htm |
| 116 | Please send comments and bug reports to c.pennycuick@bristol.ac.uk |

BOX 8.4 **Estimating fat fraction from body mass.**

**Variable definitions for this box**

$m_{min}$  Body mass of a bird with zero fat reserves, but not actually starving
$m_{max}$  Body mass of a bird with maximum stored fat

Measuring a bird's fat fraction directly is a lengthy, messy and expensive operation that involves killing birds. However, it is possible to use *Flight* to get an estimate of the fat fraction corresponding to any mass, without killing any birds. The key is to weigh a lot of birds of a given species, either from a wild population or from captive birds under different feeding regimes, and get estimates of the following two variables. $m_{max}$ is the maximum mass ever seen in a bird that is heavily loaded with fat but able to fly, and $m_{min}$ is the mass of a bird that has used up all of its disposable fat, but is not yet actually starving. One cannot get the fat mass simply by subtracting $m_{min}$ from $m_{max}$, because the difference between them is only partly due to fat. Part of it is due to changes in muscle mass, which accompany the deposition or consumption of fat. *Flight* will take account of this, and can be used to find the fat fraction at $m_{max}$ by the following method, applied here to field data on Lesser Flamingos (*Phoenicopterus minor*), which were studied by Tuite (1979) in the East African Rift Valley. This species is adapted to cope with unpredictable variations in its food supply that swing erratically between abundance and famine, so giving Tuite opportunities to estimate $m_{max}$ at 2.20 kg, and $m_{min}$ at 0.825 kg. Wing spans and areas were measured on captive birds with the help of staff at the Wildfowl and Wetlands Trust, Slimbridge.

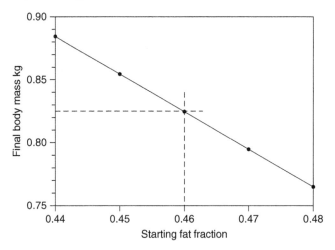

FIGURE 8.7 The final mass from five runs of the Migration calculation for Tuite's Model Flamingo (Table 8.3), starting at the estimated maximum mass of 2.2 kg. Different values of the starting fat fraction were tried, from 0.44 to 0.48. The curve shows that a starting fat fraction of 0.46 leads to a final mass of 0.825 kg, which is Tuite's estimate for the minimum mass. The fat fraction corresponding to any value of the mass between the minimum and the maximum can then be read from Table 8.3 (third block, Column G). See also Figure 8.8.

BOX 8.4 *Continued.*

The first step is to run *Flight*'s Migration calculation, with the starting mass set to $m_{max}$, and the fat fraction set to any reasonable guess (0.3, say). The output shown in Table 8.2 is from one of several runs, each starting with the same all-up mass and flight muscle fraction, but a different fat fraction. For each run, the final mass is noted, when the fat is exhausted. If this is more than $m_{min}$, then the starting fat fraction is increased for the next run, or *vice versa*. The starting fat fraction (0.46 in this case) that leads to a final mass of $m_{min}$ is found by trial and error (Figure 8.7). The fat fraction corresponding to any mass between $m_{min}$ and $m_{max}$ can then be read in column G of the third block of the migration printout, against the mass in column H of the first block, and the energy height for combined fat and protein is in Column K of the third block. Both lines are plotted against all-up mass in Figure 8.8.

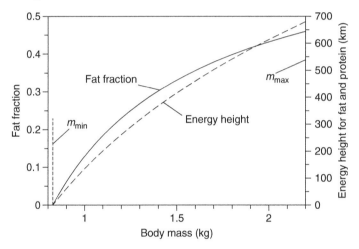

FIGURE 8.8 Graphs plotted from the Lesser Flamingo migration simulation in Table 8.3, showing the estimated fat fraction (solid line) and energy height (dashed line) corresponding to a Lesser Flamingo's measured body mass. Similar graphs can be constructed for any species for which estimates of the minimum and maximum body mass ($m_{min}$ and $m_{max}$) are available (Box 8.4), and can then be used to estimate the fat fraction and energy corresponding to any body mass between these limits, without any need for carcase analysis.

Tuite (1979) was aware that breeding concentrations Lesser Flamingos appear erratically over a vast area of Africa from Kenya to Namibia, and also that the flamingos in his study area would disappear for prolonged periods when feeding conditions were poor, and mysteriously reappear when conditions improved. Surveys failed to reveal the missing flamingos in nearby Rift Valley lakes, and Tuite speculated that the flamingos of eastern and southern Africa were actually a single population that roamed over the entire region. These would be exploratory flights in search of good feeding conditions which might or might not be found, not a predictable seasonal migration. For such a strategy to be possible, a flamingo would have to be able to

BOX 8.4 *Continued.*

make a reconnaissance flight, with enough fuel in reserve to come back if no food supply were found. The same output that is used for finding the fat fraction indicates that the range for a starting fat fraction of 0.46 would be 11,700 km. However, this is for a body drag coefficient of 0.1, which is perhaps over-optimistic for a flamingo, with its prominent head and long, trailing legs. Doubling the body drag coefficient to 0.2 does not affect the estimate of the starting fat fraction, but it does affect the distance that can be flown. The range is reduced to 8320 km, which is still more than enough to fly from Kenya's Lake Bogoria to Namibia's Etosha Pan and back again, without refuelling. Thus, the conclusion is that it would be feasible for flamingos to undertake such long exploratory flights, provided that the strategy, of which these flights would be a part, involves departing from their current location when conditions are good.

### 8.3.4 ENERGY RATES OF CLIMB AND SINK

In addition to losing energy height due to the exertions of migration, a bird also loses energy height whether it is active or not, due to basal metabolism (below). This is *metabolic sink*, and small animals sink faster than large ones. Two birds of different size with the same fat fraction are at the same energy height, and would be able to migrate the same distance if their effective lift:drag ratios were the same. However, metabolic sink effectively decreases the lift:drag ratio, and steepens the virtual "descent", more so for the smaller bird than for the larger one. If both birds sit still without migrating, then both will "descend" at their respective metabolic sink rates, and the smaller one will reach zero energy height (starve) sooner than the larger one.

Conversely, the objective of hyperphagia before migration, or during a stopover, is to gain energy height as quickly as possible, and in this direction also, it seems that small birds can achieve a higher maximum energy rate of climb than larger birds. The capabilities of different birds in this respect have been studied under the heading of "stopover ecology" (Lindström 1991; Lindström and Kvist 1995), but the results were unfortunately expressed in terms of rate of increase of fat mass, relative to "lean mass". As migrants build up protein in the flight muscles and other organs, at the same time as they build up fat mass, the "intake rate" for fat, expressed in this way, is not very enlightening. It is not possible to determine from the published results whether a bird can maintain a constant energy rate of climb until refuelling is complete, or whether (more probably) the rate of climb starts high, and decreases as the fat fraction increases. A constant rate of energy climb would mean that the rate of energy intake would have to increase as the mass builds up during the refuelling process. This would need to be clarified before drawing any conclusions about the maximum possible average migration speed, including stopovers.

## 8.4 ⬤ EFFECT OF FLYING HEIGHT ON RANGE

*Flight* allows the user to set the air density to a value corresponding to a chosen altitude in the International Standard Atmosphere (Chapter 2, Box 2.2). The effect of height on range can be seen by loading a bird, such as the Great Knot, into the programme (from the Preset Birds database) and launching it with a large initial fat fraction, first at sea level, then at a height of, say, 3000 m. This makes very little difference to the range, but reduces the flight time. Both the speed and the *mechanical* power increase by the same amount, in inverse proportion to the square root of the air density, and therefore the ratio of speed to *chemical* power, which determines the range, does not change much. However, if the bird is flying at or near maximum power at sea level, then it will not be able to meet the increased power requirement at the higher altitude, at least not until it has used up some fuel. *Flight* only considers the mechanical power requirements, not the demand for oxygen. As the partial pressure of oxygen in the atmosphere is less at higher altitudes, while the demand for oxygen increases, there is presumably some maximum height at which the bird is just able to get oxygen fast enough to meet its aerobic requirements, and a lower level at which it can fly at its maximum range speed, if it can do that at all. It is likely that bird lungs, with their "cross current" arrangement (Chapter 7, Box 7.7) are more effective than mammal lungs at saturating the blood with oxygen when the air density is low. Tucker (1968a) demonstrated this in an experiment with birds and mice in a hypobaric chamber, but there is no quantitative theory that could be incorporated into *Flight*.

From Version 1.18, the *Flight* programme allows the user to specify an optional Cruising Altitude, in addition to the Starting Altitude which is required. If a cruising altitude is specified, which is higher than the starting altitude, the programme first estimates the maximum rate of climb of which the bird is capable in its starting configuration, using the method in Chapter 7, Box 7.5. To do this, it needs the new default value for the maximum isometric stress of the muscles ($560 \, \mathrm{kN \, m^{-2}}$) from the observations of a Whooper Swan's performance in Chapter 7. The bird climbs at maximum power until it reaches the cruising altitude, then levels off, and continues at that height until its fat runs out, according to whichever speed control option has been selected. During the climb, the bird uses continuous flapping regardless of which flight style option has been selected, and the selected flight style option takes effect when it levels off at the cruising altitude.

*Flight*'s Excel output gives the total time and distance from take-off to the point where the fat runs out, at the cruising altitude, including the time and distance for the climb if one was specified. The total distance is less that it would be if the bird had been launched at the cruising altitude, because of the energy used in the climb. However, the time and distance to glide back down to the starting altitude is also listed, although it is not included in the totals. If the glide distance is added to the total, then this will usually exceed the distance for flying the whole way at the cruising level.

In terms of energy height (above), a real bird converts some of its fuel energy height into actual height in the initial climb to the cruising level, and then glides down at the end, converting height into distance without using any fuel. The distance covered in a flight profile consisting of an initial climb, a level cruise, and a final descent is little different from that covered by flying level all the way, because the gradient on which height is converted into distance is much the same for fuel energy height as for actual height, being determined by the bird's effective lift:drag ratio. Although it is true that more fuel energy is converted to potential energy in the climb than is dissipated against drag in the descent, because the bird is heavier at the beginning of the flight than at the end, this does not directly affect the total distance flown.

## 8.5 ‒ AEROBIC CAPACITY AND CLIMB

Available muscle power is likely to be limited in an aerobic climb by the declining rate at which the bird can extract oxygen from the air, whose density is progressively decreasing. Although the principle is straightforward, and was spelled out with a diagram by Pennycuick (1975a), this effect cannot be simulated in *Flight* because of the absence of quantitative information about the performance of bird's lungs at low air densities. The programme recognises that the speed and power required to fly level increase with height, and that a declining amount of excess power is available for climbing, but it does not recognise that the rate of climb may be limited to a lower level by aerobic capacity which (presumably) also declines with height. The programme has no way to determine a bird's *service ceiling*, meaning the maximum height at which a climb can be sustained, and the predicted rate of climb is likely to become over-optimistic as the climb proceeds. For example, the reader may like to try launching the Great Knot of Table 8.2 (whose starting data may be loaded directly from the Preset Birds database) with a starting altitude of zero (sea level) and a cruising altitude of 5000 m. The programme estimates the initial rate of climb as 2.48 m s$^{-1}$, and says that the bird will reach its cruising height 35 minutes after take-off, still climbing at 2.26 m s$^{-1}$ when it gets there, despite the fact that the air density at 5000 m is only 74% of its sea level value. The estimate of the starting rate of climb may be realistic, but field observations suggest that a real bird would probably be forced to level off sooner by shortage of oxygen. However, there is no basis at present for estimating the height at which that would happen.

## 8.6 ⬭ BASAL METABOLISM

An animal's basal metabolic rate (BMR) is the rate at which it consumes fuel energy when it is inactive, after allowing for any accountable items of chemical power, such as the power required for maintaining body temperature, or for digesting food. Although it is unclear exactly what the BMR is needed for, it is clear that it has no direct connection with the power required for flight, which comes mainly from the mechanical power needed to support the weight and overcome aerodynamic drag. As noted in Chapter 3, calculating the mechanical power involves taking note not only of the body mass, but also of a number of other variables not usually considered in studies of BMR, such as the strength of gravity, the air density and the bird's wing span.

The BMR is not required in the calculation of mechanical power in *Flight*, but when this is converted to chemical power, the BMR is added. The underlying assumption (very difficult to test) is that the BMR is a "maintenance overhead", which is required at the same rate whether the bird is active or not. In the Migration calculation, the BMR is recalculated at 6-minute intervals of flight time from the Lasiewski and Dawson regressions (Box 8.5), using the remaining body mass, excluding fat. These equations assume that the BMR varies with approximately the 0.75 power of the mass, whereas the mechanical power for geometrically similar birds would vary with the 7/6 power of the all-up mass. *Flight* takes account of deviations from geometrical similarity, but still shows that the ratio of the chemical power required to fly at $V_{mp}$ to the BMR is much higher in large birds than in small ones. In swans, the BMR is a trivial component of the total power requirements (like 2%), whereas estimates of this fraction are around 20% or even 25% in small passerines. This point is illustrated in

Chapter 13. Stratagems that increase the cruising speed, such as flying high, or using "bounding" flight, increase the range in small birds, by reducing the flight time, which in turn reduces the fuel energy wasted

BOX 8.5 **Basal metabolism in *Flight*.**

---

### Variable definitions for this box

$m_{empty}$    Body mass less mass of crop contents (if any)
$m_{fat}$    Mass of consumable stored fat
$P_{bmr}$    Basal metabolic rate

There is a vast empirical literature on basal metabolism, and related arbitrarily defined quantities like "existence metabolism" and so on. These quantities sometimes have the dimensions of power, sometimes those of power/mass. Summaries are commonly expressed in the form of regressions of "metabolism" on "body mass", which are themselves not always unambiguously defined. There is no theory behind any of this. The concept in *Flight* is that basal metabolism is a component of chemical power, due to the maintenance requirements of living tissues. It is seen as an "overhead" that has to be added to the mechanical power requirements, which are themselves calculated from the mechanics of locomotion (Chapter 3). The two regressions used to implement it come from Lasiewski and Dawson (1967), who measured the rates at which birds sitting quietly in a respirometer chamber consumed oxygen. They published different regressions for passerines and birds of other orders. This may reflect a real difference, and is the reason why *Flight*'s Setup screens ask whether the bird is a passerine or not.

The basal metabolic rate, being a power, gets the variable name $P_{bmr}$ (*not* the acronym BMR—see Chapter 1, Box 1.1). The "mass" to which it is referred is the difference between the empty mass $m_{empty}$ and the fat mass $m_{fat}$ (Chapter 1, Box 1.2), in other words the all-up mass excluding the fat mass and any crop mass. In the Migration calculation, it is recalculated at 6-minute intervals of flight time, as the mass declines, and added to the chemical power required to cover mechanical demands. Lasiewski and Dawson's regressions, translated into SI units (power in watts, mass in kilograms) are:

$$P_{bmr} = 6.25(m_{empty} - m_{fat})^{0.724} \text{ for passerines, and}$$

$$P_{bmr} = 3.79(m_{empty} - m_{fat})^{0.723} \text{ for non - passerines.}$$

---

No useful purpose is served by expressing chemical power as a multiple of BMR, or per unit of body mass. These practices make it difficult or impossible to retrieve the original measurements of power from published results. If the measured power is published (preferably in watts), together with the essential input data for *Flight*'s calculations (mass, wing span, wing area, air density etc.), then the results can be compared with the predicted chemical power, but this has seldom been done.

on BMR. They are less useful to large birds, whose BMR is relatively small in the first place.

One of the columns of the Migration Excel output from *Flight* lists the ratio of the chemical power to the BMR, not because this ratio is meaningful, but because some physiologists continue to express powers in this form. This ratio will be seen to change in the course of a long migratory flight, and to vary wildly between different birds. The common practice among physiologists of expressing measurements of total chemical power as a multiple of BMR serves only to make the actual power measurements (which could be compared with *Flight*'s predictions) difficult or impossible to retrieve from the published results.

## 8.7 ⬥ WATER ECONOMY

Migrating birds of many species fly for days at a time, working like marathon runners, so what about water? Birds do not sweat like mammals, but dispose of waste heat mainly by convection from the thinly insulated surfaces under the wings and on the sides of the body, which are exposed in flight to a copious stream of cooling air (Chapter 7). If the air is so hot as to require evaporative cooling, they do this through the respiratory tract, which is in any case the primary route for water loss. However, as the air temperature in clear air decreases by about 1 °C per 100 m of height, a moderate climb will usually bring relief in cooler air.

The oxidation of fat produces water as a by-product, so the question is whether enough water is produced in flight to offset the losses in the expired air. Carmi and Pinshow (1995) attempted a mathematical analysis, and concluded that a bird probably would be able to cover its losses in cool, moist air, but probably not in hot, dry air. However, the analysis contained too many hard-to-measure variables for clearcut predictions in particular cases. There are numerous instances of long-distance migration routes used by land birds crossing the sea, with no possibility of replenishing their water, which suggests that migrating birds probably can cover losses, except possibly in unusually unfavourable circumstances. The observation that swans drink copiously on completing a migratory flight has probably been misinterpreted, as these birds, being folivores, have to maintain a liquid bacterial culture in the caecum for breaking down plant cell walls. They probably empty the caecum contents to reduce the all-up mass before departure, and have to drink to replace it on arrival.

## 8.8 ⬭ SLEEP

One can only speculate whether or not migrating birds sleep while flying, as it is difficult to determine whether a warbler over the middle of the Sahara is or is not wide awake. Actually, it is far from clear why animals need to sleep, or exactly what happens when they do. Maintaining level flight and a constant heading are straightforward functions that can be managed in aircraft by a simple autopilot, and it is possible that birds have an autopilot function, which allows prolonged straight-and-level flight with some other brain functions disabled. Common swifts are generally believed never to land except when nesting, and have been observed by radar maintaining an into-wind heading when flying at night (Bäckmann and Alerstam 2002). This would imply that they can observe their drift relative to the ground, and adjust the heading to minimise it. Whether or not they need to be awake to do that is, however, unclear.

# 9

# ACCELERATED FLIGHT AND MANOEUVRING

Flight is controlled by changes of wing shape that lead to angular accelerations about the three axes of a bird's body (pitch, roll and yaw). If these lead in turn to accelerations in the dorso-ventral direction, the effect is to alter the value of gravity, together with anything that depends on gravity, including the power required to fly. Small birds exploit this by "bounding" flight, in which they use increased gravity to increase their cruising speeds. Structural strength limits the gravity increase that can be tolerated in manoeuvres such as tight turns.

The power curve of Chapter 3 and the glide polar of Chapter 10 are performance curves for a bird in "unaccelerated" flight, that is, flying in a straight line at a constant speed. Actually only straight gliding flight is truly unaccelerated. Flapping flight can only be unaccelerated in the sense that there is no change of speed or direction over one or more complete wingbeat cycles. Within each wingbeat cycle, a bird in horizontal flapping flight accelerates upwards during the downstroke of each wingbeat, and downwards during the upstroke, and these are not trivial accelerations. It does this because the aerodynamic force exerted by the wings on the body has to be greater during the downstroke than during the upstroke, in order to overcome drag. If this force

Modelling the Flying Bird

drops to zero during the upstroke, then the body's upward acceleration is momentarily equal to $-1g$, where $g$ is the acceleration due to gravity. If the upstroke and the downstroke are of equal duration, then $+1g$ is required during the downstroke, to keep the flight path horizontal on average. The bird actually feels an acceleration that is offset by $1g$, and oscillates at the wingbeat frequency between zero (free fall) and $+2g$, because the wings have to generate an average upward force equal to the weight, to keep the flight path horizontal. This might be considered an extreme case, but actually the range of acceleration observed in each wingbeat cycle of level flight can be somewhat greater than $0g$–$2g$ (Pennycuick et al. 2000). Pilots who know what $0g$ and $2g$ feel like will recognise that the apparently untroubled flight of migrating birds involves enduring this constant juddering, at a frequency of several Hz, for hours or days at a time. However, there is no reason to believe that the power required to fly horizontally by flapping is either more or less than that required from an engine driving a steadily rotating propeller.

## 9.1 ⬤ INTERMITTENT FLIGHT STYLES IN FLAPPING FLIGHT

Cyclic changes of speed, with a period longer than the wingbeat period, are also considered to be "unaccelerated" by *Flight*, provided that the average speed remains constant from one cycle to the next. Some unnecessary confusion has been introduced into the study of *flight styles* of this kind by obscure and misleading terminology. The cyclic accelerations are caused by the bird flapping during only a part of the cycle (the *power phase*). Two different styles can be distinguished, depending on what the bird does during the other phase, when it is not flapping (Figure 9.1). In *flap-gliding*, which is common in large gliding birds like eagles, pelicans, etc., the power phase alternates with a *gliding phase*, resulting in cyclic changes of height, or speed, or both. The term "undulating flight" for this type of motion is misleading and should be avoided, because a flap-gliding bird does not necessarily undulate. It may go along perfectly level, increasing and decreasing its speed in each cycle. The obscure term *bounding* has become entrenched in the literature for the other intermittent style, in which the power phase alternates with a *ballistic phase*, so called because the bird wraps its wings tightly around the body while it is not flapping, and becomes in effect a wingless, streamlined body that develops no lift, and follows a ballistic trajectory. Bounding is often (but not always) used by small passerines, and by a few larger birds, notably woodpeckers.

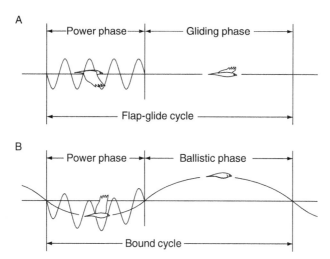

FIGURE 9.1 (A) A flap-gliding cycle consists of a power phase and a gliding phase. The flight path does not necessarily "undulate". The bird either climbs or increases its airspeed during the power phase, or does some combination of the two, and reverts to its original height and airspeed during the gliding phase. The height and airspeed return to their original values after each complete cycle. (B) In bounding, the bird is in free fall during the non-flapping phase, because its wings are closed and wrapped round the body. It follows a downward-curving, ballistic trajectory, and compensates by curving the flight path upwards during the power phase. This increases the value of gravity, against which the wings have to work, to $g/q$, where $g$ is the acceleration due to gravity, and $q$ is the power fraction, defined as the duration of the power phase, divided by that of the complete cycle. Wingbeat frequency in either flap-gliding or bounding is defined as the number of wingbeats in the power phase, divided by the duration of the power phase (not that of the complete cycle). After Pennycuick (2001).

An undulating flight path is obligatory in bounding, because the bird is in free fall during the ballistic phase.

The calculations in *Flight* for flap-gliding and bounding are outlined in Box 9.1, and refer to a bird which is flying steadily along in a straight line, with a regular cycle of flapping and non-flapping periods, maintaining a constant height and speed from one cycle to the next. They do not apply to a bird that is performing an irregular sequence of short climbs and descents, turns, accelerations and decelerations (as in a swallow hawking for insects), and they definitely do not apply to a bird that is flying erratically in a wind tunnel, because it has not been trained to fly steadily. The wingbeat *frequency* in either style is measured by counting the number of wingbeats within the power phase, and dividing by the duration of the power phase, not by the period of the complete cycle. The accelerations in bounding result in an increase in the wingbeat frequency, over its value in continuous flapping (Box 9.1).

BOX 9.1 **Intermittent flight styles.**

*Flight* offers three alternative "styles" for flapping flight, all of which refer to flight that is level and unaccelerated, when averaged over a complete cycle. *Continuous flapping* is the default flight style. The other two are *flap-gliding* in which the bird regularly alternates between short periods of flapping and gliding, and *bounding*, in which the bird flaps for a few wingbeats on an upwardly curved flight path, then closes its wings and follows a ballistic trajectory. If you select either of the intermittent flight styles in the Setup screens for power curves or migration, *Flight* will ask for a value for the "power fraction", which is the duration of the flapping phase, divided by the duration of the full cycle. It will not accept a value below 0.2 for the power fraction in bounding, as this would imply that the bird has to pull more than 5g during the flapping phase, which is considered improbable. If you enter 1 for the power fraction, the programme will revert to continuous flapping flight.

**Variable definitions for this box**

| | |
|---|---|
| $g$ | Acceleration due to gravity |
| $h_b$ | Bound height |
| $P_{ind}$ | Induced power |
| $P_{pro}$ | Profile power |
| $P_{par}$ | Parasite power |
| $P_{mech}$ | Total mechanical power |
| $q$ | Power fraction |
| $\tau$ | Bound period |

**Flap-gliding**

As explained in the main text of this chapter, the average power over a complete flap-glide cycle is the same as in continuous flapping, but the work is done in a fraction of the cycle period. The mechanical power during the gliding phase is zero, but the power during the flapping phase has to be higher than in continuous flapping flight, to make up for energy losses during the gliding phase. Flap-gliding may be necessary in some birds with high wingbeat frequencies, to raise the specific work during flapping to a level at which the muscle can work efficiently (Chapter 7, Box 7.1).

**Bounding**

The term "bounding" (not "flap-bounding", whatever that means) refers to a style with alternating flapping and ballistic phases. Passerines in bounding flight fold the wings tightly against the body in the ballistic phase, making the whole bird into a streamlined body of approximately circular cross section. A body of this shape may, if it has a tail, develop a moment that aligns its head into the relative wind, as in a feathered arrow, but it is not capable of developing aerodynamic lift. Wings do that, but they are removed from the airflow during the ballistic phase of bounding, by wrapping them around the sides of body. The notion that "body lift" affects the trajectory in the ballistic phase appears to have arisen from a mistaken belief that

BOX 9.1 *Continued.*

"lift" acts upwards, which is not necessarily so. The body does, of course, have drag, and the assumption in *Flight* is that the drag of the body with wings folded is the same as that of the wingless body, as used for calculating parasite power (Equation 1, Box 3.2). In that case, parasite power is not affected by the bounding action, and is required at the same rate as in continuous flapping.

Induced power is zero during the ballistic phase, but higher in the flapping phase than in continuous flapping. It is only required for a fraction $q$ of the time (where $q$ is the power fraction), but the value of gravity against which the weight has to be supported is $g/q$ instead of $g$, because of the upward acceleration in the flapping phase. Likewise, the profile power is required only during the flapping phase, but is found from the "absolute minimum power" as in Box 3.3, which becomes a function of $g/q$ instead of $g$. The total mechanical power $P_{mech}$ is found by adding together the induced power ($P_{ind}$), the profile power ($P_{pro}$) and the parasite power ($P_{par}$) as before:

$$P_{mech} = q[P_{ind}(g/q) + P_{pro}(g/q)] + P_{par}. \qquad (1)$$

This shorthand notation says that $P_{ind}$ and $P_{pro}$ are functions of $g/q$, and that both of them are multiplied by the power fraction ($q$), whereas $P_{par}$ is calculated in the same way as in continuous flapping. *Flight* computes the power curve in the same way as usual, by incrementing the true airspeed in steps of 0.1 m s$^{-1}$ and calculating the power at each step. Figure 9.2 shows curves of mechanical power and effective lift:drag ratio versus airspeed, calculated by *Flight* for a chaffinch in bounding flight. Each graph shows curves for values of the power fraction from 0.2 to 1.0 (continuous flapping). 0.3 would be a typical power fraction for a chaffinch in level flight. Figure 9.2A shows the power increases, gradually at first and then ever more strongly, as the power fraction is reduced, while Figure 9.2B shows the effective lift:drag ratio decreasing in a similar manner. The characteristic speeds $V_{mp}$ and $V_{mr}$ increase progressively as the power fraction is reduced.

The increased power requirement in bounding may explain the anomaly that small passerines have similar or even higher flight muscle fractions than larger birds, despite the scaling argument of Chapter 7, which indicates that their power margin in that case should be unnecessarily large. Bounding raises the specific work from a very low value to a level where more efficient energy conversion would be anticipated. The extra muscle then allows small birds to fly efficiently at higher airspeeds than would otherwise be possible, and this must be a major factor in their ability to penetrate against head winds. Predictions of migration range with a large initial fat fraction (0.4) may show slightly increased range, despite the reduction in the effective lift:drag ratio. This is due to reduced wastage of fuel for basal metabolism, because the higher speed shortens the flight time for a long flight. *Flight* will simulate bounding migration in swans, but no such advantages are seen in this case. The specific work is already near the upper limit in swans, and is raised by bounding to a wholly impractical level (Chapter 7).

BOX 9.1 *Continued.*

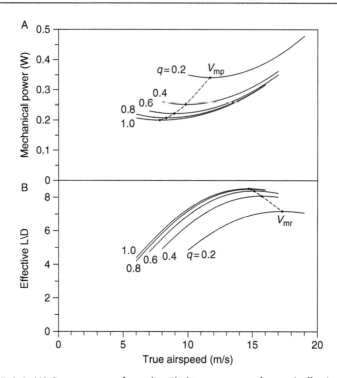

FIGURE 9.2 (A) Power curves from the *Flight* programme for a chaffinch in level bounding flight, for values of the power fraction from 0.2 to 1 (continuous flapping). At a power fraction of 0.2, the bird would be pulling 5g during the power phase, which is probably more than the practical upper limit. As the power fraction is reduced the entire power curve moves up the graph (more power) and to the right (faster). The dashed line connects the values of the minimum power speed $V_{mp}$ in each curve. (B) Effective lift:drag ratio from the same set of power curves. These curves come down with decreasing power fraction, while the maximum range speed $V_{mr}$, defined as the speed for maximum effective L/D, increases. After Pennycuick (2001).

**Wingbeat frequency in bounding**
A bird's wingbeat frequency in cruising flight is a function of gravity (Chapter 7 Box 7.3). Specifically, it varies with $\sqrt{g}$, where $g$ is the acceleration due to gravity. In the flapping phase of bounding, $g$ is increased to $g/q$ (above). If no allowance is made for the increased gravity, observed wingbeat frequencies are higher than predicted by Equation (3) of Box 7.3, but if $g/q$ is used in place of $g$ in the formula, this discrepancy is resolved (Figure 9.3) (Pennycuick 2001).

**Bound height**
If a bird in bounding flight starts a new bound cycle at the same height as it started the previous one, then its flight path can be said to be horizontal on average. However, in the course of each cycle it rises above the mean flight

BOX 9.1 *Continued.*

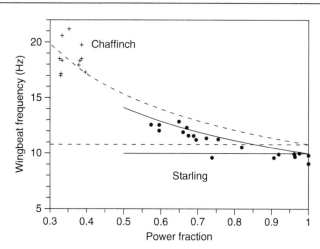

FIGURE 9.3 Wingbeat frequencies for bounding flight in migrating chaffinches and starlings, measured from video sequences in which the power fraction was also measured. The horizontal lines show the estimated wingbeat frequency in cruising flight from Equation (3) of Box 7.3 (Chapter 7), and the curves show the corrected frequency which takes account of the gravity increase caused by bounding (Box 9.1). After Pennycuick (2001).

path during the ballistic phase, and drops below it during the flapping phase. The "bound height" ($h_b$) is the vertical distance from the highest point of the ballistic phase to the lowest point of the flapping phase. It depends on the power fraction ($q$), and also on the square of the bound period ($\tau$):

$$h_b = \frac{[g\tau^2(1-q)]}{8}. \tag{2}$$

Even for short bound periods, bound heights are quite large. For example for a bound period of 1 second and a power fraction of 0.5 (half a second flapping, and half a second ballistic), the bound height is 0.61 m. Thus reports in the literature that a bird was bounding in a wind tunnel with a test section only 50 cm high need to be read with caution. If the bird does not have enough headroom for the full bounding motion, it will partially open its wings during the ballistic phase to avoid hitting the floor. A test section around 4 m high would be needed to train typical passerines to bound without undue constraint, and no bird wind tunnel in current use has that.

## 9.1.1 POWER REQUIREMENTS IN FLAP-GLIDING

Intermittent flight styles do not save energy, any more than a cyclist saves energy by alternately pedalling and freewheeling. All of the work done by a bird's flight muscles is used eventually either to support the weight or to overcome drag, and those two things have to be done for the whole cycle, whether the muscles work continuously or intermittently. In the case of flap-gliding, the bird slows down in the gliding phase, and/or loses height, and it has to compensate for this during the power phase, by accelerating and/or climbing. In energy terms, some of the work that the bird does during the power phase is used to overcome drag at the same time, as in continuous flapping, but the muscles have to do additional work, which is converted into potential energy by climbing, and/or into kinetic energy by speeding up. As the bird descends and/or slows down in the gliding phase, the energy that was stored as in increment of height and/or speed during the power phase is reused to overcome the gliding drag.

Both potential and kinetic energy are high-grade forms of energy which can, in principle, be stored and recovered without loss. Of course, some losses are inevitable in practice, but as there is no basis for calculating them, *Flight* assumes that the losses are small and neglects them. In that case, the work done over a complete flap-glide cycle is the same as in continuous flapping flight. The average power is the same as in continuous flapping, but the power during the power phase is higher, because the work for the whole flap-glide period has to be done in a fraction of that time (the power fraction). The power during the power phase is used to calculate the specific work in the flight muscles. The reason why many raptors do a lot of flap-gliding may be that their wingbeat frequencies are mostly rather high (an adaptation for high-powered manoeuvres in chasing prey), so that if they were to flap continuously in level flight, the specific work might be too low for efficient conversion of fuel energy into work (Chapter 7 Box 7.2).

### 9.1.2 POWER REQUIREMENTS IN BOUNDING

In the case of bounding flight, the wings are fully closed in the ballistic phase, and the bird is essentially a streamlined body in free fall. It follows a flight path that curves downwards, accelerating earthwards at 1*g*. During the power phase, it has to compensate by curving the flight path upwards and "pulling *g*" as pilots say (a notion which is discussed further below). As far as the mechanics of flight are concerned, the effect of curving the flight path upwards is identical with that of an increase in the strength of gravity. The amount of the increase is the reciprocal of the power fraction. If the power phase lasts for half of the bound period, then the bird has to pull 2*g* during the power phase, while if the power fraction is one-third, it has to pull 3*g*, and so on. The calculation of induced and profile power during the power phase is the same as in Boxes 3.1 and 3.3 of Chapter 3, except that gravity is increased in inverse proportion to the power fraction. Parasite power is unaffected by gravity, and continues at the same rate throughout the bound cycle. Unlike flap-gliding, bounding always results in a substantial increase in the average power required, as compared to continuous flapping, but the speeds for minimum power and maximum range are also increased. Bounding allows small birds, whose airspeeds would otherwise be unduly low in relation to commonly encountered wind speeds, to fly faster while maintaining an acceptable level of muscle efficiency. It is also very effective at increasing the specific work required in the flight muscles, which would otherwise be very low in small passerines, forcing them to use only a part of the muscles, or else to run the whole muscle very inefficiently. This is a result of the scaling relationship that limits the size of large birds because of lack of muscle power, and conversely leads to excess muscle in small birds (Chapter 7, Box 7.4).

## 9.2 ⬤ MANOEUVRING FRAME OF REFERENCE: FLIGHT CONTROLS

The view from the cockpit is different from that of a ground observer who perceives an aircraft manoeuvring, relative to the landscape. The controls available to a pilot (or bird) cause angular accelerations about three axes in a frame of reference that is fixed to the bird or aircraft, not to the ground (Box 9.2). These accelerations are in *pitch* about the transverse axis, in *roll* about the longitudinal axis and in *yaw* about the dorso-ventral axis (Figure 9.4), and in an aircraft the control moments about these axes are generated by separate and distinct controls, the elevator, ailerons and rudder respectively. The principle is the same in birds, but the adjustments of the wings that produce control moments are more complicated, and not always easy to disentangle from one another (Figure 9. 5). The aerodynamic forces on the wings that produce the control moments themselves depend on air flowing over the wings, that is the controls require some airspeed to operate. Airspeed is motion relative to the air, which is not necessarily, or usually, the same as motion relative to the ground, because the air is itself usually in motion relative to the ground (wind). The lift force developed by the wings is perpendicular to the direction from which the airflow is coming (by definition), but that direction is itself defined in the bird-centred coordinate system of Figure 9.4.

Linear accelerations are limited to two directions relative to the bird, along the flight path, and perpendicular to it in the dorsal direction,

**BOX 9.2 Frames of reference and flight controls.**

Performing aerobatics is a challenging art, not least because it requires the pilot to think in three different coordinate systems at once. A spectator on the ground, watching an aerobatic display, sees the aircraft climbing, looping, diving and rolling relative to a frame of reference that is fixed to the local earth's surface. If the observer is tracking the aircraft, its changing position will most likely be monitored by assigning numbers to three rectilinear coordinates, of which two are horizontal distances from a north-south and an east-west axis, and the third is the height above some datum. Besides keeping track of the aircraft's position and orientation relative to the earth's surface, the pilot also needs to be aware of its three-dimensional speed relative to the air, which may be moving invisibly along, relative to the landscape (wind). To manoeuvre the aircraft, the pilot has to apply forces that make it accelerate either along the flight path or perpendicular to it, and to do that, he has to think in a third coordinate system, which is fixed to the aircraft, and loops and rolls with it. In principle, there are six ways in which the pilot can make the aircraft accelerate, relative to itself, although not all of them are equally useful. Three of them are angular accelerations, about the three mutually perpendicular axes of pitch, roll and yaw (Figure 9.4). The flight controls produce moments about these three axes (Figure 9.5), and these in turn lead to linear accelerations, as follows.

**Angular acceleration about the pitch axis**

The pitch axis is the transverse axis that runs through the wings from one side to the other. Conventional aircraft create a pitching moment by deflecting the tailplane, which is a small, movable, auxiliary wing, some distance aft of the main wing. Birds do not have a tailplane, and produce a pitching moment in a different way, by sweeping the wings forwards or back (Figure. 9.5A). This is essentially the same method as the weight-shift control of hang gliders. The effect of pitching in the nose-up direction (relative to the aircraft or bird) is to increase the angle of attack of the wings, which in turn increases the lift force. A nose-up pitching moment therefore results in a linear acceleration in the dorsal direction, perpendicular to the flight path. The lift force produces curvature of the flight path, but does not necessarily act upwards. The direction depends on how the wings are oriented, relative to the ground. If the bird wants to bend the flight path in some other direction than upwards, it first has to "roll" until the wings are oriented to generate lift in the required direction.

**Angular acceleration about the roll axis**

The roll axis is the longitudinal axis of a bird's body or an aircraft's fuselage. A rolling moment (about the roll axis) is created by increasing the lift on one wing, and reducing it on the other. Birds do this by simply rotating the whole wing at the shoulder by a small amount, one wing in the nose-up sense, increasing its angle of attack and the lift that it develops, and the other nose-down (Figure 9.5B). In manoeuvring, the purpose of rolling is to rotate the bird until its dorsal direction is aimed in the direction in which a manoeuvring force is required. The force that bends the flight path is then produced by applying a pitching moment (above), so increasing the angle of attack of both wings.

BOX 9.2 *Continued.*

### Angular acceleration about the yaw axis

The yaw axis passes dorso-ventrally through the mid-point of the wings, and a yawing moment causes an angular acceleration in the plane of the wings, swinging the nose left or right. This is effected in conventional aircraft by a movable rudder, which is a small wing some distance behind the main wing, whose plane is perpendicular to that of the main wing. Yawing produces only a minor sideways force, and the effect is that the aircraft proceeds in nearly the same direction as before (relative to the air) but sideslips. The function of the rudder is not to steer the aircraft (the wings do that), but to balance and cancel "asymmetric drag", which is a side effect of creating a rolling moment (above). The up-going wing experiences increased drag as well as increased lift, while drag is reduced on the down-going wing. This produces an unwanted yawing moment, and the rudder is used to cancel this, and avoid sideslip. Birds do not need a rudder, as they have other ways in which they can adjust the shape of the wings asymmetrically, so as to produce a rolling moment without adverse yaw (Figure 9.5C).

### Linear accelerations

Manoeuvring is effected by linear accelerations perpendicular to the flight path. As only the wings can generate a large enough force for this, curvature of the flight path is invariably controlled by pitching, which in turn varies the angle of attack of the wings (above). Sideways accelerations, towards one wing tip or the other, are not significant, as there is no lifting surface that can produce a large sideways force. Longitudinal accelerations are produced by thrust or drag, and are used to control airspeed, but the accelerations are small compared to those produced by lift on the wings.

### Inverted flight

Most small aircraft can perform a loop, but this manoeuvre does not involve inverted flight as usually defined, although the aircraft is upside down as it passes over the top. In a properly executed loop, the wings are lifting towards the dorsal side, that is towards the centre of the loop, throughout the manoeuvre. The pilot's weight acts towards the seat as usual, and his coffee does not spill. Inverted flight means that the wings are lifting towards the ventral side of the aircraft. Momentary inverted loading occurs in level flight when an aircraft enters a strong down-draught, causing the angle of attack to become negative. When this happens in an airliner, unsecured objects, including passengers, gravitate towards the cabin roof, as the aircraft accelerates downwards. Special aerobatic aircraft are designed to be capable of sustained inverted flight and manoeuvring, but most aircraft are not.

Among birds, only hummingbirds have wings that are adapted to withstand inverted loads, and an elevator muscle comparable in size to the pectoralis. Hummingbirds do not fly around upside down, or perform inverted manoeuvres, but their wings develop inverted lift (directed upwards) during the upstroke in hovering. A hovering hummingbird's body has approximately zero airspeed, and the wings beat back and forth in a

BOX 9.2 *Continued.*

> nearly horizontal plane, inverted as they swing towards the dorsal side. Other birds develop minor amounts of inverted lift on the primary feathers during the upstroke in very slow flight, but the wing as a whole is not capable of withstanding inverted loading. The same is true of bats' wings, and (probably) those of pterosaurs. The dorsal direction is the *only* direction in which large forces can be generated for manoeuvring in any flying animal, including hummingbirds.

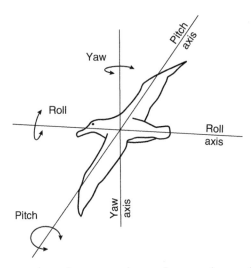

FIGURE 9.4 Flight controls apply moments that produce angular accelerations about the three axes of the bird-centred co-ordinate system, pitch, roll and yaw.

and they are produced as secondary effects of the rotations produced by the controls. For the discussion of linear accelerations, we assume that the bird has some airspeed, and that the relative airflow comes from ahead, approximately parallel to the bird's longitudinal axis. The relative airflow does not have to be horizontal, or in any particular three-dimensional direction relative to the earth's surface.

## 9.2.1 ACCELERATION ALONG THE FLIGHT PATH—CONTROL OF AIRSPEED

Airspeed can be increased by simply angling the flight path downwards. This increases the component of the weight that acts forwards along the flight path. The drag builds up as the airspeed increases, until the net force along the flight path returns to zero, and the airspeed

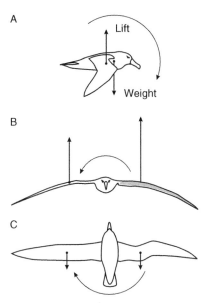

FIGURE 9.5 (A) Sweeping the wings back produces a nose-down pitching moment which leads to an increase of airspeed, while sweeping them forward causes a nose-up pitching moment and slows the bird down. (B) Small rolling moments are produced by rotating the wings in opposite directions at the shoulder joints, nose-up to increase the lift on one side, and nose-down to decrease it on the other. Larger rolling moments can be produced by shortening the down-going wing, by flexing the elbow and wrist joints. (C) Shortening one wing moves the centre of drag nearer the body, and generates a yawing moment. Birds do not have a rudder.

settles at a new value, faster than before. A gliding bird initiates a change of airspeed by essentially the same method as a hang-glider pilot, by moving the centre of lift of the wings forwards or back, relative to the centre of mass. The difference is that the geometry of a hang-glider's wings is fixed, and control depends on moving the pilot's mass forwards or back, relative to the structure, whereas a bird can reorganise the shape of its wings, using the shoulder, elbow and wrist joints to move the whole wing, or particular parts of it, relative to the body. Moving the wings forwards results in a nose-up pitching moment, whose immediate effect is to increase the angle of attack of the wings. This produces a momentary upward acceleration accompanied by increased drag, which slows the bird down until it is once again in equilibrium, but at a lower speed than before.

The effect of adding power by flapping in horizontal flight depends on the airspeed. A bird that is flying horizontally at a speed below its minimum power speed is unstable if it is generating just enough power to maintain that speed, in the sense that if it speeds up by a small amount, the power required is less than before. If the bird continues to produce the same amount of power, it continues to accelerate, until it meets the rising part of the power curve at a stable speed (see Chapter 3, Figure 3.7). Both birds and aircraft normally accelerate through $V_{mp}$ after take-off for this reason, and settle at some higher speed. Some birds and bats that need to fly very slowly because of their feeding methods are adapted to fly at speeds below $V_{mp}$, but this is not something that is easy for them to do. As a general rule, a bird that is flying steadily along can be assumed to be flying at $V_{mp}$ or faster.

A gliding bird can decelerate by pulling up, or by setting the wings into a configuration that generates extra drag, or (very commonly) by extending the feet as air brakes. Glider pilots usually perceive air brakes as a device for controlling the angle of descent rather than the speed, while pilots of fast aircraft such as airliners use them either for slowing down or for rapid descent, as required. A bird that slows down below $V_{mp}$ prior to landing has to increase power as it does so, by increasing the amplitude and/or frequency of the wingbeat.

## 9.2.2 ACCELERATIONS TRANSVERSE TO THE FLIGHT PATH—PULLING $g$

A large acceleration requires a large force, and that can only be produced by the wings, and only in the dorsal direction. The lift force on the wings is controlled by changing their angle of attack, that is the angle between the incident wind and the plane of the wings. Figure 9.6A shows a powered aircraft flying along horizontally, which implies that the angle of attack has been adjusted so that the lift equals the weight. The pilot sits on the wing, and feels gravity ($1g$) just as he would if he were sitting on the ground. If the angle of attack is increased so that the lift exceeds the weight (Figure 9.6B), then the resultant of the lift and the weight is an upward force ($R$), which bends the flight path into a curve in the dorso-ventral plane. If this plane is vertical, and the flight path is initially inclined downwards, the resultant force leans forwards, causing the aircraft to accelerate along the flight path as it comes down to the level position, and then leans back as the flight path bends upwards, causing the aircraft to decelerate. In Figure 9.6C the lift is less than the weight, and the resultant force causes the flight path to curve downwards. The pilot feels more than $1g$ in the first case, and less than $1g$ in the second. If the lift force is adjusted to zero, the pilot feels weightless, and loose objects float about in the cockpit. The flight path is a downward-curving parabola, but its

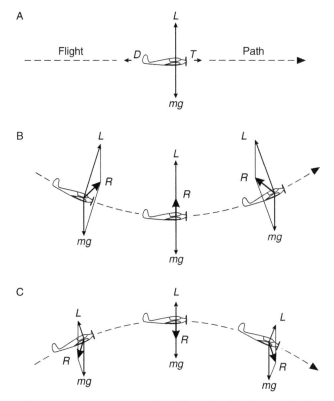

FIGURE 9.6 (A) In a powered aircraft in level flight, the lift $L$ (perpendicular to the flight path) balances the weight $mg$ (downwards), while the propeller thrust $T$ (forwards), balances the drag $D$ (backwards). The resultant force on the aircraft is zero, and it proceeds in a straight line at a constant speed, not accelerating in any direction. (B) If the lift exceeds the weight but the wings are level, and thrust and drag remain balanced, there is a resultant force $R$ in the dorsal direction, which bends the flight path into an upward curve. The pilot feels increased gravity, in proportion to the lift and in the same direction. The weight now has a forward component which makes the aircraft accelerate as it comes down to the bottom of its arc, and a backward component which makes it decelerate as it starts to climb. (C) If the lift is less than the weight, the pilot feels reduced gravity, and the flight path bends downwards.

shape is only directly apparent to a ground observer. The pilot has to refer to outside visual references to detect this, or to artificial sensory aids, which birds do not have (Chapter 11).

The process of increasing the lift on the wing is usually referred to as "pulling $g$", because the effect, as perceived by a bird or by a person in an aircraft is indistinguishable (by any physical test that does not involve observing the ground) from that of a change in the strength of the local gravity field. "Pulling $2g$" means increasing the angle of attack until the lift is twice the aircraft's all-up weight. An object inside the aircraft then has twice its weight in level flight, relative to the aircraft. The weight of such an object, as measured with a spring balance, is directly proportional to the lift on the wings, and is not affected by the external gravity field. If the aircraft executes a loop, and the wings pull 1$g$ as it passes over the top, meaning that they exert a force equal to the all-up weight towards the aircraft's dorsal side, then the weight feels the same as in right-way-up level flight, although the aircraft is inverted, and the flight path is curved downwards by the combined effect of gravity and the (downward-directed) lift. There is no practical way to deduce the orientation of an aircraft, solely from the forces experienced by the occupants, because it is not possible to distinguish the effect of lift from that of gravity, without additional information from visual observation of the horizon, or from gyroscopic instruments (Chapter 11).

### 9.2.3 MAXIMUM $g$ AND THE FLIGHT ENVELOPE

The maximum $g$ that a bird or aircraft can pull depends on the maximum lift coefficient at low speeds, and on the strength of the structure at high speeds. This is described by the *flight envelope*, a graph of maximum $g$ (both positive and negative) as a function of speed. Constructing a flight envelope is an essential step in the design process for any aircraft, but it is difficult to construct one empirically for birds, hence the example in Figure 9.7 should be regarded as somewhat conjectural. For a fixed-wing aircraft or a gliding bird, the maximum $g$ that the wing can develop starts at zero when the speed is zero, and then increases with the square of the speed, passing through 1$g$ at the stalling speed (the minimum speed for unaccelerated flight), and through 4$g$ at twice the stalling speed. At some speed, the maximum $g$ that could theoretically be pulled reaches the maximum that can be safely resisted by the structure. The point where the curve of the flight envelope breaks away from the speed-squared curve is shown as 4$g$ in Figure 9.7. In light aircraft and gliders this point is usually located somewhere between 3$g$ and 6$g$, by the simple expedient of prohibiting pilots from pulling more than the maximum permitted $g$.

It is possible to break the wings off any aircraft by diving to a high speed, and pulling too much $g$, but in birds the maximum $g$ is limited automatically by the maximum moment that the pectoralis muscles can exert about the shoulder joints. If this is exceeded, the wing rotates dorsally, overcoming the pull of the pectoralis muscles. This converts the bird into a high-drag configuration (as used intentionally by pigeons

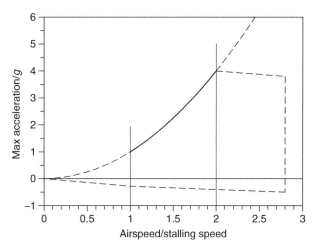

FIGURE 9.7 The maximum lift that a fixed wing can develop is proportional to the square of the airspeed, and equal to the weight when the airspeed equals the stalling speed (by definition). At twice the stalling speed, the wing is able to develop lift equal to four times the weight, that is it can "pull 4$g$", which is here assumed to be the maximum that the structure can safely withstand. Above this speed, the envelope becomes a nearly horizontal line, meaning that the lift is limited to 4$g$ or slightly less by some method other than the maximum lift coefficient of the wing. This limit is imposed in birds by the maximum moment that can be exerted isometrically by the pectoralis muscles about the shoulder joints. Most aircraft wings are designed to withstand a lesser amount of negative loading, but if birds can generate negative lift at all, the limit would certainly be less than −1$g$. Thus the lower boundary of a bird's flight envelope is conjectural, and so is the right-hand boundary, which defines the maximum speed at which the bird can fly.

dropping steeply from a high roof) that slows it down to a safe speed, without any risk of structural damage. In experiments in which pigeon wing bones were tested to destruction by applying bending and torsional moments, the humerus broke in both bending and torsional modes when the applied moment was approximately nine times the value estimated for level, gliding flight, in other words a pigeon flying at three or more times the stalling speed could theoretically pull 9$g$ before the humerus would fail (Pennycuick 1967). This is more than twice the acceleration that the pectoralis muscles can withstand before they are forcibly lengthened (about 4.2$g$). The wing bones may be said to have a "safety factor" of about 2.1, over the strength required to support the weight in a 1$g$ glide.

Peregrines can allegedly dive at over four times their stalling speed (Tucker 1998) but this does not imply that they are able to pull 16$g$. All parts of a bird's body must be strong enough to withstand the maximum acceleration that the isometric pectoralis moment permits. If a peregrine were able (and rash enough) to pull 16g while recovering from a dive, the probable effect would be to tear its viscera out of its body cavity, but in practice forcible extension of the pectoralis muscles would occur well before structural failure of any part of the body. In the absence of accelerometer measurements, 4g will serve as a current round-number "best guess" for the upper limit to the acceleration that manoeuvring birds can pull. Grodzinski et al. (2008) used the acoustic location of echolocation calls to construct three-dimensional tracks of *Pipistrellus kuhlii* bats in foraging and commuting flight, and showed that there was a sharp upper limit to the curvature of the flight path, corresponding to about 3.5g throughout the observed speed range.

## 9.2.4 FLIGHT IN A CIRCLE—THE BALANCED TURN

The simplest type of curved flight path is a turn at a constant airspeed, which can be in the horizontal plane for a powered aircraft, or at a constant rate of descent for a glider or gliding bird. Gravity always pulls towards the earth's centre whatever the bird is doing, but the lift force developed by the wings acts in the dorsal direction, and this may not be vertical. If the wings are banked (tilted to one side relative to the horizon), the resultant of the lift and the weight has a horizontal component which is unbalanced, and this causes the flight path to bend into a curve (Figure 9.8). A *balanced turn* is one which the vertical component of the lift balances the weight, which implies that the magnitude of the lift exceeds the weight, by an amount that depends on the angle of bank. The acceleration felt by the pilot depends only on the magnitude of the lift (above), and so a balanced turn can only be maintained by pulling more than 1g. The acceleration is inversely proportional to the cosine of the angle of bank (Box 9.3). As Figure 9.9 shows, it starts at 1g when the wings are level, passes 2g at 60° of bank, and thereafter rises more and more steeply to 4g at about 75° and, of course, infinity at 90°. Birds typically fly around at bank angles below about 35°, requiring accelerations up to 1.2g or so. Steeper angles of bank do not necessarily imply g according to Figure 9.9, as the turn may not be balanced. An unbalanced turn, with the wings banked near or even beyond the vertical, but pulling only 1g or less, is the least stressful way initiate a steep dive, and such "wingover" manoeuvres are often used for that purpose by both birds and aircraft.

In a balanced turn at a given angle of bank, the radius of the circle is proportional to the square of the speed, so that the ability to turn in tight circles depends on being able to fly slowly. This is important to

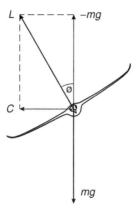

FIGURE 9.8 Seen from in front, a gliding bird that is executing a turn at a constant speed banks its wings at an angle $\varphi$ to the horizontal, so that the lift $L$ is tilted at the same angle to the vertical. For a "balanced" turn, the bird increases its angle of attack sufficiently to make the vertical component of the lift equal in magnitude to the weight, so that there is no vertical acceleration. If the lift is not adjusted to match the angle of bank, then the turn is not balanced, and the flight path bends either upwards or downwards. The horizontal component $C$ of the lift is the force that causes the bird to accelerate horizontally, in a direction perpendicular to the flight path. The effect of this continuous acceleration is to bend the flight path into a circle, whose radius depends on the angle of bank, and on the square of the speed (Box 9.3).

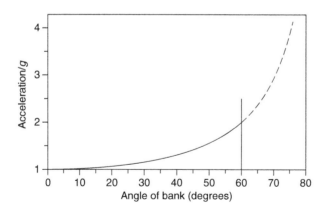

FIGURE 9.9 The centripetal acceleration required to maintain a balanced turn depends on the angle of bank [Box 9.3, Equation (2)]. A 2$g$-balanced turn is one with 60° of bank. If the angle of attack is increased to 76°, about 4$g$ is needed to maintain a balanced turn.

soaring birds that need to circle continuously within a thermal, whose horizontal extent may be limited. The same principle applies to man-oeuvres in any plane, not only to balanced turns. Manoeuvring in

BOX 9.3 **Flight in a circle.**

**Variable definitions for this box**

| | |
|---|---|
| $C_L$ | Lift coefficient |
| $C_{Lmax}$ | Maximum lift coefficient |
| $F$ | Centripetal force |
| $g$ | Acceleration due to gravity |
| $L$ | Lift force on wings |
| $m$ | Body mass |
| $r$ | Radius of turn |
| $r_{lim}$ | Limiting circling radius |
| $V$ | Velocity |
| $V_s$ | Stalling speed in a straight glide |
| $V_t$ | True airspeed |
| $S$ | Wing area |
| $\varphi$ | Angle of bank |
| $\rho$ | Air density |

It is shown in elementary calculus books that if a body of mass $m$ moves at a steady speed $V$ along a circular path of radius $r$, then there must be a steady force $F$ acting on it, directed towards the centre of the circle, where

$$F = \frac{mV^2}{r}. \tag{1}$$

This "centripetal" force makes the object accelerate continuously towards the centre of the circle, without affecting its tangential speed along the circle. In the case of a bird or aircraft flying around a horizontal circle, the centripetal force, and the acceleration it produces, act in a direction perpendicular to the flight path. Apart from aircraft with vectored thrust, only the wings can produce a controllable transverse force that is large enough to deflect the flight path into a circle. If a horizontal component of force ($F$) is produced by banking towards the centre of the circle, at an angle $\varphi$ to the horizontal (Figure 9.9), then:

$$F = mg\tan\varphi, \tag{2}$$

and the bird or aircraft follows a curved path whose radius of curvature ($r$) is:

$$r = \frac{V^2}{(g\tan\varphi)}. \tag{3}$$

These are the conditions for a "balanced turn", in which a passenger feels no sideways force, and the coffee does not spill, even though the aircraft may be banked at 30° or more. This is because "weight", as perceived in the aircraft, is the reaction to the lift force ($L$) on the wings, which is no longer vertical. The lift also has to be increased to maintain level flight in the turn, because only its vertical component balances gravity:

$$L = \frac{mg}{\cos\varphi}. \tag{4}$$

BOX 9.3 *Continued.*

If $\varphi$ is increased to 60°, then the lift has to be increased to $2mg$, to maintain a balanced turn. Pilots say that they have to "pull 2g" to hold the aircraft in the turn, meaning that the weight of the aircraft, and everything in it, including the occupants' body parts, appears to be doubled. There is actually no way to distinguish the effects of acceleration due to a gravitational field from those caused by acceleration due to a lift force. In powered flight, more power is needed to maintain horizontal flight in a turn than in straight flight, and the steeper the angle of bank, the more power is needed. In gliding flight, the rate of sink is increased in a turn.

### Circling flight with a fixed wing

Flight in steady circles is the basic manoeuvre for soaring in thermals (Chapter 10). Soaring birds typically (though not necessarily) glide when circling in thermals, and in that case circling flight is the same for a gliding bird as for a fixed-wing aircraft or glider. The following argument does *not* apply to flapping flight!

The lift force ($L$) developed by the wing can be expressed in terms of a non-dimensional "lift coefficient" ($C_L$), defined as:

$$C_L = \frac{2L}{(\rho V_t^2 S)}, \tag{5}$$

where $\rho$ is the air density, $V_t$ is the true airspeed and $S$ is the wing area, defined as in Chapter 1 Box 1.3. The lift coefficient expresses the angle through which the wing deflects the incident air to create the downwash. For a particular wing, $C_L$ is a function of the angle of attack, increasing as the wing is tilted nose-up, until it reaches a maximum value ($C_{Lmax}$). Further increase of the angle of attack causes the airflow to break away from the upper surface of the wing, with a decrease in the lift coefficient. A wing in this condition is said to be "stalled".

Inverting Equation (5) to get the airspeed at which the bird flies around the circle:

$$V_t^2 = \frac{2L}{(C_L \rho S)}. \tag{6}$$

In a balanced turn, we can substitute for the lift ($L$) from Equation (4):

$$V_t^2 = \frac{2mg}{(C_L \rho S \cos\varphi)}, \tag{7}$$

and substitute this value in Equation (3) to get the radius of turn:

$$r = \frac{2mg}{(C_L \rho S \sin\varphi)}. \tag{8}$$

If the angle of bank ($\varphi$) is zero (wings level), then the radius of turn is infinity (straight flight). For a moderate angle of bank, like 30°, the radius of turn is inversely proportional to the lift coefficient. Bearing in mind that the lift coefficient is inversely proportional to the square of the speed [Equation (5)], this is the same as saying that the radius of turn is proportional to the square of the speed, if the bank angle is held constant.

BOX 9.3 *Continued.*

If the lift coefficient is fixed at $C_{Lmax}$, then Equation (8) gives the minimum circling radius for any angle of bank, while at the same time Equation (7) gives the speed for any combination of $C_L$ and the angle of bank. If the wings are banked vertically, the speed required from Equation (7) is infinite, and the circling radius tends to a limiting value ($r_{lim}$) of:

$$r_{lim} \rightarrow \frac{2m}{(C_L \rho S)} \qquad (9)$$

There is no way to turn in a smaller circle than $r_{lim}$ *with a fixed wing*, but this limit does not apply to flapping flight. Obviously, helicopters and hummingbirds can turn at zero radius, just rotating while hovering. They can do that because they get air to flow over the wings by moving the wings relative to the body, so generating lift without having to move the bird or aircraft as a whole.

### Wing loading and circling performance

Equation (8) shows that for a particular lift coefficient and angle of bank, the radius of turn is directly proportional to the ratio of the mass ($m$) to the wing area ($S$). The "wing loading" is usually defined as the ratio of the weight rather than the mass to the wing area, but either way, it determines the minimum circling radius, and also the stalling speed ($V_s$) in gliding flight. From Equation (7) the stalling speed in straight flight is

$$V_s = \sqrt{\left[ \frac{2mg}{(\rho S\, C_{Lmax})} \right]}, \qquad (10)$$

and this is increased by a factor $\sqrt{(1/\cos \varphi)}$ in a balanced turn at an angle of bank $\varphi$. Of course, a bird's wing can stall while it is flapping, but in that case the "stalling speed" refers to the relative speed between the flapping wing and the air flowing past it, not to the forward speed of the bird as a whole. The ability to fly slowly, and turn in small circles, is critical for exploiting thermals, and therefore the wing loading is a useful indicator of a gliding bird's capacity for this activity. The section of *Flight* that calculates glide polars gives an indication of circling performance by calculating the minimum circling radius in gliding flight, for a bank angle of 24°, a representative value observed in the field for large soaring birds (Pennycuick 1972). This depends on the bird flying at the particular speed (in gliding) which results from setting the lift coefficient to the value corresponding to minimum sink in the glide polar calculation. A gliding bird cannot fly any slower than its minimum gliding speed, which is set by $C_{Lmax}$, but no such limit applies in flapping flight. If it flaps its wings, and has enough muscle power, the bird can reduce its speed and also its turning radius to zero, regardless of its wing loading. The wing loading does not have any special significance for performance in powered flight. There is no theoretical basis for using wing loading as a morphological variable, against which to plot measurements of powered flight performance.

pursuit of prey is a compromise between being able to fly fast enough to overtake the prey, and flying slowly enough to be able to follow its twists and turns.

## 9.2.5 RATE OF ROLL AND MANOEUVRABILITY

Initiating a change of direction is a two-stage process. First, the bird has to roll, that is rotate about its longitudinal axis until the wings are appropriately oriented to pull the flight path in the direction required, meaning that the resultant of the lift force and the weight lies in the required plane of curvature. Asymmetry of the wings does not in itself result in a turn, but is needed temporarily, to rotate the wings into the required orientation, before the actual turn is initiated. Then both wings pull together, so as to produce the required curvature of the flight path.

At any particular speed, the *manoeuvrability* of a bird or aircraft depends on two characteristics, the maximum *g*, discussed above, and the maximum rate of roll, which determines the time needed to roll the wings into the required orientation before pulling *g*. Most fixed-wing aircraft use ailerons to initiate a roll. These are hinged surfaces at the trailing edges of the wings, arranged so that when one is deflected downwards, the other is deflected up. This increases the lift on the side with the down-going aileron, and decreases it on the other side, so producing a rolling moment about the longitudinal (roll) axis. This in turn causes angular acceleration about the roll axis. As the angular velocity about the roll axis builds up, the rolling moment for a given aileron deflection dwindles to zero, because the angle of attack of the up-going wing decreases, while that of the down-going wing increases. The asymmetry is only maintained for a short time, until the plane of the wings has rotated into the orientation required for the turn. Then, symmetry of the wings is restored, and both wings pull together, to deflect the mass of the aircraft into a curved trajectory.

Birds do not have ailerons, but they can rotate the entire wing nose-up or nose-down at the shoulder joint, which is even more effective in producing a rolling moment. Birds and bats can achieve very high rates of roll by flexing the elbow and wrist joints of one wing, while keeping the other wing fully extended. This moves the centre of lift of the shortened wing in towards the body, so that the rolling moment produced by the fully extended wing is unbalanced, resulting in a high angular acceleration. The maximum rate of roll determines the time needed to initiate a turn, while the curvature of the flight path in the turn itself depends on the airspeed and the maximum *g*.

## 9.3 ⬣ TRANSIENT MANOEUVRES

### 9.3.1 TAKE-OFF

Take-off is the transition from being supported by something that is essentially part of the earth's surface to being supported entirely by the air. Since the weight is supported by aerodynamic forces in flight, and these depend on air flowing over the wings, take-off is the process of acquiring "flying speed", which refers to the relative speed between the wing and the air flowing over it, not the speed relative to the ground or water. Birds up to pigeon-size can take off from level ground in still air by simply jumping, and at the same time flapping the wings downwards and forwards through their maximum arc. The speed of the wing relative to the air generates a force upwards and forwards, sufficient to support the weight, even though the bird's body is hardly moving yet. Once the bird starts to move forwards, the power required to fly drops as the airspeed increases (Chapter 3). The acceleration continues with subsequent wingbeats until the airspeed of the whole bird reaches the minimum power speed ($V_{mp}$). This is the speed at which the bird has the most excess muscle power for accelerating, climbing or turning. In larger birds, $V_{mp}$ is higher, and less muscle power is available, relative to that required to fly at very low speeds. A swan, for example, cannot fly at all until it has attained an airspeed near $V_{mp}$, and requires a take-off run to gain enough airspeed to lift off the ground or water. Once airborne, it continues accelerating to $V_{mp}$ before initiating a climb. Take-off is a dangerous manoeuvre for many birds that are subject to predator attack, because a bird that has just taken off has little or no excess power available for manoeuvring to evade a predator, until it has accelerated to $V_{mp}$. A flamingo ambushed in the shallows can be caught by even a hyaena or baboon as it tries to accelerate to flying speed, while a grouse that has been spotted on the ground by a falcon has no excess power to take evasive action in the first few seconds after it explodes into the air.

The time and effort required for the initial acceleration is greatly reduced if the bird can take off along a path that slopes downwards. It is much easier to acquire flying speed by dropping from a branch or an elevated rock than by accelerating upwards from a level surface. Launching down a slope requires only minimal initial effort, and seabirds take off whenever possible down the windward slope of a wave. By taking off against the wind (if there is any), the bird starts with an airspeed equal to the wind speed, before it even begins to move over the ground. This greatly reduces the effort required to take off and accelerate to $V_{mp}$. Conversely, taking off downwind is at best hazardous and often impossible.

## 9.3.2 LANDING

Landing is the reverse transition, whereby the weight is transferred from the wings to some solid object, or to a water surface. A variation on this is seen in the larger alcids (guillemots and razorbills), which have small wings adapted for dual use in air and water, and consequently fly faster than other birds of similar size, and have smaller power margins. These birds spend most of their lives at sea, but nest on sea cliffs, and use a ballistic technique for landing on cliff ledges. The bird approaches the cliff in a shallow dive, levelling out below the landing ledge with excess airspeed, and then pulling up into a near-vertical climb. If the climb is well judged, the bird's ground speed drops to zero just above the landing ledge, and it drops on to its feet, but its air speed is too low to correct any but minimal errors. The only way to correct larger errors is to regain air speed by diving away from the ledge, then climb up a short way out from the cliff, and initiate another approach. Essentially the same method is used by cliff-nesting vultures, and variants are seen in many birds that nest on vertical surfaces, such as house martins nesting on a wall below an overhanging roof.

Landing against the wind reduces the ground speed at which the bird has to touch down, often to zero. A pigeon can approach directly towards a window ledge, in level flight and no wind, and arrive slowly enough to absorb the impact with its legs, and not crash into the window, but larger or faster birds have to resort to a climbing approach to land on a cliff ledge (above), and need some help from a head wind to drop on to their feet on level ground.

# 10

# GLIDING FLIGHT AND SOARING

A gliding bird expends energy at much the same rate as one that is flying horizontally, but it gets it by consuming potential energy (losing height) rather than by doing work with its muscles. Soaring is behaviour that replaces this energy by exploiting movements of the atmosphere. Storks, vultures and pelicans use thermals to gain potential energy, while albatrosses replenish their kinetic energy by pulling up through the detached boundary layer in the lee of ocean waves. "Energy height" allows either method to be analysed on the same basis as powered flight.

All flight involves mechanical work being done against aerodynamic drag. In powered flight this work comes from fuel energy, which is converted into work by an engine (Chapter 7). The definition of *gliding* flight is that no work is done by the engine (if there is one). The work comes instead from depleting the bird's potential energy, or its kinetic energy or some combination of the two. A gliding bird must either lose height (relative to the air) or slow down, or both. Kinetic and potential energy can also be exchanged with each other. A bird can pull up at the expense of losing speed, or gain speed in a dive at the expense of losing height, but work continues to be done against drag whatever the bird does. The energy dissipated against drag has to be replenished, either from fuel energy or from some other source. The term *soaring* refers to adaptations or behaviour whereby the bird replenishes its potential or its kinetic energy by exploiting movements of the atmosphere, in such a way that gliding flight can be continued for prolonged periods, without using fuel energy.

## 10.1 ● GLIDING PERFORMANCE

### 10.1.1 THE GLIDE POLAR

In steady gliding flight, the bird loses height (relative to the air) and the resulting rate of loss of potential energy has to account for the power required to overcome drag. The basic performance curve for a bird or glider in straight flight is the *glide polar*, which is a graph of the downward vertical component of velocity (the sinking speed or *sink*), plotted as a function of the forward speed (Figure 10.1). There is a well-defined minimum speed for gliding, the *stalling speed* (Box 10.1), defined as the lowest speed at which the wing is capable of generating enough lift to support the weight. As a gliding bird slows down, the sink increases sharply as the speed is pulled back to the stall, so that continued flight is only possible at a very steep angle of descent. Conversely, as a gliding bird increases its speed on recovering from a stall, the sink first decreases, but as speed is further increased, the sink soon levels off, and then increases at still higher speeds. $V_{ms}$ is the speed for minimum sink. As the speed is increased above $V_{ms}$, the sink increases, but not by much at first, so that the ratio of forward speed to sink (the *glide ratio*) continues to increase. The glide ratio reaches its maximum value at the best-glide speed $V_{bg}$. At still higher speeds the glide ratio declines ever more steeply.

The approach used in *Flight* to calculate a glide polar for a particular bird (Box 10.1) follows Welch et al. (1977), and is similar to that used for a power curve (Chapter 3), except that the calculation takes the form of calculating drag forces rather than components of power. As in flapping flight, a gliding bird's weight is balanced by the rate at which downward momentum is added to the air, in the downwash region behind the wing. The *induced drag* results directly from generating this downwash, and is the component of the total drag that is due to supporting the weight in air. The second component is the *profile drag* of the wings, and the third is the *parasite drag*. This last component is the same as in flapping flight, being the drag of the body excluding the wings. For any particular speed, these three components of drag are calculated and added together to get the total drag, which is

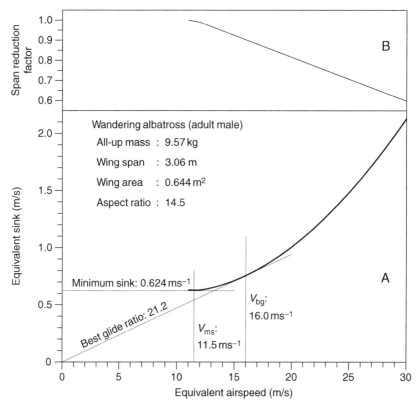

FIGURE 10.1 (A) Glide polar calculated by *Flight* for the Wandering Albatross in the Preset Birds Database. Both airspeed and sink are *equivalent* speeds, meaning that the polar has been calculated for sea-level air density (Compare Figure 10-5). The minimum sink and the best glide ratio are shown, together with the characteristic speeds $V_{ms}$ and $V_{bg}$ at which they occur. $V_{bg}$ is the speed at which a tangent drawn from the origin meets the polar. (B) Under the "Linear" wing span reduction option (the default), the wing span reduction factor starts at 1 at the stalling speed, and decreases linearly to about 0.6 at 30 m s$^{-1}$. If the alternative "Minimum Drag" option were selected, the wing speed reduction factor would stay level initially at 1, and then break sharply downwards.

multiplied by the speed to get the rate at which potential energy is lost. This is then divided by the weight to get the sink.

*Flight* will calculate a glide polar for any bird, and display a summary and a graph if required, in the same way as it does for a power curve. *Flight* plots both the power curve and the glide polar with power (or sink) increasing upwards on the *Y*-axis, regardless of whether the power comes from muscular work or from gravitational potential energy. This convention makes the two curves look generally similar, reflecting the fact that they have much in common. However, it is

BOX 10.1 **Computing the glide polar.**

---

**Variable definitions for this box**

| | |
|---|---|
| $B$ | Wing span |
| $B_{stop}$ | Wing span reduction constant |
| $C_{Db}$ | Body drag coefficient |
| $C_{Dpro}$ | Wing profile drag coefficient |
| $C_{L}$ | Lift coefficient |
| $C_{Lmax}$ | Maximum lift coefficient |
| $C_R$ | Reaction force coefficient |
| $D$ | Total drag |
| $D_b$ | Body drag |
| $D_i$ | Induced drag |
| $D_{pro}$ | Wing profile drag |
| $g$ | Acceleration due to gravity |
| $k$ | Induced drag factor |
| $L$ | Lift |
| $m$ | All-up mass |
| $R$ | Aerodynamic reaction |
| $V_{bg}$ | Speed for best glide ratio |
| $V_{ms}$ | Speed for minimum sink |
| $V_s$ | Stalling speed |
| $V_t$ | True airspeed |
| $V_z$ | Sinking speed |
| $S_b$ | Body frontal area |
| $S_w$ | Wing area |
| $\beta$ | Wing span reduction factor |
| $\beta_{opt}$ | Optimum wing span reduction factor |
| $\delta$ | Planform slope |
| $\varepsilon$ | Wing area reduction factor |
| $\rho$ | Air density |

**Gliding equilibrium**

A bird gliding in a straight line at a constant true airspeed ($V_t$), follows a path that slopes downwards, so that its airspeed has a downward vertical component ($V_z$), called the "sinking speed", or just *sink*. The *glide polar* is a graph of the sinking speed against the airspeed (Figure 10.1). It is the basic performance curve for gliding flight, and is the gliding counterpart of the power curve for powered flight. The drag force ($D$) is not horizontal in this case, but is directed backward along the flight path, while the lift force ($L$) is directed, as always, perpendicular to the flight path. The total reaction ($R$) is the resultant of the lift and the drag, and in straight gliding flight at a constant speed it acts vertically upwards, balancing the weight. It can be converted into a dimensionless coefficient by comparing it to a reference force made up from the dynamic pressure and an area, in this case the wing area ($S_w$):

$$C_R = \frac{2R}{\rho V_t^2 S_w}. \tag{1}$$

**BOX 10.1** *Continued.*

$C_R$ is the *reaction coefficient*. In steady gliding it has to balance the weight ($mg$), so Equation (1) becomes

$$C_R = \frac{2mg}{\rho V_t^2 S_w}. \tag{2}$$

If the lift:drag ratio is reasonably high, the magnitude of $R$ is nearly the same as that of the lift ($L$), and it is customary to express Equation (2) in terms of the lift coefficient ($C_L$), although this is only approximately correct:

$$C_L = \frac{2mg}{\rho V_t^2 S_w} \tag{3}$$

The lift coefficient is related to the downwash angle. The slower the gliding bird tries to fly, the higher the value of $C_L$ required (Equation 3) and the larger the angle through which the wing must deflect the incident airflow, in order to generate enough downwash to support the weight. When this angle becomes too large, the airflow separates from the wing surface, lift is lost, and the wing is said to *stall*. Gliding flight at a speed below the stalling speed is only possible (if at all) at a very steep angle of descent. The wing stalls when the lift coefficient reaches some maximum value ($C_{Lmax}$), whose value is believed to be in the region of 1.8 for the wings of birds like eagles and gulls, which are adapted for gliding flight. If we turn Equation (3) around, then the maximum lift coefficient defines the *stalling speed* ($V_s$):

$$V_s = \left[\frac{2mg}{(\rho S_w C_{Lmax})}\right]^{1/2} \tag{4}$$

**Calculating sinking speed**

At true airspeeds above the stall ($V_t > V_s$), the power equilibrium condition for steady gliding flight says that the power expended against drag ($DV_t$) must equal the rate of loss of potential energy ($mg\,V_z$), where $m$ is the mass and $g$ is the acceleration due to gravity. This allows the sink ($V_z$) to be expressed in terms of the forward speed.

$$V_z = \left[\frac{DV_t}{(mg)}\right] \tag{5}$$

Equation (5) involves the drag ($D$), which is itself a function of the speed ($V_t$). The calculation proceeds in much the same manner as that of the power curve, but in this case it calculates three components of drag (rather than power), and adds them together to get the total drag, which is then inserted in Equation (5) to get the sink.

**Induced drag**

Downwash has to be continuously created to balance the weight, as in powered flight, and this results in a drag force on the wing, the induced drag ($D_i$), given by

$$D_i = \left[\frac{2k(mg)^2}{(V_t^2 \pi B^2 \rho)}\right], \tag{6}$$

BOX 10.1 *Continued.*

where $k$ is the induced drag factor. It was noted in Box 3.1 of Chapter 3 that the induced power for a pair of flapping wings approximated by an actuator disc can be divided by the speed, so as to represent it as a virtual drag, and the formula that results from doing that is the same as Equation (6) for a fixed wing.

### Parasite drag

The parasite drag is the drag of the body, not including the wings ($D_b$), and is calculated in the same way as for flapping flight:

$$D_b = \frac{(\rho V_t^2 S_b C_{Db})}{2} \tag{7}$$

The default value in *Flight* for the parasite drag coefficient ($C_{Db}$) is 0.1, but this may not be appropriate for all birds, as noted in Chapter 3, Box 3.3.

### Profile drag

The profile drag of a gliding wing is easier to represent than the profile power of a flapping wing. The wing can be assigned a profile drag coefficient ($C_{Dpro}$), which is characteristic of the profile shape of the wing. The profile drag ($D_{pro}$) can then be found from the speed and the wing area, in the same manner as the parasite drag in Equation (7):

$$D_{pro} = \frac{(\rho V_t^2 S_w C_{Dpro})}{2} \tag{8}$$

where $S_w$ is the wing area. The default value used in *Flight* for $C_{Dpro}$ is 0.014, which is based on experimental measurements (Pennycuick et al. 1992).

### Wing span reduction

We are not yet quite ready to estimate the drag, as gliding birds have a complication all their own, not found in gliders or in the power calculations for flapping flight. They reduce both the wing span and the wing area as speed increases, by flexing the elbow and wrist joints. The definitions of wing span and area require these to be measured with the joints fully extended (Chapter 1), but a gliding bird only holds its wings at full stretch when flying actually at the stalling speed. As soon as it speeds up, even slightly, it "cracks" the elbow and wrist joints. The wing tips move in slightly, reducing the wing span, and the flight feathers begin to close up fanwise, reducing the wing area. In a fast glide, the wing planform, seen from below, acquires a distinctive M-shape, with the leading edge of the hand wing swept strongly back. The effect of shortening the wings is to flatten the glide polar, that is, to reduce the sink at high speeds, as compared to a bird that keeps its wing span constant at all speeds, like a glider. This is an option that is not available to gliders which, being unable to shorten their wings, have to resort to other, less effective methods of reducing drag at high speeds.

Shortening the wings is represented in *Flight* by a *wing span reduction factor* ($\beta$), whose maximum value (only used at the stalling speed) is 1. As the bird speeds up, $\beta$ is progressively reduced, perhaps down to 0.5 or even below. *Flight* also provides a constant called the *planform slope* ($\delta$), which

BOX 10.1 *Continued.*

determines the variable factor ($\varepsilon$), by which the wing area is reduced, when the wing span is reduced by the factor $\beta$, as follows:

$$\varepsilon = [1 - \delta(1 - \beta)] \tag{9}$$

The default value (which can be changed) is $\delta = 1$, which results in the area being reduced by the same factor as the span, so that the mean chord is not affected by shortening the wings. The value of the wing span ($B$) entered into the programme is measured with the wings at full stretch, but the value used for calculating the drag at each speed is not $B$ but $\beta B$, and the wing area used for calculating profile drag $\varepsilon S_w$ instead of $S_w$. The induced drag is modified from Equation (6) to:

$$D_i = \frac{2k(mg)^2}{(V_t^2 \pi (\beta B)^2 \rho)}, \tag{10}$$

and the profile drag from Equation (8) becomes

$$D_{pro} = \frac{(\rho V_t^2 \varepsilon S_w C_{Dpro})}{2} \tag{11}$$

**Calculating the glide polar**

Before *Flight* starts to calculate a glide polar, the user must select a "law" for making the wing span reduction factor ($\beta$) a function of the true airspeed ($V_t$). Two options are provided, *Linear* and *Minimum drag* (of which more below). It then calculates the stalling speed ($V_s$), and selects a slightly lower speed to start the polar. It increments the airspeed in steps of $0.1$ m s$^{-1}$, first setting $\beta$ at each step according to the chosen law, then calculating the induced drag from Equation (10), the body drag from Equation (7) and the wing profile drag from Equation (11), and adding the three components together to get the total drag ($D$). Then it finds the sink from Equation (5). The sink declines at first as the speed increases, then levels off and starts increasing. When the sink levels off, the programme notes the speed as the *minimum sink speed* ($V_{ms}$). This corresponds to the minimum power speed in flapping flight. At each speed step, the programme also calculates the lift:drag ratio ($L/D$) from the approximation

$$\frac{L}{D} = \frac{V_t}{V_z} \tag{12}$$

This at first increases at each speed step, then levels off and starts to decline. When it levels off, the programme identifies the speed as the *best glide speed* ($V_{bg}$). This is the speed at which the bird covers the greatest air distance for a given loss of height. The programme also does a MacCready cross-country speed calculation at each step, and uses this to determine when to terminate the polar (Box 10.2).

**Alternative wing span reduction laws**

At low speeds, induced drag predominates, but at high speeds induced drag is small, and profile drag predominates. The effect of a value of $\beta$ less than 1 is to increase the induced drag in inverse proportion to $\beta^2$, but at the same time to reduce the profile drag by a factor of $\beta$. By making $\beta$ a decreasing

BOX 10.1 *Continued*.

function of speed, profile drag is reduced at higher speeds, at the expense of an increase of induced drag (which is small at high speeds anyway). At any particular speed, there is an "optimum" value of $\beta$ that makes the sum of induced and profile drag a minimum. This can be found by differentiating the wing drag ($D_w$), which is the sum of the induced drag from Equation (10) and the profile drag from Equation (11), with respect to $\beta$.

$$D_w = D_i + D_{pro}$$
$$= [(2km^2g^2)/(V_t^2\pi\beta^2B^2\rho)] + [(\rho V_t^2\varepsilon S_w C_{Dpro})/2]. \tag{13}$$

Substituting for $\varepsilon$ from Equation (9) and differentiating,

$$\frac{dD_w}{d\beta} = -\left[\frac{(4km^2g^2)}{(V_t^2\pi\beta^3B^2\rho)}\right] + \left[\frac{(\delta\rho V_t^2 S_w C_{Dpro})}{2}\right]. \tag{14}$$

Setting this expression to zero gives the optimum value of $\beta$, for minimum wing drag at a given speed:

$$\beta_{opt} = \left[\frac{(8km^2g^2)}{(\delta\pi\rho^2B^2 C_{Dpro}S_w V_t^4)}\right]^{1/3} \tag{15}$$

If the "Minimum drag" option is selected for "Wing span reduction" in the Glide Polar Setup screen, the programme begins the calculation at each speed step by setting $\beta$ to $\beta_{opt}$, unless Equation (15) makes $\beta_{opt}$ more than 1, which it does at low speeds. In that case, $\beta$ has to be limited to 1 (maximum wing span). The result can be seen by viewing the graph from the Glide Polar Summary screen, which also shows a graph of $\beta$ *versus* speed. The line remains level at $\beta = 1$ up to quite a high speed, then suddenly breaks downwards. This does not appear to be what gliding birds actually do. Instead they start reducing the wing span directly above the stalling speed, and then reduce it linearly as the speed increases.

Instead of using Equation (15), the "Linear" option for wing span reduction sets $\beta$ according to this formula:

$$\beta = \frac{\left[B_{stop} - \left(\frac{V_t}{V_s}\right)\right]}{(B_{stop} - 1)} \tag{16}$$

$B_{stop}$ is a constant whose default value is 5. Equation (16) sets $\beta$ to 1 at the stalling speed (maximum wing span), and reduces it linearly to zero at $B_{stop}$ times the stalling speed. If $B_{stop} = 5$, no ordinary bird is going to glide fast enough to require a wing span approaching zero. The law is based on measurements by Rosén and Hedenström (2001) of a jackdaw gliding in a tilting wind tunnel, generalised to other birds by assuming, in effect, that the same value of $\beta$ occurs at the same lift coefficient. It might have to be modified in the future if data become available for wing span selection in birds that actually do glide at high multiples of the stalling speed, such as (possibly) peregrines. Meanwhile, linear span reduction is currently the default option for glide polar calculations in *Flight*. "Minimum drag" was the only method provided in early versions of the programme, and remains available as an alternative option.

conventional in the gliding literature to plot glide polars with sink increasing downwards, and some polars are drawn in this way in the discussion of soaring performance later in this chapter, for reasons of tradition.

## 10.1.2 CONTROL OF SPEED IN GLIDING

The very idea of a glide polar implies that a glider or a gliding bird is able to control its speed. A glider pilot does that by adjusting a tailplane, but birds use a different method, which works even if the tail is damaged or missing. A gliding bird controls its speed by adjusting the angles of the shoulder, elbow and wrist joints. In a most elegant adaptation, the adjustment that changes the speed is the same as the one that trades induced against profile drag to suit the new speed (Pennycuick and Webbe 1959). At low speeds, the bird spreads its wings to the maximum wing span and area, and this movement also produces a nose-up attitude which reduces the speed. Conversely, the bird moves the centre of lift back to increase speed, by flexing the elbow and wrist joints, and this movement also reduces the span and area to suit the higher speed. Bats can only do this to a very limited extent, because sweeping back the outer part of a bat's wing releases tension in the sail, without which the wing cannot function. Pterosaurs, on the other hand, would have been able to adjust their wing shape nearly as effectively as birds, if their patagium was elastic as proposed in Chapter 6.

## 10.1.3 WING SPAN REDUCTION

The three major components that make up a gliding bird's drag all depend strongly on the airspeed (Box 10.1). Two of them (induced drag and profile drag) are due to the wing, while the third (parasite drag) is due to the body. If the bird's wing geometry were fixed, then the induced drag would decrease strongly with speed (inversely proportional to the square of the speed), while the profile drag would increase directly with the square of the speed. However, gliding birds change the wing geometry by progressively flexing both the elbow and wrist joints as they increase speed, which has the effect of reducing both the wing span and the wing area (Chapter 5).

At any particular speed, reducing the wing span *increases* the induced drag, while the simultaneous reduction of the wing area *decreases* the profile drag (Box 10.1). The wings are fully extended at the stall, and nearly so at $V_{ms}$, where the induced drag is higher than the profile drag, and the wing span needs to be maximised. At higher speeds, the induced drag dwindles, and the penalty for reducing the wing span becomes small, while there is increasing benefit from reducing the wing area, as this gets rid of some of the (increasing) profile drag. Shortening the wings at higher speeds reduces the sink by an ever-increasing amount, relative to the sink that would have occurred at the same speed, with the wings fully extended. The effect is to "flatten" the polar, meaning that as the speed increases, the sink increases less than it would for a wing whose shape is fixed. An alternative method of flattening the polar, sometimes used in gliders, is to add area to the trailing edge of the wings by deploying "Fowler flaps" at low speeds, but this leaves the glider with more wing span than it needs for flight at high speeds. Shortening the wings, as birds do, is more effective, but has never been implemented in an aircraft, because of the mechanical difficulty of building a jointed spar, with actuators to flex and extend the elbow and wrist joints.

### 10.1.4 ALTERNATIVE WING SPAN REDUCTION LAWS IN *FLIGHT*

*Flight* provides two alternative "laws" that determine what wing span the bird selects at any given speed, and plots the wing span on the same graph as the polar. *Linear* wing span reduction is the default, because it seems to be what birds do (Rosén and Hedenström 2001.). The wing span is set to its maximum value at the stalling speed (as defined by the maximum lift coefficient—Box 10.1), and then reduced at higher speeds, as a linear function of speed. The alternative *minimum drag* law selects the wing span that minimises the sum of induced and profile drag at each speed. At low speeds, this law indicates an optimum wing span that is longer than the anatomical maximum. *Flight* truncates this, with the result that the wing span remains constant at its maximum value up to quite a high speed, and then suddenly plummets as the bird pulls its wings sharply in. While minimising drag seems to be what one would expect, this type of span reduction is definitely *not* what birds do! The linear span-reduction law does not minimise drag, but presumably it has some other advantage, perhaps to do with speed control. Figure 10.2 shows glide polars calculated by *Flight* for the same bird (a Swainson's Hawk) for the above two span-reduction laws, and also for a fixed-span wing.

## 10.2 ⬤ SOARING

*Soaring* is behaviour that results in the bird extracting potential and/or kinetic energy from the atmosphere, and using it to overcome drag. A flying bird possesses stored energy in three forms, fuel energy,

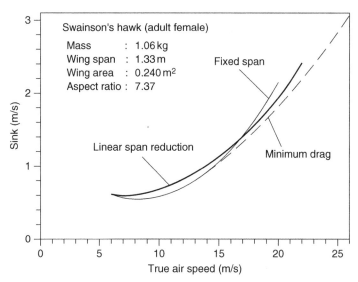

FIGURE 10.2 Three polars for a Swainson's Hawk from the Preset Birds Database, with different span-reduction assumptions. *Flight's* minimum drag option coincides with a fixed-wing polar (no span reduction) at low speeds, because the span remains at its full value ($\beta = 1$) up to about 13 m s$^{-1}$. At higher speeds, the minimum drag option flattens the polar dramatically, by reducing the wing area, and the profile drag. The default linear option is also better than the fixed-wing polar at speeds above about 16 m s$^{-1}$, but noticeably worse at lower speeds and worse than the minimum drag option at all speeds. It is *Flight's* default because it resembles observed bird behaviour more closely than the minimum drag option.

potential energy and kinetic energy, and the amount of each can be represented as a virtual *energy height*. The concept of fuel energy height has been introduced in Chapter 8 (Section 8.3 and Box 8.3) as the height to which the bird would be lifted, if the whole of its fuel energy were to be converted into work (reducing the mass in the process), and used to raise it against gravity. The bird's *potential* energy height is the same as its actual height above some datum (usually the land or water surface), and its *kinetic* energy height is the height to which it can pull up at the expense of reducing speed, converting speed to height as pilots say. Birds that start at the same virtual or actual height go the same distance if they are equally efficient, regardless of the size of the bird or of the type of energy represented by the energy height.

In round numbers, a migrating bird whose all-up mass is 20% fat would have a fuel energy height of about 200 km, it might climb in thermals to a height of 2 km, and it might have enough kinetic energy to pull up to 20 m. Practical heights for potential and kinetic energy are respectively around two and four orders of magnitude smaller than fat energy heights. However, these small energy stores are more significant than they might appear, because, unlike fuel energy, they can be repeatedly used to overcome drag, and replenished directly from energy that is freely available in the atmosphere, without eating. Alaskan Bar-tailed Godwits allegedly cross the Pacific Ocean from north to south in flapping flight, from Alaska to New Zealand. They do this by fuelling up once before departure, to a vast fat energy height of about 730 km, and then flying non-stop for a week, coming "down" on an average gradient of about 14:1 (Pennycuick and Battley 2003). Albatrosses of various species fly similar distances about the Southern Ocean, but they do it by alternately consuming part of their *kinetic* energy, and replenishing it at intervals of seconds or minutes, from the wind discontinuity where the atmospheric boundary layer separates from the crest of a wave. Since the energy height that an albatross can gain each time it does this is 10,000 times smaller than a godwit can gain by putting on fat, the albatross has to replenish its miniscule energy store 10,000 times to go the same distance that a godwit can fly by refuelling once. However, albatrosses know how to do that, and in consequence they can roam the windier regions of the world's oceans, almost free of metabolic cost. Migrants that replenish their potential energy by climbing in thermals are intermediate in the frequency with which they need to replenish their height, typically gaining up to 2 km in each of a hundred or so climbs, in the course of a migration from the temperate regions to the tropics.

## 10.2.1 SOARING BY REPLENISHMENT OF POTENTIAL ENERGY

### 10.2.1.1 Linear soaring

The way in which a gliding bird uses potential energy to overcome drag is shown by the glide polar (Figure 10.1), in which rate of sink (the downward component of speed) is plotted against the forward speed. If the bird flies in air that is rising at the same speed (relative to the ground) as the bird is sinking (relative to the air), then its potential energy loss is offset by energy supplied by the rising air. It maintains height, without having to do any muscular work. That is the simplest method of soaring. Birds and gliders *slope-soar* for hours at a time in places where a steady wind blows against a slope, simply by tacking back and forth along the slope, so as to stay in the zone where the air is deflected upwards. The term "lift" is used by glider pilots to refer to a zone of air that is rising fast enough to climb in, independently its original meaning as the lift force on a wing. If the slope lift exceeds the bird's minimum sink, then it can either gain height, or alternatively maintain height at a faster speed, at which its sink is equal to the upward air velocity.

Slope lift along the windward side of a slope is not the only kind of linear lift. In appropriate conditions, standing waves may form *downwind* of a hill. Sometimes, the wave crests are marked by linear "lenticular" clouds, which lie across the wind, continually forming along the windward edge and dissolving along the downwind edge, so that the clouds remains stationary over the ground. Wave systems often remain usable up to many times the height of the hill that triggers them, unlike slope lift, which typically extends 300 m above the top of any slope, whether it is an alpine wall or a modest English sea cliff. In parts of the world where strong prevailing winds blow across mountain ranges, as in the Southern Alps of New Zealand, and the Andes of Argentina and Chile, persistent wave systems often provide strong, linear lift in which a glider can fly fast and straight for hundreds of kilometres, at a height far above the mountain tops. Distance flights of over 3000 km in a day have been made by shuttling back and forth along an Andean wave system. Migrating Canada geese have been observed by glider pilots using a wave system in Colorado, but little is known about the extent that birds are able to exploit waves.

## 10.2.1.2 Soaring in thermals

A "thermal" is a vortex structure in the atmosphere that drifts along with the wind, powered by heat fed into the atmosphere from below by contact with a warm land or water surface (Figure 10.3). Strong solar heating of the surface, as in hot, arid areas, usually triggers columnar thermals, or *dust devils*, in which a column of air rotates about a vertical axis. The pressure in the centre of the column is reduced because of the rotation, and this causes air to converge inwards at the bottom of the column, where the rotation is retarded by friction with the ground. The only way the air can escape is upwards. The core of a dust devil may contain strong lift, but is usually narrow. Birds that rely heavily on this type of lift have to be capable of gliding in circles of small radius, with a low rate of sink, and have adaptations that permit this (Box 10.2). A columnar thermal powered by a persistent source of heat, such as a grass fire, may extend from ground level up to 2000 m or more above ground level (AGL), but low-level dust devils powered by solar heating of the surface more often extend upwards no more that a few hundred metres. They may then peter out, or trigger

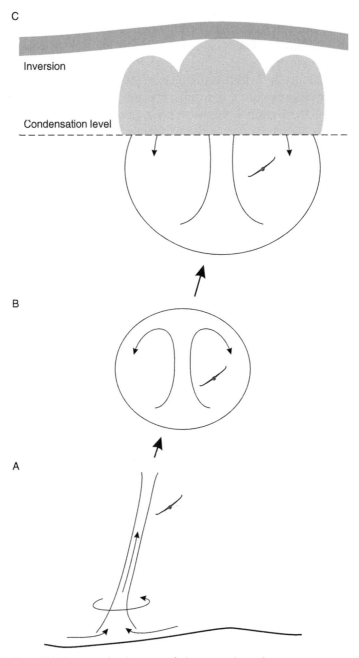

C

Inversion

Condensation level

B

A

FIGURE 10.3 **(A)** Strong solar heating of dry ground produces a steep temperature gradient near the surface, and often results in columnar thermals, which may be visible as dust devils. The reduced pressure in the rotating column sucks air in at the bottom, producing a strong but narrow thermal core, embedded in a wider region of weak and turbulent lift. **(B)** If the air above is unstable, low-level columnar thermals may trigger

BOX 10.2 **Soaring performance in thermals.**

If a soaring bird simply needs to stay airborne for a prolonged period with minimum exertion, as in a vulture patrolling over a wildebeest herd, then it requires adaptations that enable it to fly in small enough circles, with a low enough rate of sink, to maintain or gain height in the smallest and weakest thermals that it needs to use. If it needs to travel across country, in addition to remaining airborne, then there are additional requirements which conflict to some degree with those for staying up.

**Variable definitions for this box**

| | |
|---|---|
| $C_L$ | Lift coefficient |
| $m$ | All-up mass |
| $r$ | Radius of turn |
| $r_{lim}$ | Limiting radius of turn |
| $S$ | Wing area |
| $V_{bg}$ | Best-glide speed |
| $V_c$ | Achieved rate of climb in thermals |
| $V_{it}$ | Inter-thermal air speed |
| $V_{opt}$ | Optimum value for $V_{it}$ that maximises $V_{xc}$ |
| $V_{xc}$ | Average speed across country |
| $V_{zc}$ | Sink while circling |
| $V_{zmin}$ | Minimum sink in straight flight |
| $\rho$ | Air density |
| $\varphi$ | Angle of bank |

**Thermal profile and the circling envelope**

Gliding in circles increases the rate of sink, relative to the air, and the tighter the circle, the greater the sink. Circling performance can be expressed by a graph of sink versus radius of turn, when flying in steady circles at the same lift coefficient that results in minimum sink in straight flight, at different angles of bank. This curve is called the *circling envelope* for a particular bird or glider. To calculate it, we first note from Chapter 9, Box 9.3 that the radius ($r$) of the circle is related to the angle of bank ($\varphi$) by

$$r = \frac{2m}{(C_L \rho S \sin\phi)}, \tag{1}$$

where $m$ is the all-up mass, $C_L$ is the lift coefficient, $\rho$ is the air density, $S$ is the wing area, and $\varphi$ is the angle of bank. As the bank is increased, the

vortex-ring thermals, which are toroidal vortex structures that lose contact with the ground. Lift is strong and smooth in the core, but surrounded by a ring of sink. As a vortex ring continues to rise, it grows by entrainment of the surrounding air. (C) A vortex-ring thermal expands and cools as it rises, and may give rise to a cumulus cloud, based at the *condensation level*, where the moisture content of the rising (and cooling) air becomes sufficient to saturate it. The height of cumulus tops is often limited by an *inversion*, a layer of air (shown pink) in which the temperature is warmer than that of the air below, or at least decreases at a lower lapse rate.

BOX 10.2 *Continued.*

radius decreases, but the speed has to increase so that the upward component of the lift still supports the weight at the steeper angle of bank, while
keeping the lift coefficient constant. As the wings approach vertical bank,
the speed approaches infinity, and the radius approaches a limiting value
which is not zero, but $r_{lim}$, where

$$r_{lim} \rightarrow \frac{2m}{(C_L \rho S)}. \tag{2}$$

This is strictly a *fixed-wing* argument, because it assumes that the airspeed
over the wing is the same as the forward speed of the whole bird or aircraft. It
does not apply to birds or bats manoeuvring in flapping flight (or to helicopters), as they can increase the lift by moving their wings relative to the
body. However, it does apply to those birds, and one or two bats, that exploit
thermals by gliding in steady circles, with the wings essentially fixed.

The output of *Flight*'s Glide Polar calculation gives the lift coefficient for
minimum sink in straight flight. Holding the lift coefficient constant at this
value, the turning radius for different angles of bank ($\varphi$) can be found from
Equation (1), and the sink while circling ($V_{zc}$) is given by the formula:

$$V_{zc} = \frac{V_{zmin}}{\sqrt{(\cos^3 \varphi)}}, \tag{3}$$

where $V_{zmin}$ is the minimum sink in straight flight (Haubenhofer 1964). The
circling envelope is the graph of the sink against the radius, for different
values of the bank angle. Two examples for different birds are shown on
the same graph in Figure 10.4. Each curve approaches the minimum sink
in straight flight as the radius approaches infinity, while in the other direction the radius approaches $r_{lim}$ as the bank approaches 90°. The curve is
called an envelope (i.e. a boundary) because the bird can circle at combinations of radius and sink to the right of the curve and below it, by selecting a
lower lift coefficient, but it does not have access to the region above the
curve and to the left of it.

Equation (1) shows that the radius at any particular combination of bank
and lift coefficient is directly proportional to the ratio $m/S$, also known as
the *wing loading*. The upper part of Figure 10.4 shows the profile of a hypothetical weak, narrow thermal, with an upward air velocity about 0.58 m s$^{-1}$
in the centre of the core, falling to zero at a radius of about 18 m. The frigatebird is able to circle at a radius of about 9 m (vertical dashed line) at a
bank angle of about 26°, where its sink is marginally less than the upward
air velocity in the thermal, while the albatross, with a higher wing loading,
would not be able to maintain height at any radius below its circling envelope. Frigatebirds, with their very low wing loadings, as adapted to soaring
in weak narrow thermals over the trade-wind zones of the oceans, whereas
albatrosses also soar over the oceans, but in an entirely different way that
does not involve flying in small circles. For them, a low wing loading is
not an advantage, nor is it necessarily an advantage for a bird that uses thermals to travel long distances, as opposed to simply remaining airborne in
weak conditions.

BOX 10.2 *Continued.*

### Cross-country speed in thermals—the MacCready theory

The speed at which a glider can fly across country using thermals is central to the sport of gliding, and also to those birds that use thermals for migration, or for covering distance on foraging trips. Simplifying a little, a cross-country flight consists of a series of climbs in thermals at a rate of climb ($V_c$) separated by straight inter-thermal glides through stationary air, in which the pilot is free to select an *inter-thermal speed* ($V_{it}$). The average *cross-country speed* ($V_{xc}$) over one or more cycles of climb and glide depends on the rate of climb in thermals, on the pilot's choice for $V_{it}$, and on the particular glider's glide polar.

Glider pilots everywhere use the MacCready theory of cross-country speed, incorporated into mechanical or electronic calculating devices, as an in-flight guide to selecting the speed to fly between thermals, according to the conditions. Figure 10.5 shows a graphical form of the theory applied to two glide polars calculated by *Flight* for a Rüppell's Griffon Vulture, an African species that commonly forages at distances over 100 km from the nest, using thermals to travel out and back between the nest and the foraging area (Pennycuick 1972). The polars are plotted in the traditional way with sink increasing downwards, and the $Y$-axis is extended upwards to represent rate of climb in thermals. The thin curve in Figure 10.5 is the polar for the outward journey when the vulture's crop is empty, and the thick curve is for the return journey, with a crop load of 2 kg. The best glide ratio is the same (14.4) with or without the crop load, but the best-glide speed $V_{bg}$ increases from 16.6 m s$^{-1}$ when empty to 18.8 m s$^{-1}$ when loaded. The effect of the added load is to move the curve to the right and downwards, increasing the sink at low speeds. However, the two curves cross in the vicinity of $V_{bg}$, and at higher speeds (somewhat counter-intuitively) the ballasted vulture's rate of sink is less than when it is unloaded.

The MacCready theory says that for any given rate of climb, the cross-country speed for a complete cycle of climb and glide can be found by drawing a straight line from the point on the $Y$-axis representing the rate of climb to the point on the polar representing the chosen inter-thermal speed. The cross-country speed is the point where this line crosses the $X$-axis. The maximum cross-country speed for a given rate of climb is found by drawing the line as a tangent to the polar. The tangent meets the polar at the optimum inter-thermal speed ($V_{opt}$), and crosses the $X$-axis at the maximum cross-country speed. In this example, the unloaded vulture's maximum cross-country speed of 15.2 m s$^{-1}$ increases to 16.2 m s$^{-1}$ when an extra 2 kg is added as crop load, for the same rate of climb (3 m s$^{-1}$) in thermals. The extra load increases the wing loading and hence the circling radius, and also increases the sink slightly. Both effects reduce the rate of climb to a degree that may make soaring difficult in narrow, weak thermals. However, griffon vultures normally make their unloaded outward journey early in the day, when thermals may be weak, but return with their food load in the afternoon, when thermals are usually wider and stronger, and the reduced climbing performance is not noticeable. Likewise, all high-performance soaring gliders are equipped with ballast tanks, into which

BOX 10.2 *Continued.*

pilots load large amounts of water ballast before take-off, on days when soaring conditions are expected to be good. This greatly increases cross-country speeds in strong thermals, and if conditions deteriorate, the water can be jettisoned to maximise climbing performance.

*Flight* increments the speed in steps of 0.1 m s$^{-1}$ as it calculates the glide polar, and at each step, it finds the rate of climb for which the current speed would be $V_{opt}$. It stops calculating the polar after it passes $V_{opt}$ for a climb rate of 3 m s$^{-1}$. The output includes a table of optimum inter-thermal speeds and cross-country speeds, calculated from the MacCready theory for rates of climb up to 3 m s$^{-1}$. Two values of $V_{xc}$ are given for each value of the rate of climb, the first calculated on the assumption that the bird flies at the optimum speed between thermals, while the second is a lower cross-country speed for the alternative assumption that the bird flies at the best-glide speed ($V_{bg}$) between thermals. This is a more cautious strategy that covers the greatest distance between thermals, so reducing the chance of being forced to land or resort to flapping flight. Looked at another way, flying at a speed below the theoretical optimum increases the bird's chance of finding strong thermals, so increasing the cross-country speed. Glider pilots normally fly at speeds between these extremes, depending on their assessment of conditions ahead.

*vortex-ring* thermals higher up. This type of thermal is a toroid with a vertical axis, in which air flows up the centre and down the outside. A vortex-ring thermal is a self-contained structure that is not necessarily in contact with the ground, and does not depend on continuous input of heat from below to sustain it. Provided the temperature lapse rate in the surrounding air is steeper than the rate at which a rising parcel of air cools as it expands, a vortex-ring thermal can rise through its surroundings, growing in size as it does so by entrainment of environmental air. Depending on the humidity of the air mass, and the air temperature at the surface, thermals may be marked by cumulus clouds, which form when the rising air in the thermal cools to the dew point. Thermals often remain usable for gliders inside growing cumulus clouds, but flight in cloud depends on gyroscopic instruments. It is not known whether birds have any sense that would provide the same information, but so far as is known, they do not undertake intentional, sustained cloud climbs (Chapter 11).

## 10.2.1.3 Thermal profiles

Most glider pilots would say that no two thermals are alike, but on the other hand there are certainly conditions when thermals are strong but "too narrow to climb in", and other conditions when thermal cores seem to be a kilometre or two across. The core of a thermal may be imagined to have a profile, in which the upward air velocity is strongest in the middle, and tapers off to zero at some radius from the centre. Vortex-ring thermals typically have a zone of sink around the core, caused by the downward-flowing outer parts of the ring, while columnar thermals tend to have a strong but narrow core, embedded in a larger area of weak and turbulent lift. If we assume that the bird has centred its circle accurately around the thermal core, then its *circling envelope* (Box 10.2) can be plotted on the same axes, together with the thermal profile. At any particular radius, the bird's rate of climb is the difference between the upward air velocity in the thermal and the bird's sink. The upper part of Figure 10.4 shows an imaginary thermal profile with upward air velocity (increasing upwards on the graph) as a function of radius from the centre of the core. The lower part of the figure shows calculated circling envelopes for two marine soaring birds, a Magnificent Frigatebird and a Black-browed Albatross, showing the minimum sink of each (increasing downwards on the graph) as a function of circling radius. The frigatebird would be able to climb in this particular thermal at a radius of about 9 m (vertical dashed line), but the albatross would not be able to climb at all at such a narrow radius. These two species have quite similar wing measurements, but the albatross is twice as heavy as the frigatebird. Frigatebirds roam over the trade-wind zones of the oceans, where the sky is typically filled with small, evenly-spaced cumulus clouds, each with a narrow thermal underneath it, whereas albatrosses do not normally soar in thermals, either over land or sea (Pennycuick 1983). They are specialised for an entirely different method of soaring in other parts of the same oceans (below).

## 10.2.1.4 Cross-country flying in thermals

Many birds, especially predators and scavengers, use thermals primarily as a means of staying aloft for long periods with minimum expenditure of effort. For this the bird needs a circling envelope that is good enough to maintain or gain height, in thermals of a size and strength that it typically encounters. Thermals can also be used to travel when migrating or foraging, and this imposes additional requirements, which to some extent conflict with the requirements for simply staying airborne. The basic method of cross-country flying in thermals consists of alternating two phases, a climb in a thermal in which the bird gains height but covers no distance, followed by a straight glide, in which the bird loses the height it has gained, converting it into distance travelled over the ground. A high average of rate of climb in thermals has a strong effect on the average cross-country speed, by minimising the

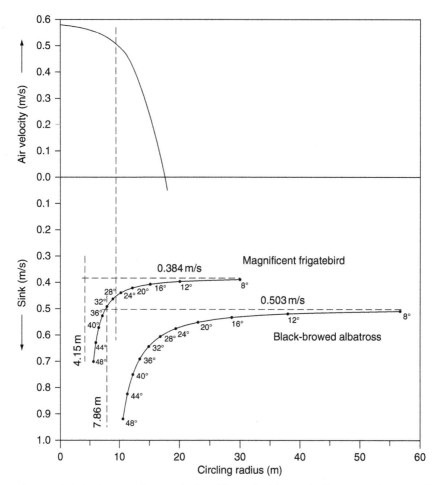

FIGURE 10.4 Circling envelopes calculated for a Black-browed Albatross and a Magnificent Frigatebird. Each bird can circle at radii and rates of sink below and to the right of its circling envelope, but not above or to the left. The horizontal asymptote is the sink in straight flight (0° bank), and the vertical asymptote is the limiting radius to which the envelope tends, as the bank angle tends to 90°. Bank angles from 8° to 48° are marked as solid points along each curve. Above is a hypothetical profile for a weak, narrow trade-wind thermal. The frigatebird, whose wing loading is 4.5 kg m$^{-2}$, can (just) maintain height in this thermal, but the albatross, with nearly twice the wing loading (8.7 kg m$^{-2}$) cannot.

time spent in climbing, when no forward progress is made. The speed at which the bird elects to fly on the straight inter-thermal glides also affects the cross-country speed, and there is an "optimum" inter-thermal speed, which maximises the cross-country speed for any particular rate of climb (Box 10.2). The Summary Screen for *Flight*'s glide polar calculation includes a "MacCready table", showing optimal

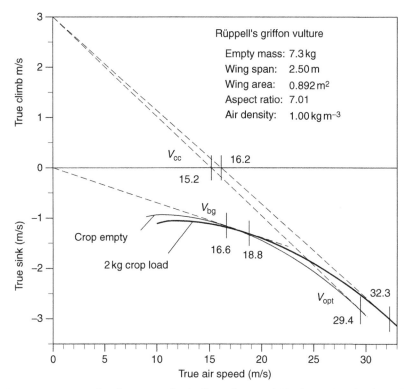

FIGURE 10.5 Two polars for a Rüppell's Griffon Vulture, with (thick curve) and without (thin curve) a 2-kg crop load. These curves are calculated for an air density of 1.00 kg m$^{-3}$, and refer to true speeds (both horizontal and vertical) at an altitude of 2060 m ASL. These polars would need to be reduced to sea level (Chapter 15, Box 15.3) before they could be compared directly to the albatross polar of Figure 10.1. The extended Y-axis represents rate of climb in thermals. A line from a point on this axis, represented achieved or antici-pated rate of climb, to some point on the polar (the inter-thermal speed) crosses the X-axis at $V_{cc}$, the average cross-country speed over one or more cycles of climb and glide. $V_{opt}$, the optimum inter-thermal speed for a given rate of climb is found by drawing a tan-gent to the curve. The reduced rate of climb due to ballast is only noticeable in weak conditions. Normally, a higher wing loading results in a faster cross-country speed.

inter-thermal speeds and cross-country speeds for rates of climb in thermals up to 3 m s$^{-1}$. Two cross-country speeds are shown in the table, one based on flying between thermals at the optimum speed ($V_{opt}$), and the other for flying at the best-glide speed ($V_{bg}$), regardless of the rate of climb. These represent extreme options in a spectrum of choices available to the bird, in which haste is balanced against caution. By flying at $V_{bg}$, the bird flies the greatest distance for a given loss of height, and scans the greatest amount of air in its search for a strong thermal. At any speed faster than $V_{bg}$, the glide angle is steeper, and the chance of finding a strong thermal for the next climb (or indeed a usable thermal) is reduced. Because thermals vary in strength, the steeper, shorter glides that result from flying at $V_{opt}$ are liable to result in the average rate of climb being less that it would have been if $V_{bg}$ had been used, and this may offset the theoretical speed advantage of flying at the higher inter-thermal speed (Pennycuick 1998b). This probabilistic element is not included in the MacCready theory, but it is well understood by glider pilots, who only fly at $V_{opt}$ between thermals if conditions are strong, and thermals ahead are well marked by cumulus clouds, with plenty of clear air between their bases and the ground. If conditions ahead look weak, cautious pilots reduce speed between thermals, so as to stretch their glides as far as possible, and maximise the chance of finding a strong thermal for the next climb.

### 10.2.1.5 Cross-country speed and wing loading

The difference in the two circling envelopes shown in Figure 10.4 is due mostly to the wing loading of the albatross being higher than that of the frigatebird by a factor of 2 or so. In the hypothetical thermal shown, the frigatebird would be able to maintain height or marginally climb, whereas the albatross would not. However, a thermal with a much wider core, as commonly found in inland tropical and temperate regions, would accommodate both birds' circles, with only a minor difference in rate of climb, due to the albatross' minimum sink being about 0.1 m s$^{-1}$ more than that of the frigatebird. If both birds were achieving a rate of climb of, say, 3 m s$^{-1}$, the albatross would go considerably faster across country. This is because a higher wing loading actually reduces the sink at higher gliding speeds. This is the reason why glider pilots who are preparing to compete in a race load up their gliders with water ballast if strong thermals are anticipated. The increased wing loading results in higher cross-country speeds if the conditions come up to expectations, while if not, the water can be dumped to restore a lower wing loading. Likewise griffon vultures, when foraging to feed their young, typically set off for the foraging area with an empty crop in the first, weak thermals of the day, and come back later in the day when the thermals are stronger, ballasted with a load of meat whose extra weight increases their cross-country speed (Figure 10.5).

### 10.2.1.6 Aircraft as platforms for observing birds

Light aircraft and gliders can be used in a variety of ways to observe birds in the air. This is a natural way to observe soaring flight, because the behaviour of soaring birds and glider pilots is essentially the same,

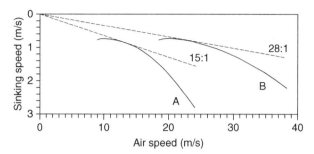

FIGURE 10.6 (A) Estimated glide polar for African White-backed Vulture. (B) Measured glide polar for Schleicher ASK-14 motor-glider used by the writer to observe African soaring birds. The glider is forced to fly faster than the vulture because of its higher wing loading, but on the other hand the glider's best glide ratio is much better than the bird's. After Pennycuick (1972).

at least over land. The basic difficulty is illustrated by the two glide polars in Figure 10.6, for an African White-backed Vulture and the Schleicher ASK-14 motor-glider which the author used to observe this species and other African soaring birds (Pennycuick 1972). The vulture's speeds are lower because its wing loading is lower, owing mainly to a simple scale effect, whereby the weight of geometrically similar machines or birds varies with the cube of the length, whereas the wing area varies with the square. The wing loading therefore varies directly with the length or with the one-third power of the mass. This speed mismatch also applies to fixed-wing powered aircraft but not, of course, to helicopters or birds in flapping flight, in which the wings can be moved relative to the body. The second major difference between the polars in Figure 10.6 is that the glider is capable of much better glide ratios than the vulture, which is mostly due to its higher aspect ratio (longer, narrower wings).

## 10.2.1.7 Observing circling flight

Figure 10.7 represents the same vulture and motor-glider circling together in a thermal. If both fly at the same lift coefficient and the same angle of bank, then the glider flies in a bigger circle than the bird, because of its higher wing loading (Chapter 10, Box 10.2). The trend can be resisted to some degree by allometry, but it would be impractical to build a glider with such huge wings that its wing loading and circling radius would match those of a vulture or stork. Figure 10.8 shows that the circling radius does indeed vary as expected with wing loading, even in very dissimilar birds. No glider can circle at a radius of 12 m with the frigatebirds, or even at 18 m with the pelicans. Consequently,

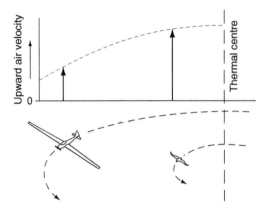

FIGURE 10.7 Geometrically similar gliders (and birds) would have wing loadings that vary with the one-third power of the mass, and the circling radius varies directly with the wing loading. Birds have the advantage over gliders in narrow thermals, because they can circle in the strongest part of the core, while gliders make wider circles in the periphery, where the lift is weaker. After Pennycuick (1972).

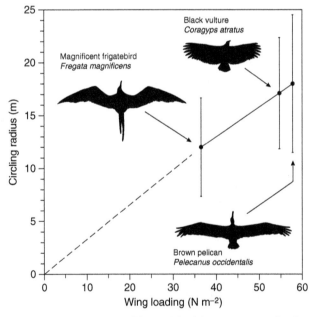

FIGURE 10.8 Circling radius observed by ornithodolite in Panama for three species of soaring birds, and plotted against wing loading. The silhouettes were traced from photographs, not directly from wings. After Pennycuick (1983).

birds outclimb gliders in narrow thermals because of their superior ability to circle in small cores, whereas gliders leave birds behind and below in straight glides, because of their higher speeds and better glide

FIGURE 10.9 A flock of white storks about 2000 m above the Serengeti Plains, photographed from a motor-glider. Photo by C.J. Pennycuick.

ratios. However, it is possible to stay with soaring birds for periods of hours, especially with species that soar in flocks, and to observe their tactics in exploiting thermals in some detail (Figure 10.9).

The wing-loading mismatch means that a glider or light aircraft that is trying to stay with a bird in straight flight is obliged to fly elongated loops around the bird, or weave from side to side behind it. Figure 10.10 is a record of a flight in a Piper PA-12 light aircraft, which belonged to the author at the time, in which a flock of cranes was followed for 125 km, as they migrated northwards over southern Sweden in April 1978 (Pennycuick et al. 1980: see also Chapter 11, Figure 11.10). Rates of climb in thermals were measured by circling around the outside of the thermal, using the engine to maintain the same rate of climb as the cranes, and the straight glides were observed by weaving behind the flock. The graph looks somewhat like a glider's barograph trace, and was used to check the speeds against the theory of cross-country soaring flight (Box 10.2), besides directly measuring the cranes' achieved cross-country speed, and observing their soaring tactics, which included intermittent flapping during the inter-thermal glides. When thermals were strong and cloudbase was high, they proceeded by pure thermal soaring like storks, but as conditions deteriorated, they flapped progressively more during the glides, and could keep going even in unsoarable conditions, by flapping steadily along like swans.

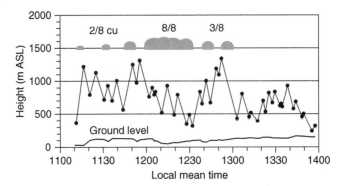

FIGURE 10.10 A record of height and time at the bottom and top of each thermal, in a flock of cranes migrating northwards in thermals over southern Sweden, observed from a powered aircraft. After Pennycuick et al. (1980). The cranes soared in thermals, with very little flapping, until about 1300, after which the thermals became progressively weaker and shallower. The cranes responded by flapping intermittently on the inter-thermal glides, until eventually they were flapping steadily along, not far above the tree tops.

## 10.2.2 SOARING BY REPLENISHMENT OF KINETIC ENERGY

The dominant soaring birds of the oceans belong to the Order Procellariiformes, comprising albatrosses and their smaller relatives the petrels, shearwaters and storm-petrels. Although they venture into some parts of the trade-wind zones of the tropics, these "tube-nosed" birds do not soar in marine thermals like frigatebirds. Their stronghold is in the windy, middle latitudes of the oceans, and they get their energy from horizontal rather than vertical movements of the atmosphere, spending their entire time when at sea within 30 m of the water surface. They soar along cliffs and slopes when they come ashore to breed, but do not climb to hundreds or thousands of metres simply to gain potential energy, as thermal soarers do. Their flight at sea is characterised, especially in the larger species, by frequent, transient "pullup" manoeuvres, in which kinetic energy is converted into potential energy, "converting speed into height" as pilots put it. Their flying heights are in the same range as their typical kinetic energy heights, namely, 10 m or less for storm-petrels up to perhaps 25 m for the larger albatrosses.

Albatrosses exploit *wind shear*, meaning differences in horizontal velocity between different layers of air. One can imagine a layer of air sliding smoothly over another layer below it, with the wind speed, measured relative to the bottom of the shear layer, increasing from bottom to top of the layer. The wind *gradient* is the rate of change of wind speed with height, measured in metres per second, divided by metres of height. The SI unit of wind gradient strength is just "per second" ($s^{-1}$). A bird is said to climb "against the wind gradient" if the air in higher layers is coming towards it as it climbs, so tending to increase its airspeed. If the wind shear is caused by retardation of the wind by friction with the surface, then the same direction is also against the wind.

### 10.2.2.1 Flight of theoretical albatrosses

The copious theoretical literature about albatross flight, reviewed by Tickell (2000), is based on the classical idea that albatrosses exploit the wind gradient in the shear layer that extends from the sea surface to a height of 20 or 30 m, caused by frictional retardation of the wind at the surface (Lord Rayleigh 1883). In a stiff breeze, the wind speed (relative to the surface) might be 15 m $s^{-1}$ at 30 m, and zero at the surface, giving an *average* wind gradient of 0.5 $s^{-1}$ over this height band. Theoretically, the gradient is not linear, but steepest near the surface, tapering off with height. This picture is quite realistic at heights of tens of metres, but gets progressively less realistic close to the surface, where the gradient is strong. This is because "height" is difficult to define above a reference level that moves, and is not flat. A level can be assigned by the practical observer to the "mean sea surface", but the actual surface at any particular point may be several metres higher or lower. The wind gradient presumably adapts itself in some way to the ups and downs of the rough ocean surface, but it certainly is not accurately described by a logarithmic increase of speed with height above a definable, flat surface, as implied by classical theories of "dynamic soaring".

If this difficulty is ignored, the albatross may be imagined climbing against the wind gradient, so that its airspeed tends to increase as it climbs, perhaps enough to offset the loss of speed due to aerodynamic drag, and to going "uphill". Then, if the albatross turns and glides down with the wind gradient, it encounters a decreasing tailwind during the descent, which also tends to increase its airspeed. If some rather insecure assumptions are made about the gliding performance of albatrosses, it can be shown by solving elaborate differential equations that it may be marginally possible for an albatross to replace its energy losses by doing endlessly repeated cyclic manoeuvres, in which it pulls up against the wind gradient, and descends with the gradient. This cycle can be an elliptical or a zigzag pattern, but since it takes place relative to the air, it always involves a net downwind displacement each time round the cycle, relative to the sea surface. It does not provide for making progress against the wind, relative to the water.

### 10.2.2.2 Flight of real albatrosses

The very first day that I saw albatrosses at sea was 26 November 1979, from the deck of the British Antarctic Survey's supply ship Bransfield, which was steaming south from Rio to South Georgia at the time in latitude 32° south, making 6 m s$^{-1}$ through the water, directly against a headwind of 8 m s$^{-1}$. Wandering Albatrosses were appearing in the distance far astern, catching up with the ship and overtaking it. They were staying close to the surface most of the time, pulling up a few metres from time to time, but going back down, and continuing their prolonged into-wind glide. They were not doing regular cyclic manoeuvres. They were not flapping their wings, and it was obvious that they could go in any direction at will, and that their performance was anything but marginal. Later observations on the same voyage showed albatrosses pulling up from near the surface to 15 m, when the wind gradient might conceivably have been strong enough to keep them going without losing airspeed up to 3 m (Pennycuick 1982). None of this behaviour was consistent with any variant of the classical theory.

### 10.2.2.3 The ocean boundary layer

The ocean surface is never flat, and the wind speed above it does not decrease in a smooth gradient. The friction at the interface between air and water drags the water surface forwards, as well as holding the bottom layer of air back. The uppermost layer of water is rolled into eddies, which become unstable, and lead to the formation of waves. In the open ocean, the waves start to form crests at wind strengths over about 5 m s$^{-1}$. The wind speed is zero at the water surface, and increases to the full wind speed through the *boundary layer*, which is a shear layer a few centimetres thick, whose top cannot be exactly defined. The boundary layer remains attached to the water surface where the wind blows up the windward slope of each wave. Where the water surface abruptly drops at the wave crest, the boundary layer separates from the surface, but continues as an identifiable, thin shear layer, arcing over the hollow on the lee side of the wave, and reattaching to the surface somewhere on the windward slope of the next wave. Underneath the separated boundary layer is a *separation bubble*, in which the wind is light. In the long swells of the Southern Ocean, a yacht's sails flap as it descends into the shelter of each separation bubble, and fill again as the next wave lifts the boat up through the boundary layer. In such conditions, *every* wave has a separation bubble on its lee side, big enough for an albatross to fly around in.

Albatrosses do not perform regular cyclic manoeuvres in the main shear zone above the boundary layer, as required by the classical theory. Instead, they gain their energy in pulses, by flying out of a separation bubble, upwards through the separated boundary layer (Pennycuick 2002). As the albatross passes through the thin layer of strong wind gradient at the separated boundary layer, it perceives a sudden change of wind strength, or *gust*. By doing the manoeuvre in a special way (below), it can exploit this to increase its airspeed, without any expenditure of muscular effort. The bird's kinetic energy depends on its airspeed, not on its speed relative to the water surface. It acquires some kinetic energy by pulling up through the boundary layer, which it can then expend against aerodynamic drag over the next few tens of seconds of flight (Box 10.3).

BOX 10.3 **Kinetic energy gained from the "Roll off the crest".**

**Variable definitions for this box**

| | |
|---|---|
| $M$ | Bird's all-up mass |
| $V_1, V_2$ | Speeds on the glider polar |
| $\Delta E_k$ | Increment of kinetic energy |

**Frame of reference for kinetic energy**

Everybody knows that a bird's kinetic energy is $\frac{1}{2}mV^2$, where $m$ is the mass and $V$ is the speed, but this statement is actually ambiguous, because it gives different answers, depending on the frame of reference to which the speed is referred. Kinetic energy in flight is interchangeable with potential energy, in that the bird can convert excess kinetic energy into potential energy by pulling up. The variometer, which is the soaring pilot's primary source of information, measures rate of climb, but only very primitive variometers do that literally. Standard instruments correct the actual rate of climb for changes of speed. They measure, in effect, the rate of change of the sum of potential and kinetic energy. A positive indication on such a "total-energy variometer" means that the glider is really climbing, and not momentarily gaining height at the expense of losing speed. The speed that is used to make this correction is the glider's airspeed, measured via the *dynamic pressure* in an open-ended "pitot tube" pointing into wind, which looks exactly like the tubular nostrils of procellariiform birds, and doubtless serves the same purpose.

It is difficult to think about a frame of reference for airspeed, as the air does not keep still, but the albatross exploits that very difficulty, by moving between layers of air that are themselves moving relative to each other. If an albatross inclines its flight path upwards and passes from a lower layer of air into an upper layer that is moving (relative to the lower layer) at a speed of $1\ \mathrm{m\ s^{-1}}$ (say) towards the albatross, then the albatross observes a "gust" whereby its airspeed increases by $1\ \mathrm{m\ s^{-1}}$. Its kinetic energy increases accordingly, even though the albatross observes no acceleration. This is basically what albatrosses do, by pulling up out of the sheltered spot in

BOX 10.3 *Continued.*

the lee of a wave, through the separated boundary layer, into the unob-structed wind above (Figure 10.11). However, the albatross does not face into wind when doing this. Instead, it rolls to a very steep angle of bank (typically 60°–70°) belly-to-wind, that is, presenting its ventral side to the direction from which the gust appears to come.

**Airspeed gain from a gust**
The effect of receiving a gust from the ventral side can be understood by considering the case of a glider flying with its wings level into a thermal. As the glider enters the core, there is a sudden increase in the upward air velocity, which is represented in aeronautical theory as a "sharp-edged gust" coming from below (Figure 10.12). In terms of the vertical component of the glider's airspeed, this is aerodynamically the same as a sudden increase in the glider's sink, but the effect perceived by the pilot is not just a transient thump. The glider tries to speed up, and pilots often comment that the surge of speed seems to continue for longer than they expect. The reason can be seen by inspecting the glide polar (Figure 10.13). The effect of the sudden increment in the glider's sink is to transfer the equilibrium position on the polar from its previous speed ($V_1$) to a higher speed ($V_2$), and if the sink is sustained, the glider will speed up until it reaches that speed. Depending on the "flatness" of the polar, the speed increment from $V_1$ to $V_2$ is much greater than the increment of sink. If the glider is allowed to speed up to $V_2$, the kinetic energy gain ($\Delta E_k$) is

$$\Delta E_k = \frac{1}{2}m(V_2^2 - V_1^2) \tag{1}$$

Normally, a thermal-soaring pilot is looking for potential rather than kinetic energy, and holds the speed down in readiness for circling in the thermal. In that case the same amount of energy is gained, but it is con-verted from kinetic to potential energy. Pilots pay little attention to this, because the amount of energy gained is trivial in comparison to the expected gain of potential energy from circling for a few minutes in the thermal. However, albatrosses keep flying by using "surge" as their sole energy source.

The energy gained from the surge does not depend on the gust being ver-tical, unless the wings are horizontal. It depends on the gust being directed against the ventral side of the wings. If the wings are vertically banked, then a horizontal sharp-edged gust produces exactly the same gain of kinetic energy, and that is what the albatross gets when it rolls to a steep angle as it passes through the separated ocean boundary layer.

## 10.2.2.4 The roll-off the crest

A conspicuous characteristic of the flight of albatrosses is their use of very steep angles of bank, close to the water surface. They *always* roll to a steep angle, typically 60°–70°, belly-to-wind, just as they pull up from the crest of a wave (Figure 10.11). The larger petrel species,

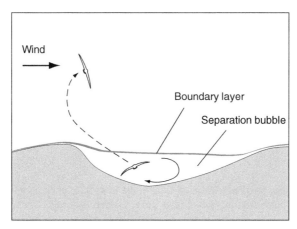

FIGURE 10.11 The "roll-off-the-crest" manoeuvre on which albatross soaring is based. The boundary layer (shown pink) is the thin layer of air within which viscous forces are appreciable, as the wind speed increases from zero at the water surface to the speed observed just above the wave crests. The boundary layer separates from the water surface at the crest of each wave, and reattaches on the upslope of the next wave, leaving a "separation bubble" in which the wind is light, along the windward side of each wave trough. The albatross flies upwards out of the separation bubble, through the boundary layer (which persists as a thin layer of strong shear) banking steeply belly-to-wind as it does so. Sufficient kinetic energy is gained from this to glide level for a minute or two in any direction including upwind. Alternatively the albatross can pull up, typically to 15–20 m, converting the kinetic energy gained into potential energy. After Pennycuick (2002).

especially Giant Petrels, do the same thing. In turbulent air, downwind of a ship or an iceberg, Black-browed Albatrosses often roll their wings past the vertical, when pulling up only two or three metres above the surface. Rather than relying on extracting energy from smooth climbs and descents through the main wind gradient, albatrosses perform this apparently obligatory roll just as they come up above a wave crest, passing upwards through the separated boundary layer. Why roll belly-to-wind when receiving the gust? The reason is that the albatross gains a much larger increment of airspeed by receiving the gust from the ventral side, perhaps by a factor of 20, than it would if it simply climbed through the boundary layer heading directly into wind. The gust is used in effect to add an increment of sink, and this leads to a much larger increment of airspeed, according to the bird's glide polar (see Box 10.3 for an explanation of how this works). It is actually the same effect (although in a different orientation) that is known to glider pilots as "surge", when they fly with wings level, directly into the core of a thermal (Figures 10.12 and 10.13).

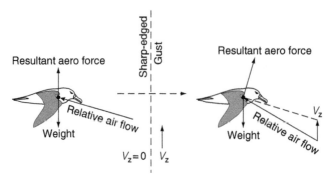

FIGURE 10.12 An albatross that flies with wings level through a sharp-edged vertical gust gets an instantaneous increment of its sinking speed ($V_z$), with no change in its horizontal speed. This causes the resultant aerodynamic force to increase, and lean forwards, causing the bird to accelerate upwards and forwards. If the bird suppresses the upward component by control inputs, the horizontal speed increases until equilibrium is restored at a new value determined by the glide polar (Figure 10.13). The speed increase depends on the gust coming from the ventral side, not necessarily from below. To extract energy from the predictable horizontal gust as it pulls up through a separated boundary layer, the bird rolls to present the ventral side of its wings to the wind. After Pennycuick (2002).

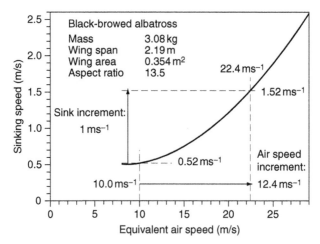

FIGURE 10.13 An albatross that pulls up through a thin shear layer with a speed difference of 1 m s$^{-1}$ gets an airspeed increase from 10 to 11 m s$^{-1}$ if it flies directly into wind, but if it rolls belly-to-wind, the 1 m s$^{-1}$ increment is applied to its sink, not to its forward speed. From the polar this sink increment results (eventually) in an air speed increase from 10 to 22.4 m s$^{-1}$. About 19 times as much kinetic energy is gained by taking the gust against the ventral side as by taking it head on. After Pennycuick (2002).

## 10.2.2.5 Covering distance

The albatross can use the kinetic energy gained from a roll-off-the-crest manoeuvre in various ways. If its objective is to make progress against the wind, it rolls back immediately after rolling off the crest of a wave, gets down as close as possible to the surface where the wind is weakest, and glides along against the wind, losing airspeed as its kinetic energy is dissipated against drag. The energy pulse from one pullup is sufficient to last a Wandering Albatross for several tens of seconds of flight in typical conditions. This straight-glide phase is terminated by another brief pullup from the shelter of another wave, and another roll off the crest, followed by another into-wind glide, and so on. If the intended direction of travel is downwind, the albatross uses the energy gained in the roll off the crest to climb to nearly its full kinetic energy height, which may be 15 or 20 m, then turns downwind and glides back to the surface, getting the benefit of the stronger tailwind higher up, and also flattening the glide a little by descending downwind through the wind gradient. Crosswind travel is achieved by a succession of pullups to an intermediate height, presenting regular flashes of white to an observer far to windward. This is the only situation in which albatrosses show regular cyclic behaviour, but not for the reason indicated by the classical theory.

Boundary layer transit was never considered by the classical theorists, presumably because it was not thought to be possible to fly under the boundary layer. Rather than gaining energy at a low rate by moving up and down through the main shear zone, an albatross gets a large pulse of kinetic energy, from each upward transit through the boundary layer. The magnitude of the kinetic energy gain is also much greater than would be calculated under traditional assumptions, because the significance of the roll-off-the-crest manoeuvre was overlooked. This manoeuvre is the basis of albatross soaring, not cyclic patterns flown in the main shear zone. The mathematical theories so ably reviewed by Tickell (2000) are irrelevant to albatross flight, because they are concerned with exploiting the wrong part of the wind gradient, by means of manoeuvres that albatrosses do not use.

## 10.2.3 SEA-ANCHOR SOARING IN STORM-PETRELS

Soaring in detached boundary layers is practical in medium-sized and large procellariiform birds, from prions to albatrosses, that normally fly around with kinetic energy heights of 10–25 m, that is, they fly at speeds that can be converted into heights of that order, by pulling

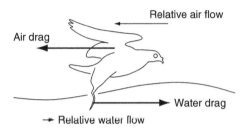

FIGURE 10.14 A feeding storm-petrel glides with its body clear of the surface, sustained by the relative airflow coming from ahead, due to the wind. The aerodynamic drag pushes the bird backwards so that its feet move slowly backwards through the water. The relative water flow comes from behind and produces forward-directed drag on the feet. When this sea-anchor drag balances the drag on the wings, the bird remains suspended, at a height where it can pick up small food items from the surface.

up. Storm-petrels (Hydrobatidae) are the smallest members of the order, and fly more slowly than the larger species, because of general scaling relationships (Chapter 13). Their kinetic energy heights at normal cruising speeds only go up to about 6 m, which is not enough to rely on the roll-off-the-crest manoeuvre as the main energy source. Storm-petrels use flapping flight for cruising, far more than the larger species, but they also use their own unique soaring method for picking up food from the surface (Figure 10.14). The bird glides head to wind with its body clear of the surface, but with its webbed feet in the water, acting as a sea-anchor. The aerodynamic drag of the wings pushes the bird backwards, so that its feet move slowly backwards through the water, creating forward-directed hydrodynamic drag. When this builds up enough to balance the aerodynamic drag of the wings, the bird hangs suspended, with its beak at a convenient height above the surface to pick up small food items.

# 11

# INFORMATION SYSTEMS FOR FLYING ANIMALS

Vertebrate senses are not necessarily reliable when contact with the earth's surface is lost, especially for monitoring orientation. Gyro instruments overcome this problem for pilots, but nothing equivalent is known in birds. Some sense organs probably provide information specific to flight, notably the middle ear as a variometer sense, and the nostrils of Procellariiform birds as an airspeed sense. Visual observation of the sun may be used for position finding, but no sense is known that could provide a route weather forecast for a long migratory flight.

Although birds have basically the same set of sense organs as other terrestrial vertebrates (eye, ear, nose etc.), the transition to flight changes the meaning of some of the information that comes in from those organs. Flight differs from other kinds of locomotion in that the weight is supported by an aerodynamic lift force, rather than by reaction from the ground, or by hydrostatic forces in water. In this situation, the inner ear no longer reliably indicates which direction is "up". Animals that evolve the power of flight have to adapt to this, as do trainee pilots.

Modelling the Flying Bird

In flight, the eye takes over from the inner ear as the primary organ for spatial orientation. Pilots also have to learn to use their eyes for an "instrument scan", which means monitoring several streams of information about measurements for which human senses are not suitable, by reading cockpit instruments that present the information in visual form. Some of these flight-specific kinds of information are thought to be available to birds, through adaptations of existing sense organs, but others apparently are not.

Our own suite of sense organs is basically the same as that of birds, and my starting assumption in this chapter is that birds do not have any magic senses. If no physical basis is known, or plausibly postulated, whereby a bird could acquire information of a particular kind, using the senses that we know it has, then the default assumption is that birds do not have access to that information, until experiments prove that they do. For example, it would appear that birds do not have sense organs of any kind that could substitute for gyroscopic instruments. According to pilot experience, that would imply that controlled flight in cloud, with no visual references, would not be a practical proposition for birds. On the other hand, the sense organs of birds are not necessarily restricted to detecting the same information for which we ourselves use the homologous organs. Some familiar sense organs, especially the middle ear and the nasal organ, can easily be adapted for new functions that are only relevant to a flying animal (Figures 11.1 and 11.2).

BOX 11.1 **The labyrinth and accelerometer sense.**

The labyrinth of the inner ear is an ancient sense organ found in all vertebrates living and fossil, even Agnathan fishes. It is a system of chambers and canals, filled with endolymph, which is a watery fluid, essentially the same as blood plasma. It is enclosed in a ventral bony box, the tympanic bulla, at the rear end of the skull, below the brain case and behind the jaw hinge. Figure 11.1 is a diagram of a typical bird inner ear (Pumphrey 1961). Within it are several different sensors, which between them provide acceleration measurement and hearing. All of these functions involve "hair cells", so called because each such cell has a projecting flagellum which responds to mechanical stimulation. When the flagellum is bent in a particular direction, the electrical potential across the cell membrane is reduced (depolarised), and this in turn stimulates a connected nerve cell terminal to generate a stream of electrical spikes (all-or-nothing impulses). The hair cell is a transducer which encodes deflection of its flagellum into the time interval between successive impulses in the output nerve fibre. The animal's brain sees streams of impulses coming in at different frequencies along different sensory nerves, and interprets these in terms of the magnitude of the signal to which each sensor responds.

BOX 11.1 *Continued.*

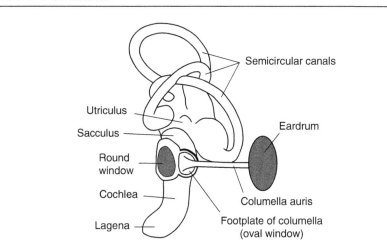

FIGURE 11.1 Diagram of a Bullfinch's labyrinth based on Pumphrey (1961). See Box11.1.

### Maculae—linear accelerometers

The three maculae (utriculus, sacculus and lagena) are inter-connected chambers within the labyrinth. Each contains an otolith made of a crystal-line substance much denser than the endolymph, which is attached to the wall of the chamber by a gelatinous layer containing the flagella of an array of hair cells. In the orientation shown in Figure 11.2A, upward acceleration of the whole organ causes the heavy otolith to be left behind, so bending the flagella downwards. The three maculae are oriented approximately at right angles to each other, and between them register linear acceleration in three dimensions. When the animal stops accelerating and moves at a steady speed, the otolith returns to the neutral position, that is, it does not register velocity, only acceleration. In a gravity field, an organ of this type indicates the direction of gravity *if the animal is restrained from falling,* by being supported by a solid or liquid surface. In flight, additional information is needed (from the eyes) to discriminate between the gravity field and the effects of acceleration relative to the earth's surface.

### Semicircular canals—angular accelerometers

The three semicircular canals are also oriented at right angles to each other, and each consists of an endolymph-filled loop of tube, which widens into a bulb at one point Figure 11.2B. The bulb contains a "cupula", which is a gelatinous projection that blocks the channel. The flagella of a group of hair cells are embedded in the cupula. If the organ is subjected to an angular acceleration in the direction shown, the endolymph inside it is left behind, and bends the cupula, resulting in a signal in the sensory nerve fibres. When the acceleration stops, the cupula returns to the neutral position. The organ does not register angular velocity (still less angular position), and cannot be used for orientation relative to the earth's surface.

BOX 11.1 *Continued.*

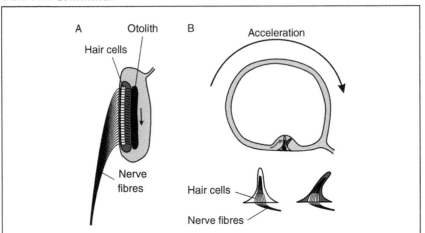

FIGURE 11.2 (A) A macula (otolith organ) is a chamber filled with endolymph (grey), containing a dense, crystalline otolith (black) supported on a gelatinous layer (red), into which the flagella of an array of hair cells penetrate. The otolith bends the flagella in the direction of the resultant of gravity and any linear acceleration that the organ may be subjected to. Three such organs, oriented mutually at right angles, can determine the magnitude of any linear acceleration, and its three-dimensional direction, but cannot distinguish the effects of acceleration from those of gravity. (B) A semicircular canal is a loop of tube filled with endolymph (grey), with a bulb in which the channel is blocked by a cupula (enlarged below) which is a flap of elastic material enclosing the flagella of a group of hair cells. Angular acceleration in the direction shown leaves the endolymph behind so that it deflects the cupula and stimulates the nerve fibres. The organ registers starting and stopping accelerations, for example when the bird rolls to initiate a turn, but does not register steady rotation or angular position. The three semicircular canals together register the magnitude and three-dimensional direction of any angular acceleration.

## The cochlea and hearing

The cochlea detects airborne sound, although it is filled with the same endolymph as the rest of the labyrinth. Sound is transmitted from the eardrum through a bony rod (the columella auris) and passes into the endolymph through a flexible membrane (the oval window) at one end of the cochlea. The area of the oval window is less than that of the eardrum by a factor of about 20–25, which increases the pressure of the sound waves by the same factor, allowing them to penetrate into the endolymph. The sound waves pass through the basilar membrane, which divides the cochlea longitudinally, and the pressure is relieved at a second flexible membrane, the round window. The basilar membrane contains a complex array of hair cells, which act as a sound detector and frequency analyser. Hearing as such does not have any special significance for flying animals, but the middle ear, which is the air-filled cavity around the columella auris is believed to be the basis of a variometer sense (Box 11.2).

# 11.1 ☞ SENSES

## 11.1.1 THE ACCELERATION SENSE AND SPATIAL ORIENTATION

The labyrinth of the inner ear is a six-channel accelerometer, and is basically the same in all vertebrates (Box 11.1). The three maculae are linear accelerometers, aligned in different directions relative to the skull, roughly at right angles to one another, while the three semicircular canals are angular accelerometers, again arranged so that they measure angular accelerations in three different planes, relative to the head. Spatial orientation is a matter of linear and angular position, not acceleration—so how does an accelerometer array come to be regarded as an organ for monitoring spatial orientation?

A linear accelerometer that is stationary on the earth's surface indicates a steady acceleration of $1g$ upwards. If you turn it upside down it indicates minus $1g$. If you drop it, it indicates zero while it is falling freely, followed by a large positive reading for a short time, as its downward (negative) velocity returns to zero on hitting the ground. If it is not broken, the reading then returns to $+1g$ and stays there. This is how the maculae of the inner ear come to serve as indicators of the direction we know as "up". A linear accelerometer indicates a steady acceleration, *if it is restrained from accelerating* in a gravitational field. If it is allowed to fall freely, the acceleration reading drops to zero, although we perceive a freely falling object as accelerating downwards, relative to the earth's surface. This is at the heart of General Relativity. It is not possible to determine from an accelerometer reading whether the measured acceleration is "real", or due to a gravitational field, or a combination of both, because that depends on what the "real" acceleration is measured relative to.

In our earthbound lives, we perceive "real" acceleration relative to the earth's surface, and if our inner ears register a steady $1g$, we conclude that we are stationary, and that we know which way is "up". A pilot in flight normally uses his eyes to determine where the earth's surface is, by locating the horizon. When the aircraft enters cloud, his acceleration sense continues to work exactly as before, but it quickly becomes apparent that it does not reliably indicate the direction of "up" as it does on the ground. Many an inexperienced glider pilot has tumbled out of the bottom of an innocuous cumulus cloud to find the familiar hills and fields of home tilted at an absurd angle, or even inverted above his head. Thus we learn to ignore our so-called "sense of balance" when flying. When the ground is hidden by cloud, other sensors are required to determine which way is "up", and these are unfortunately not included in the basic sensory outfit that we (or birds) inherited from our reptilian ancestors. Gyroscopic instruments solve the problem for aviators, providing (in the simplest case) a measurement of angular velocity, which a pilot can use for orientation in cloud. Actually there is a biological sense organ that detects angular velocity, based on an oscillating structure rather than a continuously rotating rotor as in a gyro, in the halteres of Dipteran flies (Pringle 1948). This principle has been adapted for control purposes in flying model helicopters, but no analogous sense organ is known in vertebrates. Failing the discovery of such an organ, we have to assume that sustained, controlled flight in cloud is not possible for birds.

In flight, the pilot's maculae register the resultant of gravity and any acceleration in the pitch plane, that is, in the dorso-ventral direction, relative to the aircraft. Side-to-side accelerations are normally of minor importance or interest, because neither birds nor aircraft have any structures than can produce a large sideways aerodynamic force. Fore-and-aft accelerations are only noticeable in aircraft with very powerful engines or very effective airbrakes. Only lift from the wings can generate a lift force that is large enough to produce a substantial acceleration, and this force is directed in the dorsal direction (not necessarily upwards relative to the ground), causing the flight path to curve dorsally. The dorsally directed acceleration is perceived as an increase of gravity, and indicates the magnitude of the lift force on the wings, but *not* its orientation relative to the earth's surface. Turning is achieved by banking the wings, and the semicircular canals register the transient angular accelerations when initiating and stopping a roll. They do not indicate continuous rotation (angular velocity), or the angle to which the wings have been banked, and therefore do not provide sufficient information for manoeuvring in flight (Chapter 9). Standard gyro instruments, which are found in all but the most basic aircraft instrument panels, provide the pilot with information on angular velocity and angular position, relative to the horizon, but so far as is known, neither human nor avian senses are capable of doing this.

## 11.1.2 HEARING AND THE VARIOMETER SENSE

The cochlea is another part of the endolymph-filled inner ear, which acts as a sound detector and frequency analyser in birds, as in mammals. The bird cochlea is straight, and shorter than the coiled cochlea of mammals, although it seems to be just as effective. It is obvious from the elaborate songs of birds that they must have sound analysis capabilities to match. Some owls can use sound to locate prey in total darkness, or through a layer of snow. These species have asymmetrical skulls, which results in the direction of maximum sensitivity to high-frequency sounds being different in the right and left ears, both horizontally and vertically, so enabling the owl to pinpoint the direction from which the sound comes by rotating its head (Norberg 1968).

The middle ear is the air-filled cavity between the eardrum and the inner ear, and the chain of three auditory ossicles within it, whose function is to act as a transformer that enables airborne sound waves impinging on the eardrum to move the endolymph in the cochlea (Box 11.2). However, the middle ear is one of those organs that is

BOX 11.2 **The middle ear as a variometer sense.**

Figure 11.3 is a diagram of the middle ear, which is basically the same in birds as in reptiles and amphibians. Its function is to act as a transformer that matches the low acoustic impedance of air to the much higher acoustic impedance of the endolymph. The energy of sound waves in air results from small fluctuations of pressure, accompanied by rather large movements of the air molecules to and fro, as the pressure fluctuates. If these small pressure fluctuations impinge directly on a water surface, the water hardly moves at all (being much denser than air) and most of the energy is

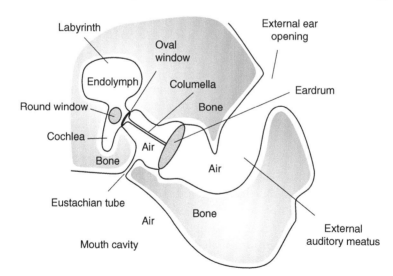

FIGURE 11.3 Diagram of a bird's middle ear, based on Pumphrey (1961). The middle ear cavity is the space between the eardrum and the oval window, which are connected by the columella auris. It contains air, whose pressure equalises with that of the mouth cavity through the Eustachian tube. Distortion of the eardrum measures the rate of change of the ambient air pressure, which can itself be used as a measure of vertical velocity in the atmosphere. The middle ear can readily be adapted to serve as a variometer, but not as an altimeter.

BOX 11.2 *Continued.*

reflected back off the surface into the air. To transmit the sound energy into the endolymph, the force collected over a large area (the eardrum) is concentrated on to a smaller area (the oval window) so increasing the pressure and producing movement in the fluid. The eardrum has to have air at the same mean pressure on both sides of it, but to make it vibrate, the fluctuating sound pressure has to act only on the outside, not on the inside. This is achieved by connecting the inside of the eardrum to the air in the mouth cavity through the Eustachian tube, which is a narrow channel that allows the average air pressure to equalise on both sides of the eardrum, while greatly attenuating the fluctuating sound pressure.

A variometer is an essential instrument for soaring over land. Its function is to register the rate of change of atmospheric pressure, which is readily interpreted by a soaring pilot or bird as an indication of vertical velocity. Glider variometers consist of an insulated or temperature-controlled cavity, connected to a source of static pressure through a thin channel, with a sensor that measures the rate of flow of air through the channel, into or out of the cavity. It is not known whether soaring birds have a sense organ that measures the rate of air flow through the Eustachian tube into or out of the middle ear cavity, but we are all aware of pressure on human eardrums induced by rapid climbs or descents. The human middle ear is a little more elaborate than that of birds, but works in essentially the same way. It gives a crude indication of rate of climb or descent despite not being adapted for that function, and would only require a more sensitive sense organ to detect distortion of the eardrum to modify it into a soaring-grade variometer.

pre-adapted for a function that is only significant for a flying animal, the measurement of rate of change of air pressure. We are all aware of popping in the ears that accompanies rapid climbs and descents, and this is due to changes of air pressure. As we descend a steep hill, the air pressure on the outside of the eardrum increases immediately, but that on the inside has to equalise through the narrow eustachian tube, which takes a little time. While the pressure is changing, there is a pressure difference across the eardrum, which distorts it. The amount of the distortion can readily be measured by mechanoreceptors, and gives an indication of the rate of change of pressure, which can be used in turn as an indication of rate of change of height. An instrument that measures the rate of flow of air through a narrow channel into or out of a reservoir is called a *variometer* in gliders, and it is the primary instrument of the soaring pilot. In powered aircraft, an instrument with the same function is known as a "vertical speed indicator". Only minor modification of the human middle ear would be needed to convert it into a variometer sense, and it is generally assumed that the middle ears of soaring birds provide this sense.

Many birds, including vultures and swifts, are extremely good at locating areas of lift (rising air), and seem to be able to do this without any nearby visual references from the ground or clouds, from which they could judge their rate of climb.

The variometer should not be confused with an *altimeter*. These are different functions, provided by different instruments in aircraft instrument panels. The most common type of altimeter used in aircraft is actually an aneroid barometer, measuring the absolute value of the air pressure, relative to a vacuum inside a sealed aneroid capsule. This pressure is then referred to a datum level, at which the pressure is known. This instrument is considered so essential that it is legally required in most countries, even for the most minimal aircraft. However, birds are not known to have an air pressure sense that could serve as a altimeter. The observation that pigeons can detect *changes* of air pressure is evidence for a variometer sense, not for an altimeter sense, as has been erroneously claimed in the literature.

## 11.1.3 A HYPOTHETICAL AIRSPEED SENSE

All pilots have to learn to be aware of their airspeed at all times, and this is something for which visual clues are not only insufficient, but also dangerously misleading. The visual impression of the landscape speeding past is due to the ground speed, which is the vector sum of the airspeed and the wind speed, whereas the forces and control moments available from the wings depend on the airspeed alone. Airspeed is measured from the *dynamic pressure*, which is the difference between *pitot pressure*, measured in a forward-pointing open tube, and *static pressure*, taken from a hole in the aircraft's skin, at a point where the pressure can be assumed to be the same as the ambient pressure in the surrounding air (Box 11.3). A "pitot tube" pointing forwards can be found on nearly every aircraft, while the static pressure source may be combined with it in a pitot-static assembly as in Figure 11.4B, or derived from a hole or slit somewhere on the surface. The airspeed indicator is a differential pressure gauge connected between the pitot and static pressure sources.

Most birds do not have an obvious pitot tube, and presumably make do with the sound of the air flowing past their ears, and the feel of the relative wind on their faces, as hang glider pilots traditionally do. The Procellariiformes, comprising the albatrosses, petrels and storm petrels, are a conspicuous exception, whose prominent, forward-facing tubular nostrils gave them the alternative name "Tubinares" (Figure 11.5). The large nasal sense organ in these birds, and the size

**BOX 11.3 The nostrils of Procellariiformes as an airspeed sense.**

**Variable definitions for this box**
$P$      Dynamic pressure
$p_{pt}$   Pitot pressure
$p_{st}$   Static pressure
$V_e$    Equivalent airspeed
$V_t$    True airspeed
$\rho$      Air density
$\rho_0$     Air density at mean sea level in the International Standard Atmosphere

**Airspeed measurement**
The airspeed indicator is an essential instrument in all aircraft, although it would be more accurate to describe the usual form of this instrument as a "dynamic pressure indicator". It measures the difference between *pitot pressure*, which is the pressure in a blind tube pointing into the relative wind, and *static pressure*, which is the local ambient air pressure, as it would be measured with a stationary barometer. Figure 11.4B shows a common arrangement in which the pitot pressure ($p_{pt}$) from a central pitot tube is connected to the inside of an aneroid capsule, while the static pressure ($p_{st}$) comes from a ring of slits in a concentric outer tube, and is connected to the instrument casing surrounding the capsule. The capsule expands according to the dynamic pressure ($p$) where

$$p = p_{pt} - p_{st} \qquad (1)$$

The dynamic pressure is itself related to the true airspeed ($V_t$) by the equation

$$p = \tfrac{1}{2}\rho V_t^2, \qquad (2)$$

where $\rho$ is the air density. The capsule drives a pointer as it expands, against a scale which is calibrated to read the airspeed obtained by inverting Equation (2):

$$V_t = \surd(2p/\rho) \qquad (3)$$

To make the instrument read the true airspeed at any height, $\rho$ would have to be set to the local air density, but this is somewhat inconvenient, and it is actually not what pilots want to know. Instead a fixed value of the air density is used so that the instrument can be calibrated once and for all. By convention, this value is $\rho_0$, the air density at mean sea level in the International Standard Atmosphere, whose value is $\rho_0 = 1.226$ kg m$^{-3}$ (Chapter 2). The instrument measures the dynamic pressure, but interprets this to read the airspeed that would correspond to the measured dynamic pressure, if the air density were equal to $\rho_0$. This is called the *equivalent airspeed* ($V_e$) and is given by

$$V_e = \surd(2p/\rho_0) \qquad (4)$$

The equivalent (not the true) airspeed determines the magnitude of the aerodynamic forces available to support and manoeuvre the aircraft.

BOX 11.3 *Continued.*

For example, a fixed-wing aircraft stalls at the same equivalent airspeed at any height, whereas the true airspeed at which it stalls is higher at higher altitudes than at sea level. Likewise, speeds that maximise particular aspects of performance, such as rate of climb, occur at fixed values of the equivalent airspeed, but not of the true airspeed.

It may be noted that there is also another type of airspeed indicator, in which the rate of rotation of a small turbine is used to measure the airspeed. Instruments of this type do not depend on the dynamic pressure, and measure the true (not indicated) airspeed (Chapter 14). Turbine airspeed indicators are used in aircraft such as hang gliders and paragliders, whose flying speeds are so low that the dynamic pressure is difficult to measure. They are not known in any flying animal.

**A pitot-static system in Procellariiform birds**

Albatrosses and their smaller relatives the petrels and shearwaters have a soaring method that depends on detecting discontinuities in the wind speed, as the air flows over the waves at sea (Chapter 10). They use this to boost their airspeed, and hence replenish their kinetic energy at intervals of a few tens of seconds, and it seems not unlikely that a fast and accurate airspeed sense would be needed to make this possible. A distinguishing feature of the group, which gives them their alternative name "Tubinares", is that the nostrils are forward-pointing tubes, which are unlike those of any other group of birds. The nostrils of petrels (Figure 11.5) look very like pitot tubes and Mangold (1946) proposed that the beaks of these birds contain a sense organ that measures airspeed. Figure 11.4A is based on his account, and shows a diagrammatic transverse section through the upper part of the beak of a fulmar (*Fulmarus glacialis*), a medium-sized petrel whose beak is similar to that of the Giant Petrel in Figure 11.5B. Mangold describes a flexible pocket on either side of the nasal septum, the inside of which is connected to pitot pressure via the nostrils, while the surrounding cavity is connected to static pressure via the mouth cavity. This requires only some mechanoreceptors to measure the expansion of the pocket, to make it directly analogous to the aneroid capsule of a standard airspeed indicator as shown in Figure 11.4B. Mangold also examined some land-soaring birds including the American Turkey Vulture (*Cathartes aura*) but did not find a similar arrangement in their nasal organs. Albatrosses (Figure 11.5A) differ from petrels in having a deep knife-shaped bill, with separate nostrils opening as forward-pointing scoops on either side. This arrangement could serve to detect side-slip, that is, a transverse component in the relative airflow. Avoiding side-slip is an important element in minimising losses against drag in any phase of flight, but perhaps especially in gust soaring.

of the associated area of the brain, have always puzzled anatomists, as some species (especially storm petrels) have a sense of smell, whereas others that have been tested apparently do not. Mangold (1946) identified an expandable pocket on either side of the nasal septum of petrels,

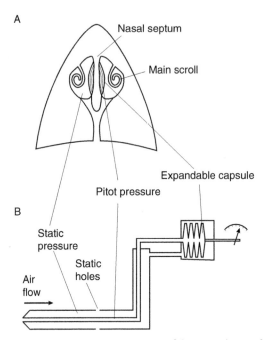

FIGURE 11.4 (A) Diagrammatic transverse section of the central part of a Fulmar's beak, passing through the second of the three nasal chambers (based on Mangold 1946). The inside of the expandable capsule on each side of the nasal septum is connected to pitot pressure from the forward-pointing nostrils (similar to those of the Giant Petrel in Figure 11.5B), while the cavity outside the capsule is connected to static pressure from the mouth cavity. The scrolls are not sensory, but concerned with adjusting the temperature and humidity of the air as the bird breathes in and out, like the turbinal bones of mammals. (B) Combined pitot-static tube as commonly used in aircraft for airspeed measurement. The expandable aneroid capsule registers the difference between the pitot pressure coming from the central tube, and the static pressure coming from slits in the wall of the concentric outer tube.

which appears to be connected to pitot pressure from the nostrils on the inside, and to static pressure via the mouth cavity on the outside, and proposed that this is a sense organ that measures airspeed, in the same way as the aneroid capsule of a conventional airspeed indicator. This arrangement is only known to occur in Procellariiform birds, which use a special form of gust-soaring that depends on exploiting discontinuities in the wind speed caused by separation of the atmospheric boundary layer from the surface of waves (Chapter 10, Section 10.2.2). Wherever the wind is sufficient to curl the crests of the waves, tube-nosed birds can replenish their kinetic energy in the lee of every wave, an unlimited and ubiquitous energy source which is, however, only available to

FIGURE 11.5 (A) Wandering albatross (*Diomedea exulans*) showing separate forward-pointing tubular nostrils on each side of the bill, typical of albatrosses (Diomedeidae). (B) Northern Giant Petrel (*Macronectes halli*) showing combined tubular nostrils on top of the bill, typical of petrels (Procellariidae). Photos by C.J. Pennycuick.

birds with the sensitive airspeed sense required to exploit it. Other sea-birds in the orders Pelecaniformes and Charadriiformes, which do not have the tubular nostrils, fly in very different styles, and are obliged to forage much closer to their nests than the tube-nosed birds.

As to the olfactory function of the nasal organ, storm-petrels are famous for detecting animal oil and homing upwind to the source, while turkey vultures can detect carrion by smell, although other New World (and also Old World) vultures apparently cannot. The kiwi is also known for its sense of smell, but that is hardly a flight adaptation. An acute sense of smell is generally regarded as unusual in birds, whereas a well-developed nasal organ is not. That may be because olfaction is not the only function of this organ (Figure 11.6).

## 11.1.4 VISION

Nearly all birds are highly visual animals, relying on their eyes as their primary source of sensory information. The eyes are commonly so large that they almost touch in the middle of the skull (Figure 11.6B), and may occupy more space in the skull than the brain. The left and right eyeballs are separated only by a thin bony septum in the midline. The eyes' optical axes are directed to the sides and obliquely forwards, more so in some birds (owls, raptors, swallows) than in others (waders). The image is formed on the retina by four optical elements, the cornea, the aqueous humour, the lens and vitreous humour. The interface between the air and the curved, outer surface of the cornea

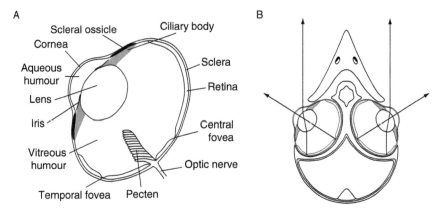

FIGURE 11.6 (A) Diagram of a bird's eye based on Slonaker (1918) and Pumphrey (1948). The eye is typically not spherical but dish shaped, so that it has limited freedom to move its socket, and cannot roll like mammal eyes. The cornea is the main refracting surface, and the sclera is reinforced by scleral ossicles to provide attachments for the muscles of the ciliary body, which distort the lens for focusing. The pecten is a structure which varies greatly in form in different birds, and is attached to the blind spot where the optic nerve enters. It is unique to birds, and its function is thought to be nutritive, maintaining the composition of the vitreous humour. The two foveas are pits in the retina associated with a high density of cones. (B) Diagrammatic horizontal section through the head of a Swallow, based on Slonaker (1918). The central foveas provide acute vision obliquely forwards and to the sides, the shape of the retina allows sharply focused and fine-grained vision to be maintained over a wider angle than is possible in the spherical eyes of mammals, with less emphasis on pointing the fovea directly at an object of interest. The temporal foveas give a fine grained, binocular image on both eyes of objects ahead of, and below the bill.

is the main refractive surface. The lens is an elastic body whose shape can be altered by the muscles of the ciliary body, so allowing the focal length of the whole array to be adjusted for accommodation (focusing).

Rather than being spherical, bird eyes have a dish-shaped retina, and consequently cannot roll around in their sockets to the same extent as mammal eyes. A bird rotates its head to look at something, in situations where a mammal would roll the eye in its socket. The exact direction in which the eye "looks" is less important in a bird than in a mammal, because the retina is fine grained, and in focus, over a wider area than a mammal retina. Many birds have a linear *area* of enhanced vision, which crosses the retina horizontally, and registers the horizon when the head is held in the normal flight position. Detailed vision along the horizon is readily understood in connection with the eye's function as the primary reference for spatial orientation in flight, and it may also have further implications in connection with navigation (below). Bird retinas typically have two *foveas* (pits), one central (centred on the eye's optical axis) and one temporal (Figure 11.6A). The central fovea provides a small area of especially acute vision to the side, while the temporal fovea is at the back of the eye and has the same forward field of view in both eyes, providing binocular vision directly ahead, and just below the bill. In waders the eyes are directly lateral, and because of the tapering shape of the head, both eyes can be seen from in front of the bird, and also from behind. This works in either direction. The curlew with its bill probing deep in the mud can still see the gyr falcon approaching it from behind.

The optic nerves of birds "decussate" fully, that is all of the nerve fibres coming in from the left eye cross over to the right side of the brain and *vice versa*, whereas in mammals decussation is only partial, with some optic nerve fibres going to their own side of the brain and some to the opposite side. This may be due to different methods of stereoscopic vision. We converge our eyes to examine nearby objects, and judge distance from the amount of convergence needed to centre the same part of the image on the foveas of both eyes. Birds, with their dish-shaped retinas, cannot generally converge their eyes to the same degree as mammals, and are presumably obliged to judge distance by observing the discrepancy in the position of the images of the same object on the two retinas. This method is easily applied to a pair of digital cameras, and may be associated with the total decussation of the optic nerve fibres.

## 11.1.5 ECHOLOCATION

Unlike the other senses, echolocation is an "active" sense, in which the animal illuminates its surroundings, not with light but with sound. A few fruit bats (Megachiroptera) and some cave-nesting birds (swiftlets) supplement their vision with limited forms of echolocation which are adequate for avoiding obstacles in deep caves, beyond the reach of daylight, but bats of the Sub-Order Microchiroptera, which includes most bat species, have more elaborate echolocation systems that make them wholly independent of ambient light. They emit very loud sounds at frequencies that are above the upper limit of human hearing, mostly in the range 25–100 kHz. The short wavelengths of these ultrasonic sounds allow insectivorous bats to detect and identify flying insects, and to pursue and catch them in flight. For example common Pipistrelles use frequencies around 50–60 kHz corresponding (at sea level) to wavelengths around 6–7 mm, and are able to detect insects of that size or larger. Bats that use higher frequencies (shorter wavelengths) can detect smaller insects, or features of large insects that serve for identification. Deriving directional information from echolocation requires that either the sound transmitter or the detector has to be directional. Many bat species have "nose-leaves" whose function is to concentrate sound emitted from the nostrils into a narrow beam, and such bats have relatively small external ears, whereas long-eared bats emit sound over a wider arc through their mouths, and detect the direction from which the echoes come with their large, mobile ears.

Insectivorous bats can generally avoid obstacles and hunt when blindfolded, but on the other hand they do have functional eyes, which presumably have been retained for some reason throughout their evolution. Many bat species migrate over distances up to a few hundred kilometres, and some forage a few kilometres from their roosts. These are short distances by avian standards, but long enough to make navigation difficult by echolocation alone, as this is an inherently short-range sense. Fruit bats, which have large eyes and no echolocation, find their way around visually, and it is possible that all bats use their eyes for orientation on the scale of the landscape, as opposed to their immediate surroundings.

## 11.2 ⬤ ORIENTATION AND NAVIGATION

### 11.2.1 THE COMPASS SENSE

The sun compass is the best known special sense of birds. It refers to the ability to maintain a compass direction by reference to the sun's azimuth, which is its direction, conventionally measured clockwise from north, as projected on the horizontal plane at the observer's position. This requires a time sense, to compensate for the sun's apparent movement across the sky, due to the earth's rotation. The classical demonstrations that birds possess a sun compass (Kramer 1952) depend on "clock-shifting" birds by keeping them under artificial lighting, which is switched on and off on a 24-hour cycle, but phase shifted from the local cycle of day and night, for example by 6 hours forward or back. Birds that have been trained to select a particular direction, or are attempting to migrate, duly select a direction that is shifted by 90°, but if they cannot see the sun, then their orientation is random.

This observation spawned an industry in which homing pigeons or wild-caught birds are tested under different experimental conditions. The measure of a bird's response is a direction, which may be the vanishing bearing of a homing pigeon as seen from the release point, or the direction in which a wild migrant tries to escape from an orientation cage. These directions usually show a large amount of scatter, and in the case of orientation cages are usually only recorded as the choice of one out of eight 45° sectors. Samples of such directions are analysed by circular statistics, with the objective of identifying the mean direction of a sample, and determining whether it differs significantly from random, or from a reference direction, or from the mean direction of another sample. Another industry has developed around the magnetic sense of birds, also based entirely on statistical analysis of the directions selected by animals, in a magnetic field manipulated by the experimenter. In this case, there is not even a plausible physical hypothesis as to how the magnetic field might be detected. There is no information on the precision with which a heading can be maintained, either by using the sun, or by a magnetic sense.

## 11.2.2 PRECISION OF ORIENTATION

Statistical experiments detect whether the bird reacts to the experimental treatment, but shed little light on whether its compass sense, whether optical or magnetic, is of any practical use for navigation. That depends on the degree of precision with which a heading can be maintained, a topic which is never discussed in the orientation literature. Any competent private pilot can hold a magnetic compass heading within 5°, while a commercial pilot who deviates by a couple of degrees from an airway can expect some pointed comments from air traffic control. A compass that only has a significant tendency to point to the right hemisphere would not be considered a practical navigational instrument, and that is all that the biological literature on the sun and magnetic compasses tells us about their performance.

Despite that, people use the sun for rough orientation, consciously or not. It works best for an observer near the north or south pole, during the six months when the sun is visible, as the sun moves almost horizontally all the way around the sky, at a steady 15° per hour. In the tropics, the sun indicates east in the morning, and west in the afternoon, but it is unusable in the middle of the day, because its azimuth becomes hard to determine when the sun is almost overhead. In the middle latitudes, the sun's azimuth changes slowly in the morning and afternoon, and faster in the middle of the day, and it can be used to determine a rough general direction, but not to steer an accurate heading. In the northern hemisphere (or more accurately, in latitudes north of the sun's current declination) the sun moves clockwise around the sky, whereas for an observer south of the sun, it moves anti-clockwise, which presumably calls for some re-programming in the case of trans-tropical migrants. The polarisation of blue sky can be used to infer where the sun is, even when the sun itself is obscured by cloud. Some historians believe that viking navigators used polarising crystals for this purpose, and it is possible that some birds may be able to do the same thing.

Sky polarisation patterns are more visible in the ultraviolet, to which many birds are sensitive, than at wavelengths visible to the human eye. The compass principle also works under the night sky, by monitoring recognisable patterns of stars rather than a single object. Many bird species normally migrate at night, and the results of planetarium experiments suggest that night migrants observe the whole star field, or whatever part of it can be seen through gaps in the clouds, so as to identify the point in the sky about which it appears to rotate. This then becomes the reference for geographical north or south (Figure 11.7).

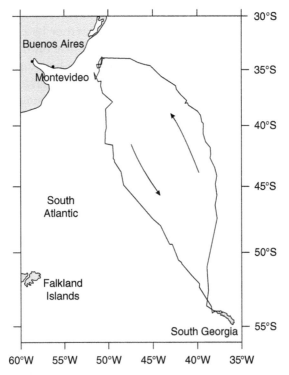

FIGURE 11.7 Track of a single foraging expedition by a female Wandering Albatross (*Diomedea exulans*) nesting on Bird Island, at the western end of South Georgia, and feeding off the coast of Uruguay, recorded by the Argos satellite system, from Prince et al. (1992). The bird covered a distance of 6479 km in 8.2 days, of which 2 days were spent in the feeding area at the north-western extremity of the track.

## 11.2.3 NAVIGATION BEYOND THE HORIZON

A sun or magnetic compass is a sense like any other, that responds to local stimuli that the animal can perceive directly. A compass is necessary for navigation but not sufficient in itself, because long-distance navigation also requires information that is beyond the range of direct perception. Satellite tracking of albatrosses nesting on small islands has shown that they forage over vast areas of open ocean, and return direct to the nesting island. The track of an 8-day foraging trip by a Wandering Albatross from Bird Island, South Georgia shown in Figure 11.7 is from Prince et al. (1992), and shows the bird first flying north, and then curving round to the west as it approaches its feeding area off the coast of Uruguay. There is a suggestion that it might have been "running down the latitude" to its destination, a technique that was used by the captains of sailing ships, in the period when they could determine their latitude from observations of the sun, but did not have chronometers that enabled them to determine longitude. However, when it had finished feeding, the albatross flew an almost straight track, apart from a few minor zigs and zags, all the way back to Bird Island, a distance of about 2200 km. Although South Georgia was a small target at this distance, there is no sign of any search pattern to locate it. A straight track when returning to the nesting island was usual in other tracks from the same albatross and others, and there was not usually any indication that the albatross needed to run down a latitude on either the outward or return journey. In the outward track of Figure 11.7, it could equally well have been searching the ocean to the east for possible feeding opportunities. Albatrosses and other birds that have been tracked in this way seem to know the direction of the destination, and can steer a course straight to it from hundreds of kilometres away, deviating only to take advantage of favourable winds.

Two methods have been proposed for position finding, that would work anywhere on the earth's surface, and do not depend on a knowledge of geographical features, which, of course, change over geological time. *Inertial navigation* was used by ships and trans-oceanic airliners until the introduction of the satellite-based Global Positioning System, and remains essential for spacecraft navigation and for submarines. It requires an array of linear and angular accelerometers, and could, in principle, be undertaken by the standard vertebrate labyrinth (Box 11.1). As noted above, a linear accelerometer, such as an otolith organ, does not indicate position, but if the position and velocity are initialised at a starting time (before take-off, say), then the velocity at some later time can be found by integrating the accelerometer reading with respect to time, and the position can be found by integrating the velocity. Likewise, direction can be monitored by double integration of the output of an angular accelerometer, such as a semicircular canal. The precision requirements for the accelerometers are very high (Britting 1971), and it seems unlikely that biological sensors could perform to the level required, but this is a topic that remains open to experimental investigation. The practical difficulties can be simplified by stabilising the platform on which the accelerometers are mounted (the bird's head) in such a way that it behaves like a pendulum with a period of 84 minutes, because the bob of such a pendulum would be at the centre of the earth, so that the platform remains parallel to the earth's surface, irrespective of any horizontal accelerations to which it is subjected. If a bio-rhythm with an 84-minute period were detected in birds, this would be a clue suggesting that inertial navigation may after all be used. However, the observation that Whooper Swans (*Cygnus cygnus*) migrating between Scotland and Iceland stop on the water when visibility is poor, but keep going over land regardless of the visibility (Pennycuick et al. 1999) suggests that they need a visual horizon over the sea, but not over land. This would not be expected if they had an 84-minute pendulum, which would, in effect, provide an artificial horizon.

The principle of *celestial navigation* is to observe the angular distance above the horizon of the sun or a star (its altitude), then take account of the earth's known rotation relative to the fixed stars, to get the observer's position relative to the earth's surface. Viking navigators a millenium ago could hold a constant latitude by observing the sun accurately enough to cross westwards from Norway to Iceland, and later direct to Greenland. Conveniently for them, this method is easiest near mid-summer, when the sun's declination changes only slowly, but not so easy for birds that migrate in spring and autumn. Determining longitude requires an accurate time reference, and remained an intractable problem for navigators until the balance-wheel chronometer was invented in the eighteenth century. Any comprehensive hypothesis about long-distance celestial navigation in animals implies that they can be assumed to have a time sense whose precision is of the order of a minute or two per day, or better. This remains a moot point despite the massive literature on biorhythms, because the *precision* of time-keeping has never interested experimenters to the same degree as the biochemical mechanisms involved.

A bird that has a sufficiently precise time sense could in principle determine the direction to a destination from simultaneous measurements of the sun's altitude and rate-of-change of altitude, as outlined in Box 11.4. This method requires the bird to be able to determine the difference between the sun's observed altitude above the horizon,

BOX 11.4 **Sun Navigation.**

---

### Variable definitions for this box

$A$ Sun's angular altitude
$B$ Crossing angle between observer-sun line and equinoctial
$D$ Sun's declination
$H$ Hour angle between sun's and observer's longitude
$L$ Observer's latitude
$α$ Rate of change of sun's altitude with respect to hour angle

### Terminology

Sun and star navigation, as developed for the use of mariners over the past several centuries, depends on an ancient fiction, the *celestial sphere* (Figure 11.8), which greatly simplifies thinking about the movements of celestial bodies. It is an imaginary sphere that is concentric with the earth, and its radius is unspecified but much larger than that of the earth. The stars are fixed to it, and the sun moves all the way around it once per year, following a path through the constellations called the *ecliptic*. Points and lines on the earth's surface can be projected on the celestial sphere by rays shining from the earth's centre. The celestial north and south poles are projected from the earth's poles by extending the earth's axis of rotation until it meets the celestial sphere. The celestial sphere rotates about the earth (or *vice versa* if you prefer) once per *sidereal day* (23.934 hours) whereas the sun lags behind by one revolution per year, and takes exactly 24 hours per revolution, on average over a whole year. The *equinoctial* is the projection of the earth's equator on the celestial sphere. It is a *great circle*, meaning one that it cuts the sphere exactly in half. The *meridians* are also great circles, which pass through both poles, and cross the equinoctial and equator at right angles. The axis of rotation lies in the plane of any meridian, and is perpendicular to the plane of the equator and the equinoctial.

Spherical trigonometry is entirely about angles, which are of two types. One type corresponds to what we normally think of as the distance between two points on the earth's surface, except that it is the angle that the two points subtend at the earth's centre. For example the distance from any point on the equator to the North Pole is 90°, in other words the *radius* of the equator (or any great circle) is 90°. Circles whose radius is less than 90° are called *small circles*. An angle measured north or south from the equator marks out a small circle on the earth's surface, parallel to the equator, whose radius is 90-$L$ degrees, where $L$ is the *latitude*. Likewise, a celestial object's *declination* is its angular distance north or south of the equinoctial. The other type of angle is measured on the surface of the sphere, where two great circles cross. For example, the angle measured at the North Pole between any meridian and an arbitrarily chosen zero meridian is called *longitude* on the earth's surface, and *right ascension* on the celestial sphere. By convention zero longitude is the meridian that passes through Greenwich, and zero right ascension is the celestial meridian through the First Point of Aries. A celestial object also has a longitude on the earth's surface, but this continually changes as the earth rotates, whereas its right ascension is fixed on the celestial sphere.

**BOX 11.4** *Continued.*

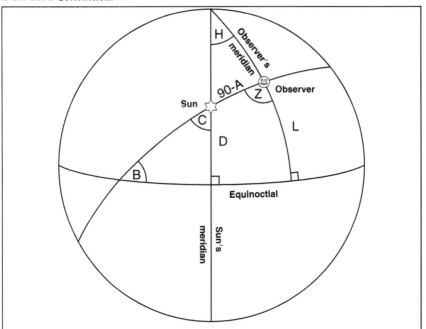

FIGURE 11.8 The observer's and the sun's positions plotted on the celestial sphere.

Figure 11.8 shows a celestial object, the sun, whose declination is *D*, and an observer at a latitude *L* on the earth's surface, whose position is projected on the celestial sphere. The longitudes of the observer and of the sun are not shown, but the difference between them is *H*, the hour angle, measured where the two meridians cross at the North Pole. *H* can be measured either in degrees or in hours, as the earth rotates at 15° per hour. The sun's altitude (*A*) is its angular distance above the horizon, as seen by the observer. The horizon (not shown) is a great circle 90° from the *zenith*, which is the point on the celestial sphere directly above the observer, and therefore the angular distance between the sun and the zenith is 90-A degrees. The basic theorems of spherical trigonometry consist of relationships between the six angles that make up a triangle formed on the surface of a sphere by the intersections of three great circles. These angles are the three corners, as in a plane triangle, and the three sides, which are also angles (above). Applying these theorems to the four spherical triangles that can be seen in Figure 11.8 results in the following relationship between the sun's altitude (*A*), the sun's declination (*D*), the observer's latitude (*L*) and the hour angle (*H*) between the observer's longitude and that of the sun:

$$\sin A = \cos H \cos L \cos D + \sin L \sin D \qquad (1)$$

### Position lines from the sun's altitude

Figure 11.9 represents the earth seen from a point directly above the North Pole using a stereographic projection, which represents any circle (great or small)

BOX 11.4 *Continued.*

either as a circle, or as a straight line if it is seen exactly edge-on, as the meridians of longitude are in this view. The sun is shown at a declination of 23.45°N, its furthest north position at the northern summer solstice. The *terminator*, where the sun's altitude is zero, divides the day side of the earth from the night side, and is a great circle, centred under the sun, with a radius of 90°. It reaches 23.45° beyond the North Pole at this season, so that the arctic regions (above 66.55° latitude) are in continuous daylight. The thick circles with the sun (red) at their centre, in the lower left quadrant of the diagram, are lines of equal sun altitude. The radius of each of these small circles is 90-$A$ degrees, where $A$ is the altitude. Conversely, if the navigator observes that the sun's altitude is $A$, and knows where on the earth the sun was overhead at the date and time of the observation, he can draw a circle on the chart with a radius of 90-$A$ degrees centred on the sun's position, knowing that his own position at the same time was somewhere on that line.

**Position lines from the rate of change of the sun's altitude**
Just as the small circles of latitude (thin lines centred on the North Pole) are intersected everywhere at right angles by the great circles of longitude radiating from the pole, so the small circles of equal sun altitude are intersected everywhere at right angles by another set of great circles, along each of which the *rate of change* of the sun's altitude is constant. These are shown in Figure 11.9 as thick lines radiating from the sun's position. Returning to Figure. 11.8, the rate of change ($\alpha$) of the altitude ($A$) with respect to the hour angle ($H$) is:

$$\alpha = dA/dH = \cos B \qquad (2)$$

where $B$ is the angle between the equinoctial and the line joining the observer's position to that of the sun. An observer standing on the equator, at the time of the spring or autumn equinox sees the sun rise above the eastern horizon and climb up to the zenith at a constant rate of $\alpha = 1$, and then continue down to the western horizon at a constant rate of $\alpha = -1$, meaning that the altitude changes by 1° for each degree change of the hour angle. These are the maximum and minimum values of the rate of change. On the summer solstice chart of Figure 11.9 the maximum rate is 0.917 along the thick dashed line. The pattern is repeated as a mirror image in the bottom right quadrant (but not shown in the figure), with the same values of the altitude, but negative values of the rate of change. The entire pattern rotates clockwise with respect to the stationary earth below. An observer at 40° latitude, say, sees the sun rise above the eastern horizon at a rate of about 0.65, which increases slowly to about 0.77 by 0900 or so, and then declines ever more rapidly to zero at noon. The altitude is then at its highest, which is 73.45°. The moment of zero rate of change is traditionally observed by navigators to determine the time of local noon (and hence the longitude) while the altitude at that moment directly gives the latitude, the sun's declination being known. After local noon, the rate of change of altitude becomes negative, as the altitude decreases back to zero, in a mirror-image of its ascent during the morning.

BOX 11.4 *Continued.*

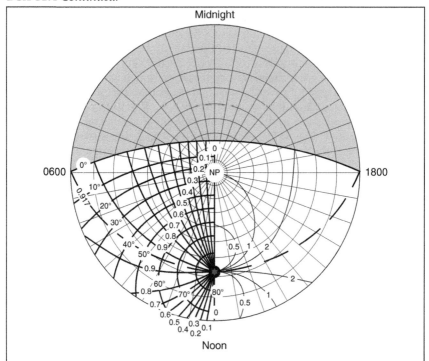

FIGURE 11.9 The earth seen from above the North Pole, plotted on a stereographic projection. The sun's declination is 23.45° N at the northern summer solstice. The family of sun-centred equal-altitude circles are plotted in the lower left quadrant, as are the family of great circles of equal rate of change of altitude. After Pennycuick (1960).

Since the hour angle ($H$) changes at a steady rate with respect to time ($t$), the rate of change of altitude with respect to hour angle is a constant multiple of the rate of change with respect to time, that is,.

$$dA/dH = \alpha = (dA/dt)/(dH/dt) \qquad (3)$$

where $dA/dt$ is measured, and $dH/dt$ is 15° per hour. Together with Equation (2), this converts an observation of $dA/dt$ into a position line that cuts the line from the sun's altitude (Sumner circle) at right angles. This is a practical method of position finding that could plausibly be implemented using a bird's eye as the observing instrument (Pennycuick 1960). It requires the bird to make a simultaneous measurement of the sun's altitude and the rate of change of altitude. The same method was proposed independently for use by surveyors by Hervieu (1960), whose account includes corrections to the rate of change of altitude for the annual motion of the sun around the sky.

**Precision of position finding**
The precision of position finding is the relationship between the probable error of the original celestial observation, and the resulting distance error on the

BOX 11.4 *Continued.*

ground, which is different for the two observations required, altitude and rate of change of altitude. The precision of the line from the sun's altitude is the same as the precision with which the altitude is observed, that is. an error of 1 arc minute in the observed altitude results in an error of 1 arc minute in position (i.e. 1 nautical mile or 1.85 km) in a direction perpendicular to the line. An error in the measured rate of change of altitude, however, causes an error of position that depends on the spacing between neighbouring lines. Where the lines are closely spaced, near the sun and near noon, a given change in the measurement corresponds to a small distance on the ground, and the precision of position finding is good, but where the lines are widely spaced, at low latitudes early and late in the day, a small error in the observation corresponds to a larger error on the ground. More explicitly, a small error ($\Delta\alpha$) in the measured rate of change of altitude results in a position error ($\Delta S$) given by

$$\Delta S = 3440Q\Delta\alpha \tag{4}$$

where the factor 3440 converts $\Delta S$ from radians to arc minutes (nautical miles), and the factor $Q$ is:

$$Q = \cos A / \sqrt{(\cos^2 D - \alpha^2)} \tag{5}$$

Lines for three values of $Q$, 0.5, 1 and 2, are shown in the lower right quadrant of Figure 11.9. The lowest values of $Q$ (highest precision) are found near the sun's position and near the noon meridian, while the precision becomes poor out on the fringes, where the altitude is low and cos $A$ is high. As an example, if $Q = 2$, a precision of about 12 nautical miles would require the bird to detect a discrepancy in the rate of change of altitude of about 1.5 arc seconds per minute. Nearer the sun, the same error of observation would lead to a lower position error in proportion to the local value of $Q$.

and the altitude where it would be, if the bird had already arrived at its destination. The response required is simple—if the sun is too low in the sky, fly towards it, and *vice versa*. A short-term time sense to measure the rate of change of the sun's altitude would provide a second, independent measurement that can be translated into a set of position lines that cross the first set (from the altitude) at right angles. If the rate of change is too high, the appropriate response is to deviate left if flying up-sun, and right if flying down-sun, and *vice versa*. A good view of the sun is needed for a few minutes, not necessarily at noon. If this sounds challenging in cloudy parts of the ocean, it should be remembered that "sun sights" (although not exactly this method) were the mainstay of marine navigation for centuries, and were the key to the successful outcome of a number of epic voyages in small boats under appalling conditions. Thus Frank Worsley (to name only one of the most famous) navigated the disintegrating lifeboat *James Caird* from Elephant Island to South Georgia in April 1916, across 800 nautical miles of mountainous seas, with only occasional glimpses of the sun, and thereby saved the stranded crew of Shackleton's ship *Endurance* from certain death.

## 11.2.4 NAVIGATION BY COMMITTEE

No goose, crane or stork willingly flies alone. Each bird in a migrating flock will modify its own preferred heading and speed as necessary, to avoid getting separated from the flock. As a result, the direction in which the flock flies is some kind of statistical average of the directions that would be selected by the individuals in it. Not all migrants fly in flocks, but in those that do, this type of committee navigation may compensate to some degree for the shortcomings of their individual compasses, and estimates of the required direction. There is evidently a random element in the navigation of some gregarious migrants, as in the flock of Cattle Egrets (*Bubulcus ibis*) that left southern Europe or North Africa in about 1930, and accidentally discovered Brazil, so initiating the invasion by that species of the Americas (Figure 11.10).

Some soaring birds also use formations as a means of locating thermals (Pennycuick 1972). Figure 11.10 shows a small flock of migrating cranes gliding along in "vee" formation. If the birds in any part of the line start to rise relative to the others, the flock will quickly converge on those that are climbing fastest, until (if the rate of climb is good) the whole flock is circling together around the core of the thermal. Storks fly along in a cluster without organising themselves into lines, and spread out laterally so as to search a wide swathe of air for thermals. Raptors, including vultures, will join other birds or gliders that are circling in a thermal, but appear to search for thermals individually, reacting to clouds or ground features as glider pilots do.

## 11.2.5 WEATHER INFORMATION

Modern pilots rely heavily on weather briefings, based on information that comes from a world-wide network of surface observers and orbiting satellites, but migrating birds have no such information. Several statistical studies have shown that migrating birds of various kinds prefer fine weather with a following wind when they make the decision to depart, but this only shows that they respond to the conditions that they can see when they depart, not that they are aware of future conditions, out of sight over the horizon ahead. It was apparent from the same study of Whooper Swans mentioned above (Pennycuick et al. 1999) that the swans had no information about the weather ahead when they departed from the south-east coast of Iceland, or from the Western Isles of Scotland, across the 800 km of open sea that separates

FIGURE 11.10 A flock of cranes (*Grus grus*) migrating northwards over southern Sweden in spring. In between thermals, the cranes spread out in echelon or vee formation, which increases the chance of finding another thermal. If one part of the line starts to rise, the others quickly converge on those individuals that are climbing fastest. Once centred in the thermal, the flock concentrates in a tight formation on one side of the circle, all circling in the same direction. White pelicans behave in a similar way, but storks circle in both directions in thermals, and form a cluster rather than a line between thermals, spreading out to the sides to increase the chance of finding thermals (see also Chapter 10 Figure 10.10). Photo by C.J. Pennycuick.

these coasts. The swans would not delay their departure because of a front or impending contrary gale along the route, so long as conditions were benign at departure. On several occasions the swans got into serious difficulties during the crossing, due to bad weather which had been correctly forecast by the meteorologist on the project (Tom Bradbury), but which they evidently failed to anticipate. As to the flights of extreme long-distance migrants such as the Alaskan Bar-tailed Godwit, which apparently takes a week or so to fly non-stop across the equator, direct from Alaska to New Zealand (Chapter 8), there is no way that even modern aviation weather services could forecast conditions over such a vast distance, so far into the future. The implication is that all migrants must depart with fuel reserves sufficient to cope with weather emergencies that are statistically likely to occur, and indeed the fuel reserves estimated by Pennycuick and Battley (2003) were much higher than the levels required for aircraft by aviation regulations.

# 12

# WATER BIRDS

Some birds (cormorants) use their feet to swim under water, but this is inefficient. Others use a flapping motion with the wings for propulsion under water, and this works best with very small wings (penguins). Wing swimmers that also fly have wings of reduced size, which increases their speed and wingbeat frequency, makes slow flight for landing difficult or impossible, and renders them incapable of flap-gliding. There are no bats that swim, for reasons that appear to apply equally to pterosaurs.

Here on earth, we are accustomed to two very distinct modes of locomotion in fluids, flight and swimming. The distinction results from a characteristic of our planet, which has extensive fluid environments of two types, air and water. Where air and water meet, the interface is so well defined that "sea level" is used as a datum for the radius of the planet. At the interface, the density of the air just above is about 800 times less than that of the water just below. Above the surface, the air density declines strongly with increasing height (Chapter 2), while below it the water density is almost constant, increasing only slightly down into the ocean depths. One can imagine planets with fluid environments that vary over a wide range of densities, or even have a gaseous atmosphere that shades imperceptibly into a liquid ocean without an identifiable surface, but on earth there are no intermediates between flight and swimming. A swimming animal's density differs only slightly (if at all) from that of the water, and consequently all, or nearly all of the energy that it expends is used directly in overcoming drag. A flying animal, on the other hand, gets negligible support from hydrostatic forces, and its energy requirements (Chapter 3) are dominated by the need to overcome drag, at the same time as supporting its weight in air, which is much less dense than itself. Gravity rules in air, but is difficult even to measure in water.

Just as different birds are adapted to live in every possible terrestrial habitat from tropical forests to arctic tundra, so other birds live in every aquatic habitat from inland streams and lakes to the open ocean. The limits to their aquatic adaptations are that all birds (like marine mammals) are confined to depths from which they can return to the surface to breathe, and that no bird has escaped from the need to nest on a solid surface (ice in the case of Emperor Penguins). In the latter respect penguins, the most aquatic of birds, are equivalent to fur seals and sea lions, but are not so fully independent of the land as whales and dolphins. The ocean surface, the interface between water and air, is the habitat of the most prolific and wide-ranging order of seabirds, the petrels and albatrosses (Procellariiformes), while three other orders, the Pelecaniformes (pelicans, cormorants, boobies, tropicbirds, frigatebirds), Anseriformes (swans, geese, ducks) and Charadriiformes (gulls, terns, auks) dominate particular parts of either the ocean or freshwater surface. Most of these birds can fly, walk and swim with varying degrees of proficiency, a degree of locomotor versatility that is shared by water beetles, but otherwise unmatched elsewhere in the Animal Kingdom. One order of birds, the Sphenisciformes (penguins) has become so specialised for swimming that none of its members can now fly, and a few species of auks, cormorants and ducks have also taken this route. Several birds, including ospreys, frigatebirds and skimmers, get their food from the water without actually entering it, and there are also bats that snatch fish from the water while flying just above the surface. However, there are no bats that swim, perhaps because a membrane wing would be unmanageable in water, whether used for propulsion or not.

## 12.1 ⬗ WATERPROOFING AND THERMAL INSULATION

The initial evolution of the bird body plan added the power of flight to a bipedal dinosaur-like animal, with little or no loss of its ability to walk and run. This versatility is due to the use of stiff flight feathers to carry the bending and torsional loads in flight, so that the leg is not part of the wing structure as it is in bats (Chapters 5–6). The contour feathers that cover the outer surface of a bird's body are mechanically similar to flight feathers, and also serve as thermal insulation by trapping air in the layer of down feathers under them. It seems that this arrangement also works in water, with almost no modification apart from the application of waterproofing compounds, whereas most marine mammals are insulated by a blubber layer, which would not be practical in a flying animal.

## 12.2 ⬤ MECHANICS OF SWIMMING

### 12.2.1 DRAG-BASED FOOT SWIMMING

Birds like ducks and gulls hardly differ from ordinary terrestrial birds, except in their waterproof outer layer of feathers. They walk on land like other birds, and float rather high in the water, owing to the buoyancy of a thick air layer trapped below the contour feathers. They propel themselves on the surface by a fore-and-aft movement of their webbed feet, spreading the toes as the foot swings back, and furling them as it swings forward (Figure 12.1A). The propulsive force comes from the drag of the foot as it moves backwards, relative to the water.

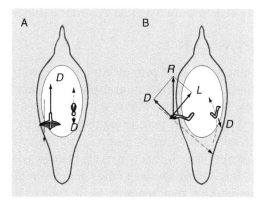

FIGURE 12.1 Two birds swimming on the surface, as seen by a diver looking upwards from below. (A) In gulls, ducks etc., the propulsive force is the drag of the spread webbed foot, as it swings backwards relative to the body (dashed arrow). The water flows forwards relative to the foot, producing a forward drag force (*D*). The foot is furled on the return stroke (right), so that it creates much less drag as it swings forwards. (B) In divers (loons) the leg articulation is modified so that the foot sweeps obliquely backwards, and towards the centre line (dashed arrow). The propulsive force is the reaction (*R*) which is the resultant of the lift (*L*) perpendicular to the local water flow and the drag (*D*) in line with it.

The forward speed of the body through the water can be no faster than the backward speed of the foot relative to the body, and work is done directly against drag, which is not an energetically efficient arrangement. On the other hand ducks stand, walk and run on land as well as other birds. Diving ducks use essentially the same foot motion when swimming under water, and so do some pelecaniform birds that forage under water, especially cormorants and anhingas.

## 12.2.2 LIFT-BASED FOOT SWIMMING

Divers or loons (Gaviidae) avoid leaving the water except when they have to climb on to their nests, which are invariably at the water's edge. They are unable to stand, and use their legs on land to push themselves along on their bellies. This is because the legs are set far back, flattened, and articulated so that their natural motion is from side to side rather than fore-and-aft. In the water, the tarsus and foot move in a direction that is oblique to the direction of swimming, so that the forward force is not pure drag as in a duck, but the resultant of a drag force parallel to the local water flow relative to the foot, and a lift force perpendicular to that direction (Figure 12.1B). This type of motion allows the body to move faster through the water than the foot moves relative to the body, and uses lift as a component of the force that balances the body drag, which is more energetically efficient than using foot drag alone. The penalty is lack of agility on land, and consequent restriction of nest sites to locations that predators such as foxes cannot easily reach.

## 12.2.3 WING SWIMMING

A far more efficient method of propulsion is wing swimming, in which the wing is rotated in the dorso-ventral plane, to produce a lift force directed forwards (Figure 12.2). Auks and diving petrels swim under water with their wings in this way, using their feet only to swim on the surface. The difficulty with this is that to produce comparable forces with the same wing in both air and water, the velocity required is smaller in water than air, by a factor similar to the square root of the ratio of the densities of water and air, that is about 28. To look at it another way, a wing that works for flying is far too big to work properly in water. Albatrosses and boobies do open their wings under water, but only to a small extent that allows a limited amount of steering. In auks and diving petrels, the wings are reduced, so that they can operate under water as propulsive hydrofoils, but they also operate in air as wings. The small wings of auks force them to fly faster than other birds of similar mass, and to beat their wings in air at a higher frequency than birds that do not swim

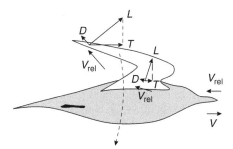

FIGURE 12.2 An auk such as a guillemot, totally submerged, and swimming with its wings. During the power stroke, the wings are partially flexed, and rotate about the shoulder joint towards the ventral side of the body. The bird moves forwards at a speed $V$, so that the relative water flow over the head is of the same magnitude as $V$, but opposite in direction. Over the wing, the relative water flow has an additional component, due to downward rotation of the wing about the shoulder joint. This component is small near the body, and progressively larger at more distal points along the wing. The resultant water flow ($V_{rel}$) over the outer part of the wing is faster, and angled strongly upwards, as compared to the relative flow over the head. The lift force on the wing is perpendicular to the local relative water flow, and is therefore angled forwards on the outer part of the wing. Its horizontal component is the propulsive thrust force ($T$). The muscles work against the drag force ($D$), which is parallel to the local flow, and much smaller than the lift. The feet of auks (and penguins) are furled back during wing swimming. They are not involved in propulsion, but serve for convective heat disposal.

with their wings. Auks flap their wings at a much lower frequency in water than in air, which causes problems in matching the flight muscles to the wings (Chapter 7, Box 7.1). They also flex the elbow and wrist joints, which reduces the wing area, but makes the wings a less-than-ideal zigzag shape. By reducing the wings further and abandoning flight, penguins avoid the need for these compromises.

The primary morphological difference between an auk and a gull of the same mass is that both the wing span and the wing area are smaller in the auk (Figure 12.3). The aspect ratio, which expresses the shape of the wing (Chapter 1, Box 1.3) is much the same in birds that use their wings for swimming and those that use them only to fly. Even penguins have similar aspect ratios to medium-sized petrels (Table 12.1). The consequences for flight performance of reducing the wing, without changing its shape, are outlined in Box 12.1. They are that an auk has to fly faster than a gull of the same mass, requires more power from its flight muscles, and has to beat its wings at a higher frequency. Another direct result of wing reduction considered in Box 12.1 is that an auk can flap, and it can glide, but it cannot flap-glide, because its minimum gliding speed is faster than its minimum power speed in flapping flight.

BOX 12.1 **Effects of reducing wing size.**

Wing swimming is barely practical for typical seabirds like petrels, gannets and gulls, because their wings are far too large to operate in water in the manner of penguin wings. Some seabirds, especially auks (Alcidae) and diving petrels (Pelecanoididae) have reduced their wings to a size where they work satisfactorily for propulsion in water, which results in fast flight with a high wingbeat frequency. The wings of wing swimmers are not greatly different in shape (aspect ratio) from those of seabirds of similar mass that do not use wing swimming, but they are a lot smaller. Table 12.1 shows the aspect ratios of some seabirds, with those that use wing swimming in bold type. The Common Diving Petrel's position at the bottom of the table (aspect ratio 7.41) might suggest wing swimming leads to a reduction of aspect ratio, but this is also by far the smallest species in the table. Flying birds also have smaller aspect ratios at smaller sizes (Chapter 13). The ultimate wing swimmer, a Gentoo Penguin, has the same aspect ratio as a Fulmar (10.7).

The effects of this type of wing reduction on flight performance can be understood by considering the two hypothetical birds shown in Figure 12.3, whose bodies have the same size and shape in both A and B. However, B has only 60% of A's wing span, although the wing *shape* (defined by the aspect ratio) is the same in both. This is a fair approximation to the modification needed to change from a bird shaped like a gull to a typical flying wing swimmer like a guillemot. We can use some equations that have been introduced earlier in the book to see the general nature of the performance implications (Pennycuick 1987b).

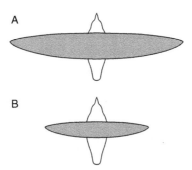

FIGURE 12.3 Birds such as auks that are specialised for wing swimming under water, but also able to fly (B), differ from other flying birds (A) in that the wing span and area are reduced relative to the size of the body, but the aspect ratio is little changed. The same trend leads to flightlessness when carried to extremes, as in penguins and the Great Auk.

BOX 12.1 *Continued.*

TABLE 12.1 Aspect ratios of some seabirds.

Wing swimmers are in heavy type

| | |
|---|---|
| **Eudyptes chrysolophus** | 10.7 |
| Fulmarus glacialis | 10.7 |
| Sula sula | 10.5 |
| Sterna fuscata | 10.4 |
| **Alca torda** | 9.44 |
| Larus ridibundus | 9.4 |
| **Uria aalge** | 9.34 |
| Pachyptila desolata | 8.62 |
| **Fratercula arctica** | 8.21 |
| **Pelecanoides urinatrix** | 7.41 |

## Variable definitions for this box

| | |
|---|---|
| $B$ | Wing span |
| $C_{Db}$ | Body drag coefficient |
| $C_{Lmax}$ | Maximum lift coefficient in gliding |
| $f$ | Wingbeat frequency |
| $g$ | Acceleration due to gravity |
| $k$ | Induced power factor |
| $m$ | All-up mass |
| $P_{am}$ | Absolute minimum power required at $V_{mp}$ |
| $P_{av}$ | Power available from the flight muscles |
| $P_{req}$ | Power required to fly at $V_{mp}$ |
| $R_a$ | Aspect ratio |
| $S_w$ | Wing area |
| $S_b$ | Body cross-sectional area |
| $V_{mp}$ | Minimum power speed in flapping |
| $V_s$ | Stalling speed in gliding |
| $\rho$ | Air density |

## Effect on flight speed

The speed at which birds normally fly around can be seen as a multiple (usually only a little greater than 1) of the minimum power speed ($V_{mp}$), which is given in Equation (1) of Box 3.4 as

$$V_{mp} = \frac{(0.807 k^{1/4} m^{1/2} g^{1/2})}{(\rho^{1/2} B^{1/2} S_b^{1/4} C_{Db}^{1/4})} \tag{1}$$

The comparison in Figure 12.3 refers to two birds of the same mass ($m$) with the same body frontal area ($S_b$) and drag coefficient ($C_{Db}$), flying in the same gravity ($g$) and air density ($\rho$) with the same induced power factor ($k$).

BOX 12.1 *Continued.*

The only variable in Equation (1) that changes is the wing span ($B$). As every-thing else in Equation (1) is constant, the minimum power speed varies inversely with the square root of the wingspan, and that is all there is to it:

$$V_{mp} \propto B^{-1/2} \tag{2}$$

No special observations are needed to see that guillemots and razorbills hurtle around the sea cliffs, much faster than gulls.

What about the wing area? The area differs between Figure 12.3A and B in proportion to the square of the span, but this does not affect the speed. The wing area does not appear in Equation (1). The apparently common sense notion that wing area affects the speed of flight is true in the case of fixed wings, and applies to gliding birds (Chapter 10), but it does not apply if the wings are moved relative to the body, as they are in helicopters, and in flapping bird flight. Biological authors who discuss flapping-flight perfor-mance in terms of wing loading (mass or weight per unit wing area) have failed to notice that this is a fixed-wing notion, which is not transferable to flapping flight. The reasons for this are explained in Chapters 3 and 4.

**Effect on power required to fly at V$_{mp}$**

The "absolute minimum" power ($P_{am}$) is given in Equation (2) of Box 3.4 as

$$P_{am} = \frac{(1.05k^{3/4}m^{3/2}g^{3/2}S_b^{1/4}C_{Db}^{1/4})}{(\rho^{1/2}B^{3/2})} \tag{3}$$

This is the sum of induced and parasite powers at $V_{mp}$. The profile power has to be added to get the minimum power, and the aspect ratio (and indi-rectly the wing area) is involved in calculating this. However, as noted above, birds that use wing swimming are distinguished from those that do not by a major difference in wingspan, but little if any difference in aspect ratio. Using the same reasoning as above, one can say that the main effect of reducing the wing span, keeping everything else including the aspect ratio constant, is that the power required to fly varies inversely with the 3/2 power of the wingspan:

$$P_{am} \propto B^{-3/2} \tag{4}$$

As the wing span shrinks, the power required to fly at $V_{mp}$ increases strongly. Whether or not this makes it difficult for wing swimmers to fly depends on how this effect on the power *required* to fly compares with any effect that shrinking the wings may have on the power *available* from the flight muscles.

**Effect on power available from the flight muscles**

The performance of the flight muscles, considered as engines, is analysed in Chapter 7 along the lines that the maximum amount of *work* that can be done in each contraction, by a given mass of muscle, is essentially fixed by the properties of the sliding filament mechanism. The maximum *power*, averaged over many contraction cycles, is this amount of work, multiplied

BOX 12.1 *Continued.*

by the contraction frequency. As the flight muscles drive the wings directly in a bird, their contraction frequency has to be the same as the wingbeat frequency, which is itself determined by the properties of the wings and the flight environment. According to Equation (1) of Box 7.3, the wingbeat frequency ($f$) in cruising flight is:

$$f \propto m^{3/8} g^{1/2} B^{-23/24} S_w^{-1/3} \rho^{-3/8} \tag{5}$$

Unlike Equations (1) and (3), Equation (5) involves both the wingspan ($B$) and the wing area ($S_w$). However, in the comparison of Figure 12.3, the aspect ratio ($R_a$) is assumed to be constant, where

$$R_a = \frac{B^2}{S_w} \tag{6}$$

In other words, constant $R_a$ implies that

$$S_w \propto B^2 \tag{7}$$

Proportionality 5 says that $f \propto S_w^{-1/3}$, which can be converted to $f \propto B^{-2/3}$ with the aid of Proportionality 7. Thus, if $m$, $g$ and $\rho$ are assumed to be constant, Proportionality 5 can be reduced to:

$$f \propto B^{-23/24} B^{-2/3} = B^{-13/8} \tag{8}$$

As the wing span is reduced, the wingbeat frequency increases dramatically, giving auks their characteristic "whirring" flight. If the flight muscle mass remains constant, the power available from the flight muscles ($P_{av}$) will increase by the same amount:

$$P_{av} \propto B^{-13/8} \tag{9}$$

whereas the power required ($P_{req}$) can be assumed to vary in the same way as the absolute minimum power, according to Proportionality 4:

$$P_{req} \propto B^{-3/2} = B^{-12/8} \tag{10}$$

The negative exponents are somewhat confusing, but can be understood from Figure 12.4, where it is assumed that there is some value of the wing span (vertical dashed line) at which the power available from the muscles is equal to that required to fly at $V_{mp}$. If the wing span is now reduced (moving left on the diagram), the power required to fly increases, along the line whose slope is −3/2, while the power available from the muscles increases along the slightly steeper line whose slope is −13/8, because that is the slope for the wingbeat frequency, which determines the power available. Although the power required to fly increases as the wing span decreases, the power available from the muscles increases slightly more, because of the increase in wingbeat frequency. Shortening the wings does not in itself prevent the muscles from meeting the power requirements, nor does the requirement for the power available to match the power required impose a limit on the amount of wing reduction that is possible.

BOX 12.1 *Continued.*

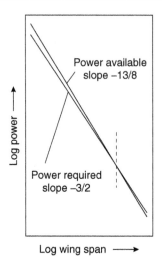

FIGURE 12.4 Wing reduction as in Figure 12.3 leads to an increase in the power required to fly in proportion to the −3/2 power of the wing span, whereas the wingbeat frequency increases at the slightly steeper slope of −13/8. The power available from a fixed mass of muscle should be proportional to the wingbeat frequency, and thus able to deliver the higher power required to fly, even in penguins.

### Effect on minimum gliding speed

Auks may not usually be thought of as gliding birds, but they do glide, and they can soar in the slope-lift along sea cliffs, in the same way as fulmars, gannets and other more typical gliding birds. The difference is that they require a stronger wind than other birds before they can do this. The minimum speed at which any bird can glide (the stalling speed $V_s$) is given in Chapter 10 Box 10.1 as

$$V_s = \left[ \frac{2mg}{(\rho S_w C_{Lmax})} \right]^{1/2} \tag{11}$$

where $C_{Lmax}$ is the maximum lift coefficient. This is a fixed-wing equation, and it says that, other things being equal, the stalling speed is inversely proportional to the square root of the wing area ($S_w$).

$$V_s \propto S_w^{-1/2} \tag{12}$$

If the aspect ratio is held constant while the wings are reduced, Proportionality 7 converts this to:

$$V_s \propto B^{-1} \tag{13}$$

This is also true for other characteristic gliding speeds, that is, the speeds for minimum sink and best glide ratio. The slope of Proportionality 13 is twice that of Proportionality 2, which describes the effect of shrinking the

BOX 12.1 *Continued.*

wings on the minimum power speed. Reducing the wing span increases gliding speeds twice as steeply as speeds in flapping (Figure 12.5).

The result of this is that although auks can glide (very fast), and also flap, they cannot flap-glide. A bird like a fulmar flying at $V_{mp}$ while it is flapping is going faster than its stalling speed. If it stops flapping, it can carry on gliding horizontally, slowing down below $V_{mp}$ until it reaches $V_s$, when it has to resume flapping or start losing height. A guillemot's $V_{mp}$ is faster than that of a fulmar, but not as fast as its own stalling speed in gliding. The guillemot has to speed up well above $V_{mp}$ before it can glide, and this makes flap-gliding impracticable.

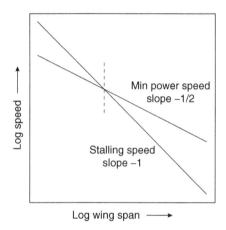

FIGURE 12.5 Wing reduction as in Figure 12.3 leads to an increase in the minimum power speed ($V_{mp}$) in proportion to the $-1/2$ power of the wing span, while gliding speeds increase in proportion to the $-1$ power of the wing span. Wing swimmers (left side of diagram) can glide, but they cannot flap-glide, because the stalling speed in gliding is higher than $V_{mp}$ when flapping.

# 12.3 ⬤ MORPHOLOGICAL TRENDS IN WATERBIRDS

## 12.3.1 THE NOTION OF THE "STANDARD SEABIRD"

Of the four families making up the Order Procellariiformes, one (the diving petrels) consists of wing swimmers which are convergent on the smaller species of auks, while the other three (storm-petrels, petrels and albatrosses) form a remarkably homogeneous sequence, ranging from tiny storm-petrels whose mass is less than 30 g to the largest albatrosses at over 9 kg. All of them are strictly pelagic, foraging at sea and coming ashore only to breed, and they all have many characteristics in common, which are not shared by other groups of seabirds. The morphological characteristics of the main procellariiform sequence (excluding diving petrels) are described in Chapter 13. The "standard seabird" is a hypothetical species, whose mass, wingspan and wing area are at the mid-point of the sequence. It is close to a White-chinned Petrel (*Procellaria aequinoctialis*), which is shown in silhouette in Figure 12.6, enclosed in a ring. Other seabirds can be derived from this starting point by changing the wing measurements. For example, adapting the standard seabird for wing swimming results in reduced wing span, with little or no change of aspect ratio (above), leading to auks and eventually to penguins. Modifications in other directions from the standard seabird follow from other adaptive requirements, but before considering some of these, I will first look at extreme wing reduction for wing swimming,

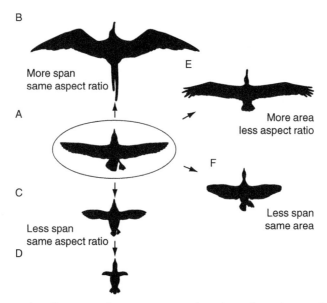

FIGURE 12.6 The silhouette in the ring (A) is a White-chinned Petrel (*Procellaria aequinoctialis*) which is close to the "standard seabird" derived from the mid-point of the procellariiform sequence (see Chapter 13). All of the six silhouettes in the figure have been scaled so that the bodies appear the same size. The differences between them reflect changes in the wings relative to the body. In a Razorbill (C), which flies but is specialised for wing swimming, the wings are reduced, with little or no change of aspect ratio. The same trend is continued to an extreme degree in the Gentoo Penguin (D), which swims in a similar manner but is flightless. The Magnificent Frigatebird (B) has increased wing span and area, also with little change of aspect ratio, which adapts it for soaring in weak, narrow thermals over the trade-wind zones of the oceans. The frigatebird's large scapular feathers and long tail make its body look bigger in the silhouette than it really is. The Brown Pelican (E) and Blue-eyed shag (F) have lower aspect ratios than the standard seabird, and different planforms that are associated with less fully pelagic lifestyles than those of those of A–D. After Pennycuick (1987a).

which leads to flightlessness, as it is not entirely clear from the arguments presented so far why penguins cannot fly.

## 12.3.2 WHY ARE PENGUINS UNABLE TO FLY?

Penguins are thought to be descended from flying birds related to petrels, and it is possible that they first became flightless as a result of selection pressure for large size. Modern penguins range in body mass from about 1–30 kg, overlapping the upper end of the mass range of flying birds, but extending well above it. The Razorbill (*Alca torda*) is an auk with a body mass around 0.62 kg, which swims with its wings under water, and flies fast in air. Its extinct congener the Great Auk (*Alca impennis*) looked quite similar, but was much bigger, and flight-less. Its locomotion in water was penguin like, and its wings were reduced to a similar degree to those of penguins. Its mass (around 5 kg) was well above that of the heaviest flying auks (about 1.5 kg), but well below the limit for flying birds (about 16 kg—see Chapter 7, Box 7.4). This suggests that the upper limit of mass for flying wing swimmers may be below the limit for birds in general, with no limits on wing morphology. However, the argument in Box 12.1 indicates that although reducing the wing span increases the power required to fly, it also increases the wingbeat frequency by a slightly larger amount, and with it the power available from a fixed mass of muscle. If this is cor-rect, then wing reduction should not in itself make a flying penguin impossible, or even call for any allometric adjustment of muscle mass.

There is a Gentoo Penguin (*Eudyptes chrysolophus*) in *Flight's* Wings Database. Its mass is 4.80 kg, its wing span is 0.510 m (a bit less than a Knot) and its aspect ratio is 10.7. The reader can verify that the power curve calculation says that its $V_{mp}$ for sea-level air density would be 33.4 m s$^{-1}$, with a wingbeat frequency of 34.5 Hz (like a hummingbird). Giving it the default flight muscle fraction of 0.17, the specific work would be 32.6 J kg$^{-1}$, similar to that of large birds like swans when flying at $V_{mp}$. These numbers do not in themselves exclude the possibility that the penguin could fly, although it might need a steep toboggan run to take off, and would certainly have trouble using its hummingbird-like flight muscles for swimming.

It is possible that the flightlessness of penguins is due to a physiolog-ical rather than a mechanical limitation, but this takes the discussion outside the *Flight* programme's scope. Wing swimmers like guillemots, that are able to fly, do so in order to travel back and forth between their nests and their foraging areas at sea, and also to move around the ocean outside the breeding season, and they have to be able to fly aerobically to meet these requirements. *Flight* says that the penguin would need nearly 5 times as much chemical power than a 5 kg Bewick's Swan to fly at $V_{mp}$, and if it were to do that aerobically, presumably its heart and lungs would have to be a lot bigger as well, perhaps too big to fit in its body. Sadly, there is no quantitative theory that says how big the heart and lungs have to be, or how much power they require for themselves, to support a given level of aerobic power in the flight muscles. Hill (1950) proposed a way in which such a theory could be developed, but there are difficulties with following this up, outlined in Chapter 13.

### 12.3.3 WING ENLARGEMENT IN FRIGATEBIRDS

Frigatebirds have aspect ratios around 12, similar to or slightly higher than those of medium-sized petrels, but both their wing spans and wing areas are considerably larger than those of procellariiform birds of similar mass (Figure 12.6B). Considered as a deviation from the procellariiform standard, this is the opposite of the wing reduction seen in wing swimmers. The long wings of frigatebirds enable them to glide very slowly, and to turn in circles of small radius. This is an adaptation to circling in the small, weak thermals that occur over the trade-wind zones of the oceans (Chapter 10). Unlike most water birds, frigatebirds do not have waterproof plumage, and they are unable to swim or even to alight on the water. As they disperse all over the tropical oceans, this must imply that they soar continuously in trade-wind thermals, day and night, for months at a time (Pennycuick 1983). They are incredibly agile in the air, and feed by snatching such creatures as hatchling turtles from the water surface, or flying fish from just above it. They are also well known for kleptoparasitism, which means intercepting other seabirds such as boobies as they return to their breeding colonies, forcing them to disgorge whatever food they are carrying, and catching it before it hits the surface. This is, of course, only a local and seasonal food source for them.

### 12.3.4 REDUCED ASPECT RATIO IN FOOT SWIMMERS

Cormorants (Figure 12.6F) and anhingas forage under water by foot swimming with their wings furled, but the legs are not modified for sideways motion as in divers. These birds can stand and walk on land, and they also perch on branches, and even on wires. They are less buoyant than other water birds, and float lower in the water, often with only the neck and head above the surface, apparently because their plumage is wettable. They are often seen standing on rocks or trees after a dive, holding out their wings to dry. Apart from the famous flightless cormorant of the Galapagos Islands, which has reduced wings, typical cormorants have wings that are shorter than those of a procellariiform bird of the same mass, but with aspect ratios of 7–8, that is, with reduced span but no reduction of wing area. Both cormorants and anhingas are capable of soaring in thermals over land, and some cormorant species migrate in flocks, flapping or flap-gliding in formation, somewhat like geese. Their wings do not seem to be directly influenced by the needs of locomotion in water, except that they may be an encumbrance under water, and are therefore no bigger than necessary when it comes to moving around in search of foraging opportunities.

The same could be said of pelicans (Figure 12.6E), which have somewhat more span than procellariiform birds of the same mass, and lower aspect ratios, with broad, slotted wing tips reminiscent of those of storks and vultures. White pelicans feed while swimming on the surface, by dipping their heads and filling their gular pouches with water, often forming groups which do this in synchrony. They make lengthy overland migrations by thermal soaring in flocks. American brown pelicans are coastal birds that feed by plunge diving. Their large wings allow them to levitate repeatedly a metre or two above the surface, and then plunge on a fish, which is drawn into the gular pouch as it inflates into a water-filled balloon.

## 12.4 ⬤ OTHER AQUATIC ADAPTATIONS

### 12.4.1 TAKE-OFF FROM A WATER SURFACE

The objective of take-off is to accelerate until the bird has sufficient air-speed to fly, and, if that is achieved at a speed below the minimum power speed ($V_{mp}$), to continue accelerating at least up to $V_{mp}$. This requires more effort in large birds than in small ones, and especially in birds with reduced wings, because both large size and wing reduction increase $V_{mp}$. Most waterbirds use their legs to run over the water surface, especially when taking off from sheltered water or in a light wind. The backward motion of the webbed feet, relative to the water, provides a forward and upward force which helps the flapping wings to accelerate the body to flying speed, and also makes the acceleration easier by raising the body above the surface, so reducing the drag caused by creating a bow wave. The skittering run over the water is often prolonged in swans and auks, and is followed by a period of almost level flight as the bird lifts its feet out of the water, and continues accelerating up to $V_{mp}$. When a flock of swans is taking off, some from a pond and others from a nearby grassy sward, those taking off from water need a longer take-off run than those taking off from solid ground.

Any bird takes off more easily in a wind than in calm air, because the objective of the take-off run is to acquire airspeed. If a bird takes off against the wind, its airspeed is already equal to the wind speed before it even starts to move, relative to the water. Petrels and albatrosses (excluding diving petrels) have sufficiently large wings to glide at speeds that are often lower than the wind speed over the open ocean. Their normal method of take-off is to face into wind and simply spread their wings, allowing the wind to lift them off the windward slope of a wave with little or no exertion on their part.

## 12.4.2 LANDING ON A WATER SURFACE

Just as a water bird cannot acquire speed at take-off by dropping, as land birds often do (Chapter 9), so conversely a bird cannot get rid of excess speed by pulling up before a water landing. However, some residual speed is acceptable when landing on water, for example divers simply slide on to the surface, with their feet trailing behind. Ducks and geese can usually drop lightly on to their feet on land, especially if there is some wind, but in water landings they prefer to come in with some residual ground speed, deploying their large webbed feet below the body and pointing forwards, where they act as water-skis. Gannets and boobies, which plunge dive at a steep angle when feeding, often do the same thing slowly when they land on the water, submerging briefly as they touch down head first. Petrels and albatrosses simply glide into wind until the speed relative to the water drops to zero, and then settle on to the surface.

## 12.4.3 VISION UNDER WATER

In the eye if a bird that operates only in air, the curved outer surface of the cornea is the main image-forming element, and the lens serves mainly for focusing (Chapter 11). When the eye is immersed in water, the cornea ceases to operate as a refracting surface, because its refractive index is close to that of water. Human eyes likewise will not provide a focused image under water for this reason. Divers have to wear a mask with a flat glass face plate, and an air space behind it in which the eyes can operate normally. In his classic book *The Vertebrate Eye*, Walls (1942) describes three ways in which different aquatic birds have solved this problem. Cormorants have unusually large ciliary muscles, and an unusually compliant lens, and simply distort the lens by a massive amount, sufficient to allow it to form a focused image either in air or in water. In auks, the nictitating membrane or "third eyelid", which sweeps from front to back across the eye in all birds to clean it, is kept closed under water, as it contains a supplementary lens which allows the eye to focus in water. Kingfishers, which need to keep the image of a fish in focus as they plunge from air into water, do this by passing the image from the central fovea, which focuses in air, to the temporal fovea, which focuses in water because of the unusual elongated shape of the eye.

## 12.4.4 HEAT DISPOSAL IN WATER

The legs of penguins are used for swimming in a duck-like fashion only on the surface. When the penguin is submerged and swimming with its wings, the feet trail behind, and dispose of waste heat through their large surface. Convective heat disposal is far more effective in water than in air, and is controlled by *retia mirabilia*, which are counter-current heat exchangers, situated at the base of the leg, inside the insulating layer of feathers. If heat needs to be conserved, the outgoing blood in the artery passes heat directly into the incoming blood in the vein, so cooling the blood before it leaves the thermally insulated interior of the body, but when excess heat has to be disposed of, the blood passes out by a different route, so that it is cooled by the water flowing past the foot.

# 13

# ALLOMETRY

There is a statistical trend for larger birds to have higher aspect ratios than smaller ones, although the wing area varies as expected with the mass. This chapter uses original measurements (not trawled from the literature) of 220 bird species from the Wings Database of the *Flight* programme, and a subset of 44 species for which there are data on flight muscle mass, to make allometric plots of variables calculated by the programme. These include the minimum power speed, wingbeat frequency, specific work in the muscles and many others.

Allometry is the study of differences in shape between different animals (like birds) that are all built on the same plan. This commonly takes the form of looking for deviations from a null hypothesis, which postulates that all animals in a particular set or taxon are "geometrically similar". If a miniature aircraft model is described as a "1/72 scale model" of a real aircraft, this means that it is geometrically similar to the original, and that every linear measurement on the model, such as the wingspan, the length of the fuselage and so on, is exactly 1/72 of the corresponding measurement on the original aircraft. If that is so, then it follows that the ratio of any area on the model to the corresponding area on the original is $(1/72)^2$, and the ratio of any pair of corresponding volumes is $(1/72)^3$. With such models, that is as far as it goes, but the same idea can be carried further with additional assumptions. For example in the case of birds of different sizes, it may reasonably be assumed that the density of a bird is constant, and therefore that the mass of any bird is directly proportional to its volume. In that case, the null hypothesis states that for geometrically similar birds, any area "varies with" or "scales with" the length squared, while any volume (and also any mass) scales with the length cubed. The conventional shorthand for this is the proportionality sign "$\propto$". For the basic geometrical relationships, we can write:

Area $\propto$ Length$^2$

Volume $\propto$ Length$^3$

These can be read "Area varies with length-squared" and "Volume varies with length-cubed". With the additional assumption of constant density, we can also write:

Mass $\propto$ Length$^3$

Since mass is the easiest variable to measure, these statements are usually turned around, to express the way in which different types of measurements are expected to vary (or "scale") with the mass, under the assumptions of geometrical similarity and constant density. Any length is expected to scale with the one-third power of the mass, any area with the two-thirds power, and any volume directly with the mass, thus:

Length $\propto$ Mass$^{1/3}$

Area $\propto$ Mass$^{2/3}$, and

Volume $\propto$ Mass

Deviations from the expected relationships indicate "allometry" as opposed to "isometry".

TABLE 13.1 Species codes for the birds in Figure 13.1.

| Code | Scientific name | English name |
|------|-----------------|--------------|
| WAN | *Diomedea exulans* | Wandering Albatross |
| BBA | *Diomedea melanophris* | Black-browed Albatross |
| GHA | *Diomedea chrysostoma* | Grey-headed Albatross |
| STY | *Phoebetria palpebrata* | Light-mantled Sooty Albatross |
| MAC | *Macronectes giganteus/ M. halli* | Giant petrel |
| WCP | *Procellaria aequinoctialis* | White-chinned Petrel |
| CAP | *Daption capensis* | Cape Pigeon |
| PRN | *Pachyptila desolata* | Dove Prion |
| WIL | *Oceanites oceanicus* | Wilson's Storm-petrel |

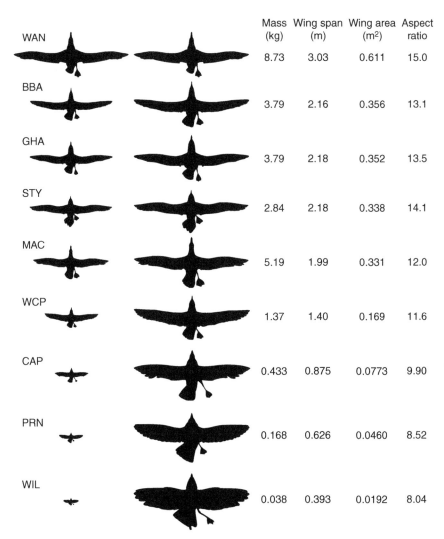

|  | Mass (kg) | Wing span (m) | Wing area (m²) | Aspect ratio |
|---|---|---|---|---|
| WAN | 8.73 | 3.03 | 0.611 | 15.0 |
| BBA | 3.79 | 2.16 | 0.356 | 13.1 |
| GHA | 3.79 | 2.18 | 0.352 | 13.5 |
| STY | 2.84 | 2.18 | 0.338 | 14.1 |
| MAC | 5.19 | 1.99 | 0.331 | 12.0 |
| WCP | 1.37 | 1.40 | 0.169 | 11.6 |
| CAP | 0.433 | 0.875 | 0.0773 | 9.90 |
| PRN | 0.168 | 0.626 | 0.0460 | 8.52 |
| WIL | 0.038 | 0.393 | 0.0192 | 8.04 |

FIGURE 13.1 Silhouettes of nine procellariiform species from Bird Island, South Georgia, with measurements from Pennycuick (1982). The three-letter species codes are identified in Table 13.1.

## 13.1 ● ALLOMETRY OF MORPHOLOGICAL VARIABLES

### 13.1.1 ALLOMETRY IN A SMALL SAMPLE OF CLOSELY RELATED BIRDS

Figure 13.1 shows a set of silhouettes made from wing tracings of nine bird species belonging to the Order Procellariiformes that nest on Bird Island, South Georgia (Pennycuick 1982). These are all pelagic birds that feed at sea on fish, squid, krill or zooplankton, and come to land only to nest. In the left-hand column, the size of each silhouette is proportional to the bird's actual size, while in the next column the smaller species have been enlarged by different amounts, so that all the silhouettes have the same wing span on the page. This group covers a wider range of body mass than any other order of birds (over 300:1). If they were all geometrically similar, then one would expect that the wing span (a length) would vary with the one-third power of the mass, and the wing area (an area) with the two-thirds power of the mass. The aspect ratio, being the dimensionless ratio of two lengths, would be independent of the mass. In the terms introduced above, the expected relationships are that

Wing span $\propto$ Mass$^{1/3}$

Wing area $\propto$ Mass$^{2/3}$, and

Aspect ratio $\propto$ Mass$^{0}$

To test whether the data do in fact vary in these ways, the numbers listed in Figure 13.1 have been plotted in Figures 13.2–13.4 on double-logarithmic graphs. The principle behind this is explained in Box 13.1, and it is more than just a device for spreading the smaller data points apart, and squashing the larger ones together (although it does that). If the birds are indeed geometrically similar, and one plots the logarithm of the wing span against the logarithm of the mass, then the result will be a straight line, with a slope of one third. It is unnecessarily obscure to look up the logarithms and plot them on the graph, as

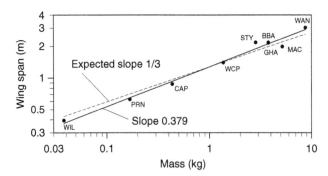

FIGURE 13.2 Allometric plot of wing span $(Y)$ versus mass $(X)$ for the birds in Figure 13.1. The logarithmic scales are the same on both axes in this graph (and in all the allometric graphs in this chapter), meaning that a factor of 10 is represented by the same distance in both the $X$ and $Y$ directions. Expected slope 0.333. Reduced major axis slope 0.379. $N = 9$ (number of points). $r = 0.991$ (correlation coefficient).

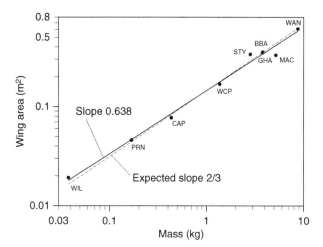

FIGURE 13.3 Allometric plot of wing area (*Y*) versus mass (*X*) for the birds in Figure 13.1. Expected slope 0.667. Reduced major axis slope 0.638. *N* = 9. *r* = 0.995.

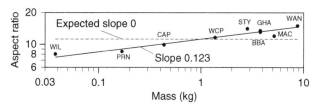

FIGURE 13.4 Allometric plot of aspect ratio (*Y*) versus mass (*X*) for the birds in Figure 13.1. Expected slope 0. Reduced major axis slope 0.123. *N* = 9. *r* = 0.952.

some authors do. A better way is to label the axes with the original mass and wing span numbers, but to position these logarithmically along the axes. This means that any fixed distance along the *X* scale represents a fixed *ratio* between the numbers at its ends, not a fixed difference as on a linear scale. For example, the distance from 0.1 to 1 kg along the mass (*X*) scale in Figures 13.2–13.4 is the same as that from 1 to 10 kg. If the scales are the same on both the *X* and *Y* axes, that is if a given distance represents the same ratio on both scales, then the slopes of lines can be measured directly on the page. If the wing span does indeed vary with the one-third power of the mass, then the fitted line will go up one centimetre for every three centimetres along the body mass axis. If the slope differs from that, there is an allometric relationship between wing span and mass.

BOX 13.1 **Double-logarithmic plots.**

Although every biology textbook is full of graphs showing the logarithms of physical measurements such as mass, one should be aware that a logarithm is a transformation that can be applied only to a pure number, not to a physical measurement. However, if we plot the ratio of a bird's mass to a "reference mass", which happens to be exactly 1 kg, the numbers stay the same as before but are detached from their physical dimensions, thus allaying any scruples we might have about plotting "log wing span" against "log mass". The logarithms themselves are not shown on the axes of the graphs in this book. The original numbers are shown, but the scales are distorted so that a fixed distance along the scale represents a constant multiple, not a constant difference as in an ordinary linear graph.

**Variable definitions for this box**

| | |
|---|---|
| $a$ | Intercept of fitted line with $Y$–axis |
| $b$ | Slope of fitted line |
| $b_{lr}$ | Slope of linear regression line |
| $b_{rma}$ | Slope of reduced-major-axis line |
| $N$ | Number of data points |
| $r$ | Correlation coefficient |
| $X$ | $X$ value of data point |
| $X_{mean}$ | Mean value of $X$ for all data points |
| $x$ | Deviation $X - X_{mean}$ for a particular data point |
| $Y$ | $Y$ value of data point |
| $Y_{mean}$ | Mean value of $Y$ for all data points |
| $y$ | Deviation $Y - Y_{mean}$ for a particular data point |

If you fit a straight line through the logarithms of an original set of points $(X, Y)$, either by eye or by calculation, this implies that the line expresses a functional relationship between $X$ and $Y$. For instance, suppose the result of a regression calculation on the logarithms of the original $X$ and $Y$ numbers is

$$\log(Y) = \log(a) + b \log(X), \tag{1}$$

where $a$ and $b$ are constants. This represents a straight line, whose slope is $b$, while $\log(a)$ is the intercept of the line on the $Y$ axis. Reverting to the original numbers, the same equation can also be written

$$Y = aX^b. \tag{2}$$

In this form, the Equation (2) says that $Y$ is proportional to $X$ raised to the power of $b$, while Equation (1) says that $b$, being the slope of the line, can be read directly off the graph. All the graphs in this chapter have been drawn with equal scales on the two axes, meaning that an increment of 1 in $\log(X)$ is represented by the same distance as an increment of 1 in $\log(Y)$. In terms of the original variables, a factor of 10 on the $X$ axis is represented by the same distance as a factor of 10 on the $Y$ axis. This sometimes results in graphs of letter box or chimney shape, but it also means that the exponent $b$ is the same as the gradient of the line, as it is drawn on the page. With practice, exponents can be judged quite accurately by eye from these graphs.

BOX 13.1 *Continued.*

## Linear versus reduced-major-axis regressions

The familiar linear regression is based on the assumption that the $X$ variable is determined without error, while the $Y$ variable, whose value is subject to error, is a function of the $X$ variable. It is commonly used in situations where one variable $(X)$ is clearly independent, and the other $(Y)$ is dependent on it, for example time might be the independent variable $(X)$, while $Y$ is the height of a growing plant, which is measured at fixed time intervals. The linear regression calculation (below) is inherently asymmetrical, and results in a different line, depending which variable is defined as $X$, and which as $Y$, unless all the points happen to fall exactly on a straight line. At the other extreme, if the points are completely random (zero correlation coefficient) the slope is zero, meaning that the line is parallel to the $X$ axis. In this case, if the variables are switched so that the $X$ axis becomes the $Y$ axis and vice versa, the new line is parallel to the new $X$ axis, perpendicular to its previous orientation.

This is not satisfactory if there is no reason to select either variable as independent, for example where wing span is to be plotted against wing area, which could equally well be done either way round. This is the usual situation in allometry, and the usual solution is to calculate a "reduced-major-axis" line, which passes through the mean values of $X$ and $Y$, as in a linear regression, but whose slope is calculated differently, in such a way that the same line results (relative to the data points) whichever variable is chosen to be $X$. If the data are completely random, the slope is 45°, whichever way round the data are plotted.

## Fitting a straight line

The algorithm for fitting a straight line through a given set of data points proceeds in two stages, first the accumulation of sums, and then the calculation of the constants $a$ and $b$ that determine the offset and slope of the line. If a double-logarithmic line is to be fitted (either linear regression or reduced-major-axis) the first operation when each data point is entered is to transform the original values of $X$ and $Y$ into their logarithms. From this point on, the symbols $X$ and $Y$ refer to the logarithms, not to the original numbers. The first stage is to accumulate the following six sums, where $X$ and $Y$ refer to the logarithmically transformed input numbers:

$\sum X$:  The sum of the values of $X$ for all the data points.
$\sum X^2$:  The sum of the squares of the $X$ values.
$\sum Y$:  The sum of the values of $Y$ for all the data points.
$\sum Y^2$:  The sum of the squares of the $Y$ values.
$\sum XY$:  The sum of the products $X$ times $Y$ for all the data points.
$N$:  The number of data points.

All six of these sums are set to zero before commencing the accumulation. The sums are then incremented, one data point at a time, as follows:

$$\sum X := \sum X + X$$

$$\sum X^2 := \sum X^2 + X^2$$

BOX 13.1 *Continued.*

$$\sum Y := \sum Y + Y$$

$$\sum Y^2 := \sum Y^2 + Y^2$$

$$\sum XY := \sum XY + XY$$

$$N := N + 1$$

These are assignment statements, not equations. The ":=" sign can be read as "becomes". For example the value of $\sum X$ is set to zero before accumulation begins and then, as each data point is entered, the new value of $\sum X$ becomes the existing value, plus the value of $X$ for the current data point. Having finished accumulating the sums, the mean values of the transformed $X$ and $Y$ can be calculated from the equations:

$$X_{mean} = \frac{\sum X}{N}, \text{ and} \tag{3}$$

$$Y_{mean} = \frac{\sum Y}{N}. \tag{4}$$

The deviations ($x$ and $y$) of each data point from the means are:

$$x = X - X_{mean}, \text{ and} \tag{5}$$

$$y = Y - Y_{mean} \tag{6}$$

The sums of the squares and product of these deviations are required for the fitted line, and these can be found directly from the accumulated sums, without first calculating the means:

$$\sum x^2 = \sum X^2 - \left[ \frac{(\sum X)^2}{N} \right] \tag{7}$$

$$\sum y^2 = \sum Y^2 - \left[ \frac{(\sum Y)^2}{N} \right] \tag{8}$$

$$\sum xy = \sum XY - \left[ \frac{(\sum X \sum Y)}{N} \right] \tag{9}$$

It should be noted that Equations 7 to 9 each find the difference between two numbers, which are often rather long and nearly the same, differing only in the last few decimal places. It is advisable to use double-precision numbers when accumulating the five sums that appear on the right-hand sides of these three equations.

The slope for the required type of line can now be found, either $b_{lr}$ for a linear regression line, or $b_{rma}$ for a reduced-major-axis line:

$$b_{lr} = \frac{\sum xy}{\sum x^2}, \text{ or} \tag{10}$$

BOX 13.1 *Continued.*

$$b_{\text{rma}} = \sqrt{\left(\frac{\sum y^2}{\sum x^2}\right)} \qquad (11)$$

If $X$ and $Y$ are interchanged, the slope of the reduced-major-axis line from Equation (11) is simply the reciprocal of its former value. That is not the case for the slope of the linear regression line, because Equation (10) will have a different sum of squares of deviations in its denominator, if the variables are interchanged. Whichever version of the slope ($b$) is used, the $Y$-intercept ($a$) is

$$a = Y_{\text{mean}} - bX_{\text{mean}}, \qquad (12)$$

and the correlation coefficient ($r$) is:

$$r = \frac{\sum xy}{\sqrt{(\sum y^2 \sum x^2)}}. \qquad (13)$$

The straight lines in Figures 13.2–13.4 are calculated reduced-major-axis (rma) lines, not linear regression lines (see Box 13.1). The advantage of the rma line is that it gives the same line (relative to the data points), regardless of whether wing span is plotted against mass, or mass against wing span, whereas a linear regression does not, in general, do this. The rma calculation (but not the linear regression) changes the slope of the line to its reciprocal if the axes are transposed. If the slope of wing span versus mass is one third, then the slope of mass versus wing span will be three. Figure 13.2 is a log-log plot of wing span versus mass for the nine birds in Figure 13.1, with a fitted rma line, whose calculated slope is 0.379, a little more than the expected slope of one third (dashed line). Likewise, Figure 13.3 is a log-log plot of wing area versus mass for the same birds, and the slope of the rma line is 0.638, a little less than the expected slope of two thirds. Can these small differences in slope be significant, with only nine data points?

A plot of the aspect ratio (span-squared/area) against the mass contains no further information, as we may anticipate from the first two graphs that its slope will be $(2 \times 0.379) - 0.638 = 0.120$. The slope actually comes out to be 0.123 (Figure 13.4). Not only does this slope differ more obviously from the expected slope (zero), but the three points at the left end of the graph are all below the expected line, whereas the cluster of points at the right end are all above it. To see what this deviation from the expected slope means in terms of wing morphology, we have to go back to the second column of silhouettes in Figure 13.1. The aspect ratio in the last column decreases from 15 to 8, as the birds in the first column get smaller. Inspection of the second column of silhouettes certainly does not suggest that the aspect ratio is independent of the mass. The shapes change progressively from the slender-winged albatrosses at the top to the storm-petrel at the bottom, which has little more than half their aspect ratio. All nine species are closely related in a taxonomic sense, but despite this the storm-petrel looks in silhouette more like a crow than an albatross.

## 13.1.2 FIDUCIAL LIMITS FOR SLOPES OF FITTED LINES

Anyone with a statistical outlook, including most biologists, will want a test to determine whether the slope of a fitted regression or reduced-major-axis line is significantly different from an expected value. The linear regression calculation is based on minimising the sum of the squares of the deviations of the data points *in the Y direction*, above and below the line, and straightforward formulae, based on this sum of squares, are given in statistical textbooks (e.g. Bailey 1995) to estimate the standard error of the slope of a linear regression line. In the case of the reduced-major-axis line, it is the deviations perpendicular to the line that are of interest, not those in the $Y$ direction, and these are not so easy to calculate. Also the effect of the logarithmic transformation of the data points is far from clear, as it may weight the deviations at one end of the line relative to those at the other, so biasing the uncertainty of the slope. Rayner (1985) goes into these difficulties, and provides ways of estimating the standard errors of the slopes of different types of lines. It is possible to derive fiducial limits from Rayner's formulae, but the validity of these depends on assumptions about the distributions of the deviations, which are unfortunately impossible to verify. Therefore I do not give fiducial limits for slopes in this book. My graphs are annotated with the constants that define the fitted line, the number of data points, and the correlation coefficient, and that is as far as I go. The correlation coefficient can be used as evidence that a trend is present, but not to test whether the slope differs from an expected value. There is evidence (above) that the slope of the reduced-major axis line of aspect ratio versus mass is significantly different from zero, but it does not come from statistical tests on the slopes of the three lines in Figures 13.2–13.4.

## 13.1.3 MEASUREMENTS FROM A LARGER DATA SET

The nine species in Figures 13.1–13.4 are a small but homogeneous data set, in that they are closely related species with similar ecology, and all were measured in the same locality (Bird Island, South Georgia) by the same observer (myself). Other much larger databases have been compiled by trawling published data from the literature, but these are of limited use because many of the original field observers had their own definitions for "wing span" and "wing area", and some neglected altogether to say what they meant by these terms, or how the measurements were made. The best known database of this type is by Greenewalt (1962), who was frank and explicit about the shortcomings of his sources. As in other more recent compilations that are based on literature trawling rather than original observation, the data suffer from the limitation that their quality and homogeneity have not been, and cannot be checked, however impressive the number of species that are included.

In the following pages, a number of allometric plots are presented that include 220 bird species from the "Wings" database which comes with the *Flight* programme. The wing measurements in this database are not based on literature trawling, but were made either by myself or by others personally known to me, using the wing measurement procedures described in Chapter 1. Body masses are also taken from the database where birds were weighed in good condition, but in some cases where the wings were measured on birds that were found dead, masses have been taken instead from Dunning (1993) who has trawled a large number of mass measurements from the field ecology literature. Body masses vary far more than wing measurements within most bird species, but most field observers can at least be relied on to mean more or less the same thing by "mass" when they weigh a bird. Some of the following allometric plots are restricted to a subset of 44 species for which measurements of flight muscle mass are available, and these are all my own measurements.

## 13.1.4 ALLOMETRY OF MORPHOLOGICAL VARIABLES FROM THE WINGS DATABASE

Figures 13.5–13.7 are similar to Figures 13.2–13.4, but show 220 bird species from the Wings Database. These represent 18 orders, which are identified by different point styles, according to the key in Table 13.2. In Figure 13.5, the wing span has been plotted (as $Y$) against the body mass (as $X$), as in Figure 13.2. The slope of the reduced-

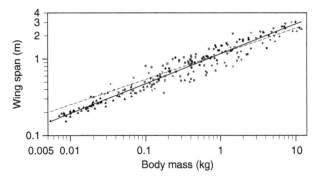

FIGURE 13.5 Allometric plot of wingspan (Y) versus mass (X) for 220 bird species from the Wings Database. The point symbols identify the orders to which the birds belong, according to the key in Table 13.2. Expected slope 0.333. Reduced major axis slope 0.390. N = 220. r = 0.968.

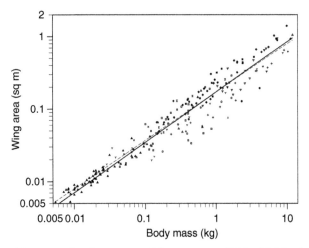

FIGURE 13.6 Allometric plot of wing area (Y) versus mass (X) for 220 bird species from the Wings Database. Expected slope 0.667. Reduced major axis slope 0.694. N = 220. r = 0.967.

major-axis line is almost the same as the slope of Figure 13.2, 0.377 as opposed to 0.379 (expected 0.333) despite the much greater diversity of this data set. This graph differs from the small procellariiform data set in that, although the overall allometric trend is the same, there is more scatter above and below the line, caused by variations of span in birds of similar mass, but belonging to different orders, and adapted for different flight requirements. The lower correlation coefficient (0.968 instead of 0.991) reflects this.

In Figure 13.6 the slope of the rma line for wing area *versus* mass is 0.694, marginally above the expected slope of two-thirds, unlike the

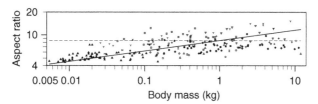

FIGURE 13.7 Allometric plot of aspect ratio (Y) versus mass (X) for 220 bird species from the Wings Database. Expected slope 0. Reduced major axis slope 0.136. N = 220. r = 0.612.

TABLE 13.2 Key to orders

| | |
|---|---|
| + Anseriformes | Ducks Geese Swans |
| × Caprimulgiformes | Nightjars |
| □ Charadriiformes | Gulls Terns Skuas Waders Auks Sheathbills |
| ■ Ciconiiformes | Storks Herons Ibises |
| ◇ Columbiformes | Pigeons |
| ☆ Coraciiformes | Kingfishers |
| ★ Cuculiformes | Cuckoos |
| ◆ Falconiformes | Falcons Hawks Eagles Vultures (Old and New World) |
| ○ Galliformes | Game birds |
| ● Gaviiformes | Divers (loons) |
| △ Gruiformes | Bustards Cranes Limpkin |
| ▲ Passeriformes | Songbirds Crows |
| ▽ Pelecaniformes | Pelicans, Cormorants Anhingas Frigatebirds |
| ▼ Phoenicopteriformes | Flamingos |
| ⊢ Piciformes | Woodpeckers Flickers |
| ⋏ Podicipediformes | Grebes |
| ⋎ Procellariiformes | Petrels Albatrosses |
| ⊥ Strigiformes | Owls |

procellariiform series of Figure 13.3, in which the slope is slightly below two-thirds. The slope of both lines is close to isometric. When wing area is combined with wing span, whose slope is higher than expected, the slope of the line for aspect ratio (Figure 13.7) is 0.136 (expected to be zero). The left-hand end of the line is weighed down by a cluster of solid triangles, which represent small songbirds with short, stubby wings. It seems that only small birds have aspect ratios of 5 or below, hence the slope of the line.

One might imagine that the flight muscle fraction would be higher in bigger birds, because of the scale effects described in Chapter 7 Box 7.4, which cause the power required to fly to increase more steeply than the power available from the muscles. However, Figure 13.8 does not show any such trend. Instead it shows a weak negative slope, albeit with a lot of

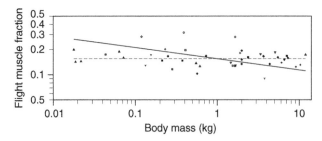

FIGURE 13.8 Allometric plot of flight muscle fraction (Y) versus mass (X) for the 44 bird species in the Wings Database for which flight muscle mass is recorded. Expected slope 0. Reduced major axis slope −0.133. $N = 44$. $r = −0.196$.

scatter and a low correlation coefficient. This has to imply that large birds operate their muscles at a higher specific work than small ones (below).

## 13.2 ⬥ ALLOMETRY OF CALCULATED VARIABLES

The primary variables recorded in the Wings Database are the mass, wing span, wing area and (in some cases) flight muscle mass of individual birds. The means of these for each species are the variables which have been examined for allometric trends in Figures 13.5–13.8. One of the variables which has already been discussed, the aspect ratio (Figure 13.7) is actually a secondary variable, derived from the primary measurements of wing span and wing area. Its allometry results from the allometry of the primary variables from which it is calculated. The *Flight* programme's output lists estimates of many other secondary variables, such as the minimum power speed, the wingbeat frequency and so on, whose values follow directly from the bird's mass, wing span and wing area, provided that all other variables are set to their default values. The remaining allometric graphs in this chapter were constructed by first running a power curve calculation for each of the 220 species in the database, and then going through the Excel output files and picking off the mass of each species, and the value of some variable to be plotted. If this procedure reveals an allometric trend in a secondary variable, it must be due to allometry in the primary data, because the mass, wing span and wing area are the only information that the *Flight* programme has about any species. Other variables that would affect the result are set to fixed values for all species, specifically the air density (set to 1.23 kg m$^{-3}$ for sea level in the International Standard Atmosphere) and the acceleration due to gravity (set to 9.81 m s$^{-2}$). The expected slopes of the lines for different secondary variables (Box 13.2) are calculated on the assumption, which may be

BOX 13.2 **Expected slopes of log-log lines.**

The expected slopes of the reduced-major-axis (rma), double-logarithmic lines for different variables as functions of body mass, are based on the assumption of isometric scaling, specifically that wing span scales with the one-third power of the mass, and wing area with the two-thirds power of the mass. External variables like gravity and the air density scale with the zero power of the mass, that is they are regarded as constants that are not affected by the mass.

**Variables for this box with their expected scaling exponents**

| | |
|---|---|
| $B \propto m^{1/3}$ | Wing span |
| $C_{Db} \propto m^0$ | Body drag coefficient |
| $c_m \propto m^{1/3}$ | Wing mean chord |
| $f_{red} \propto m^0$ | Reduced frequency |
| $g \propto m^0$ | Acceleration due to gravity |
| $k \propto m^0$ | Induced power factor |
| $m$ | All-up mass |
| $m_{musc} \propto m$ | Flight muscle mass |
| $N_{mech} \propto m^0$ | Effective lift:drag ratio based on mechanical power |
| $S \propto m^{2/3}$ | Wing area |
| $S_b \propto m^{2/3}$ | Body frontal area |
| $P_{am} \propto m^{7/6}$ | Absolute minimum power |
| $P_{mech}$ | Mechanical power output of the flight muscles |
| $q_m \propto m^{1/3}$ | Specific work in the flight muscles |
| $Re \propto m^{1/2}$ | Reynolds number of wing |
| $V$ | Airspeed |
| $V_{mp} \propto m^{1/6}$ | Minimum power speed |
| $\rho \propto m^0$ | Air density |
| $\zeta \propto m^0$ | Mitochondria fraction in the flight muscles |
| $\nu \propto m^0$ | Kinematic viscosity of air |

**Minimum power speed**

The formula for the minimum power speed ($V_{mp}$) is given in Chapter 3, as Equation (1) of Box 3.4.

$$V_{mp} = 0.807 k^{1/4} m^{1/2} g^{1/2} \rho^{-1/2} B^{-1/2} S_b^{-1/4} C_{Db}^{-1/4}. \tag{1}$$

There are six variables on the right-hand side of Equation (1) (in addition to the mass itself), two of which, the wing span $B$ and the body frontal area $S_b$, would vary under isometric scaling with the one-third and two-thirds powers of the mass respectively, as indicated in the table of variables above. The numerical constant is, of course, independent of the mass. The air density ($\rho$) and the acceleration due to gravity ($g$), are set to constant, sea-level values for all species, and are therefore independent of the mass as far as the rma line is concerned. Likewise, the induced power factor ($k$) and the body drag coefficient ($C_{Db}$) are set to constant values for this calculation, and are therefore proportional to $m^0$. We can now turn Equation (1) into a proportionality (ignoring the numerical constant), and substitute the power of $m$ with which each variable is assumed to vary:

BOX 13.2 *Continued.*

$$V_{mp} \propto (m^0)^{1/4} m^{1/2} (m^0)^{1/2} (m^0)^{-1/2} (m^{1/3})^{-1/2} (m^{2/3})^{-1/4} (m^0)^{-1/4},$$
$$= m^{1/2} m^{-1/6} m^{-1/6} = m^{1/6}. \tag{2}$$

One-sixth is approximately 0.167, and that is the expected slope of the rma line for minimum power speed versus mass.

### Minimum mechanical power

The absolute minimum power ($P_{am}$) is given in Chapter 3 as Equation (2) of Box 3.4:

$$P_{am} = 1.05 k^{3/4} m^{3/2} g^{3/2} S_b^{-1/4} C_{Db}^{-1/4} \rho^{-1/2} B^{-3/2}. \tag{3}$$

The minimum power is the total mechanical power required to fly at $V_{mp}$, and this is obtained by adding the profile power to $P_{am}$. This depends on the aspect ratio, but as isometric birds all have the same aspect ratio, the slope of the expected line can be found in the same way as for the minimum power speed:

$$P_{am} \propto (m^0)^{3/4} m^{3/2} (m^0)^{3/2} (m^{2/3})^{1/4} (m^0)^{1/4} (m^0)^{-1/2} (m^{1/3})^{-3/2})$$
$$= m^{3/2} m^{1/6} m^{-1/2} \tag{4}$$
$$= m^{7/6}.$$

Seven sixths is approximately 1.17. If the slope were 1, this would mean that the minimum power required to fly is proportional to the mass of the bird, and therefore also (in geometrically similar birds) to the mass of flight muscle that provides the power. A slope greater than 1 means that *each gram of muscle* has to produce more power in a larger bird than in a smaller one.

### Effective lift:drag ratio

This is defined in Equation (3) of Box 3.4:

$$N_{mech} = \frac{mgV}{P_{mech}}. \tag{5}$$

If all the birds fly at $V_{mp}$ (or at some other characteristic speed such as the maximum range speed $V_{mr}$), then Proportionality 2 shows that the speed is expected to vary with the one-sixth power of the mass, and Proportionality 4 that the mechanical power is expected to vary with the seven-sixths power of the mass. $N_{mech}$, the effective lift:drag ratio based on mechanical power, should therefore vary as:

$$N_{mech} \propto m m^0 m^{1/6} m^{-7/6}$$
$$= m^0. \tag{6}$$

### Wingbeat frequency at $V_{mp}$

The wingbeat frequency when cruising at $V_{mp}$ is discussed in Chapter 7, and calculated from Equation (3) of Box 7.3:

$$f = m^{3/8} g^{1/2} B^{-23/24} S^{-1/3} \rho^{-3/8} \tag{7}$$

BOX 13.2 *Continued.*

Substituting as before, we get:

$$f \propto m^{3/8}(m^0)^{1/2}(m^{1/3})^{-23/24}(m^{2/3})^{-1/3}(m^0)^{-3/8}$$
$$= m^{3/8}m^{-23/72}m^{-2/9} \qquad (8)$$
$$= m^{-1/6}.$$

Bigger birds flap their wings at lower frequencies than smaller ones. This is *not* the slope expected for an individual bird, when its mass increases due to laying down fat, without any change to its wing measurements. Equation (7) might suggest that the slope in that case would be 3/8, but actually it is 1/2. The 3/8 exponent of mass in Equation (7) includes 1/8 for the wing's moment of inertia, which does not change when the bird takes on ballast (Pennycuick 1996).

**Specific work in the flight muscles**
Neglecting complications due to mitochondria in the flight muscles, the specific work ($q_m$) in the flight muscles, that is the work done in one contraction by unit mass of muscle, is given in Chapter 7, Box 7.5 as:

$$q_m = \frac{P_{mech}}{(m_{musc}f)}, \qquad (9)$$

where $P_{mech}$ is mechanical power output, $m_{musc}$ is the mass of the flight muscles, and $f$ is the wingbeat frequency. Substituting the exponents calculated above for $P_{mech}$ and $f$,

$$q_m \propto \frac{m^{7/6}}{(mm^{-1/6})} \qquad (10)$$
$$= m^{1/3}.$$

**Reduced frequency**
Reduced frequency ($f_{red}$) is a dimensionless number that describes the geometry of the wingbeat in a way that is related to the type of airflow to be expected around the wing (Chapter 4, Box 4.3). It is defined as:

$$f_{red} = \frac{\pi f c_m}{V}, \qquad (11)$$

where $f$ is the wingbeat frequency, $c_m$ is the mean chord of the wing, and $V$ is the airspeed. Substituting as usual:

$$f_{red} \propto \frac{(m^{-1/6}m^{1/3})}{m^{1/6}} \qquad (12)$$
$$= m^0.$$

Reduced frequency is expected to be independent of body mass.

**Wing Reynolds Number**
The Reynolds number ($Re$) is another dimensionless number, indicating the scale of the flow, and the relative importance of inertial and viscous forces

BOX 13.2 *Continued.*

(Chapter 3, Box 3.6). It has to be defined in terms of a specified linear measurement, which by aeronautical convention is the mean chord ($c_{\mathrm{m}}$) of the wing:

$$Re = \frac{Vc_{\mathrm{m}}}{\nu}, \tag{13}$$

where $\nu$ is the kinematic viscosity of the air. In terms of the body mass:

$$Re \propto \frac{m^{1/6}m^{1/3}}{m^0} \tag{14}$$

$$= m^{1/2}.$$

The wing Reynolds Number of birds cruising at $V_{\mathrm{mp}}$ is expected to vary with the square root of the mass.

regarded as a null hypothesis, that all the birds in the database are geometrically similar.

## 13.2.1 MINIMUM POWER SPEED

The expected slope for the minimum power speed ($V_{\mathrm{mp}}$) is one-sixth (0.167), and the actual slope (Figure 13.9) is marginally less (0.153). The calculated and expected lines are so close together on the graph that the difference is difficult to see. It is due to the higher-than-expected slope of wingspan versus mass (Figure 13.5), with a minor contribution from the slope of wing area versus mass (Figure 13.6) which is also slightly higher than expected.

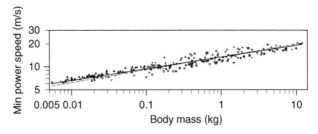

FIGURE 13.9 Allometric plot of minimum power speed (Y), computed by the *Flight* programme, versus mass (X) for 220 bird species from the Wings Database. Expected slope 0.167. Reduced major axis slope 0.153. $N = 220$. $r = 0.947$.

## 13.2.2 MINIMUM MECHANICAL POWER

$V_{\mathrm{mp}}$ is the speed at which the least power is required to fly horizontally (by definition), and so the power required to fly at $V_{\mathrm{mp}}$ is the minimum that will suffice for level flight. A slope of 1 would imply that this power

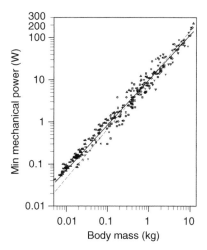

FIGURE 13.10 Allometric plot of minimum mechanical power (Y), computed by the *Flight* programme, versus mass (X) for 220 bird species from the Wings Database. Expected slope 1.17. Reduced major axis slope 1.07. N = 220. r = 0.986.

is proportional to the mass, but because of the expected positive slope of $V_{mp}$ (above), the expected slope for the power is 7/6 (1.17). The computed slope (Figure 13.10) is less than this (1.07), because of the increasing wing span trend, but it is still greater than 1, meaning that the power *required* from each gram of muscle is higher in larger birds than in smaller ones.

## 13.2.3 MAXIMUM EFFECTIVE LIFT:DRAG RATIO

The effective lift:drag ratio at any given speed is the weight times the speed, divided by the power. It directly determines the distance flown per unit mass of fuel consumed (and hence the range), and it reaches its maximum value (by definition) at the maximum-range speed $V_{mr}$. Other things being equal, geometrically similar birds should all have the same maximum effective L/D, that is the expected slope for the rma line in Figure 13.11 is zero, but the computed line deviates strongly above this, with a slope of 0.175. The main cluster of points for small passerines starts at around 7, and there are plenty of points above 20 for medium-sized or larger birds. As these are computed (not measured) values, the trend is entirely due to the allometry of wing span and wing area, and not to differences of Reynolds number, although those might be expected to accentuate the trend.

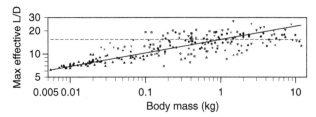

FIGURE 13.11 Allometric plot of effective lift:drag ratio ($Y$) at $V_{mr}$, computed by the *Flight* programme, versus mass ($X$) for 220 bird species from the Wings Database. Expected slope 0. Reduced major axis slope 0.175. $N = 220$. $r = 0.766$.

## 13.2.4 WINGBEAT FREQUENCY

Wingbeat frequency declines as mass increases, and determines the power available from unit mass of flight muscle. The expected slope is minus one-sixth (–0.167), but the computed slope in Figure 13.12 is steeper than this (–0.256). This means that the power *available* from each gram of muscle is less in larger birds than in smaller ones. The discrepancy between power available and power required is the basis of the scaling argument in Chapter 7, Box 7.4. Allometry of the wings does not get rid of the problem. There is still an upper limit to the size and mass of birds that are able to fly horizontally.

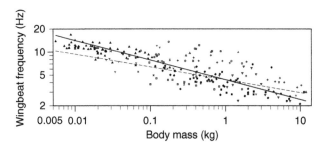

FIGURE 13.12 Allometric plot of cruising wingbeat frequency ($Y$), computed by the *Flight* programme, versus mass ($X$) for 220 bird species from the Wings Database. Expected slope –0.167. Reduced major axis slope –0.256. $N = 220$. $r = -0.820$.

## 13.2.5 SPECIFIC WORK IN THE FLIGHT MUSCLES

*Flight* calculates the power required to fly at $V_{mp}$, and divides it by the wingbeat frequency to get the work done in each contraction. Then it divides by the flight muscle mass to get the specific work, which is the work done by unit mass of muscle in each contraction. The rma line can be computed for the subset of 44 species for which

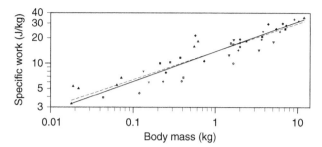

FIGURE 13.13 Allometric plot of specific work in the flight muscles (Y) at $V_{mp}$, computed by the *Flight* programme, versus mass (X) for the 44 bird species in the Wings Database for which flight muscle mass is recorded. Expected slope 0.333. Reduced major axis slope 0.356. N = 44. r = 0.924.

measurements of muscle mass are included in the database (Figure 13.13). The slope of the line (0.356) is close to the expected value (0.333). This gives another view of the reason why there is an upper limit to the mass of birds that are capable of level flight. According to the argument in Chapter 7, the upper limit to the specific work when the muscles are generating maximum power is 0.3 times the isometric stress (560 kN m$^{-2}$), times the active strain (0.26), divided by the density of muscle (1060 kg m$^{-3}$). This works out to 41 J/kg of muscle, which is only marginally above the amount that the largest birds in Figure 13.13 (swans) need for level flight at $V_{mp}$. The estimate for the isometric stress used here is higher than most estimates derived from experiments on isolated muscles, and it comes from the observation that Whooper Swans can in fact fly horizontally (Chapter 7, section 7.3.7).

## 13.2.6 REDUCED FREQUENCY

This is a dimensionless number that basically represents the distance that the wing tip moves up and down for each unit of distance moved horizontally. If the wing tips move up and down very steeply (high wingbeat frequency, low speed) the sharp transitions between upstroke and downstroke tend to produce unsteady aerodynamic effects, whereas if the motion is mostly horizontal, with not too much up-and-down movement, the flow can be considered "quasi-steady", which makes it more amenable to calculation. As a rough rule of thumb, the flow can be considered quasi-steady if the reduced frequency is below about 0.2. The scatter in Figure 13.14 extends from about 0.1 to 0.25, with a computed slope near –0.1. The expected slope for geometrically similar birds is zero.

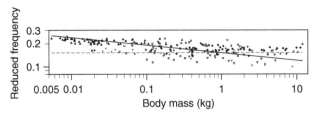

FIGURE 13.14 Allometric plot of reduced frequency (Y) at $V_{mp}$, computed by the *Flight* programme, versus mass (X) for 220 bird species from the Wings Database. Expected slope 0. Reduced major axis slope –0.0991. N = 220. r = –0.617.

### 13.2.7 WING REYNOLDS NUMBER

Reynolds number is another dimensionless number, which describes the scale of the flow. It is often seen as expressing the relative importance of inertial forces, as compared to those due to viscosity. The Reynolds numbers of bird wings (based on the mean chord) run from about 10,000 to 500,000 when flying at $V_{mp}$, increasing with mass at an expected slope of 0.5. The computed slope in Figure 13.15 is slightly less (0.449). Experience with model aircraft, which occupy essentially the same range of Reynolds number as birds, indicates that problems are to be expected in keeping the boundary layer attached to curved surfaces. However, feathered surfaces seem to be remarkably resistant to separation of the boundary layer, for reasons which are not understood (Chapter 3, Box 3.6).

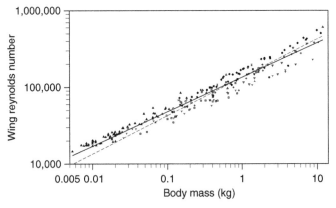

FIGURE 13.15 Allometric plot of wing Reynolds number (Y) at $V_{mp}$, computed by the *Flight* programme, versus mass (X) for 220 bird species from the Wings Database. Expected slope 0.5. Reduced major axis slope 0.449. N = 220. r = 0.977.

## 13.2.8 RATIO OF MINIMUM CHEMICAL POWER TO BASAL METABOLIC RATE

The chemical power is the chemical cost of generating the mechanical power, with an added allowance for the basal metabolic rate (BMR). The BMR varies empirically with the 0.75 power of the mass, whereas the mechanical power varies with about the 1.07 power of the mass (above). The chemical power is not therefore expected to show a straight-line relationship with the mass on a double-logarithmic plot. That would not deter a physiologist from creating such a plot, but it does mean that there is no basis for expecting any particular value for the slope of the line, still less for the slope of a line in which the ratio of the chemical power to the BMR is plotted against the mass. Nevertheless, many physiologists attach great importance to this ratio, to the extent that when they measure a chemical power, they often express it as a multiple of BMR, and suppress the original observations of power. The implication seems to be that the ratio of the chemical power in flight (or "flight metabolism") to the BMR is expected to be the same for any bird, regardless of its wing measurements, or the air density, or even the strength of gravity. This nonsensical expectation is depicted in Figure 13.16, along with a log-log plot of *Flight*'s estimate of the ratio of minimum chemical power to BMR. This ratio increases from less than 3 in some small passerines up to about 50 in some large birds. The reason for the strong positive trend is obvious from the difference in the slopes of the two main components of the ratio (above).

Neither the mechanical nor the chemical power required in flight have any direct connection with the BMR, which is a separate component of

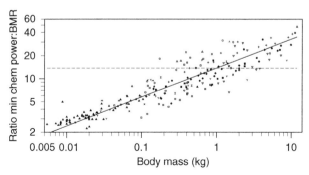

FIGURE 13.16 Allometric plot of the ratio of minimum chemical power to the basal metabolic rate (Y), computed by the *Flight* programme, versus mass (X) for 220 bird species from the Wings Database. Expected slope 0. Reduced major axis slope 0.378. N = 220. r = 0.916.

TABLE 13.3 Regression coefficients from Kirkpatrick (1994).

| | Birds | Bats | Expected exponent |
|---|---|---|---|
| Wing area | $0.134\,B^{1.78}$ | $0.124\,B^{1.83}$ | 2 |
| Wing moment of inertia | $9.23 \times 10^{-4}\,B^{5.08}$ | $1.02 \times 10^{-3}\,B^{5.11}$ | 5 |
| Humerus 2nd moment area | $6.02 \times 10^{-11}B^{4.19}$ | $2.68 \times 10^{-11}B^{4.42}$ | 4 |

Kirkpatrick (1990) also gives the following reduced-major-axis regression for the mass of one wing of a bird ($m_w$) *versus* the body mass ($m_b$):

$$m_W = 9.74 \times 10^{-2} m_b^{1.10}.$$

Note that in this context $m_b$ includes the mass of the wings.

chemical power that is required for reasons that are poorly known, but unconnected with flight. The practice of expressing the chemical power as a multiple of BMR obscures the actual values of the measured power, which could be compared with predicted values from *Flight*, and used to test the hypotheses that underlie the programme. The ratio of power to BMR cannot be used for this.

# 13.3 ▬ VARIATIONS ON ALLOMETRY

## 13.3.1 CHOICE OF INDEPENDENT VARIABLE

Allometric graphs are most commonly drawn with the body mass as the independent variable, but actually any variable can be used in this role. For example, Kirkpatrick (1990, 1994) published a set of allometric equations covering structural measurements of a variety of birds and bats, in which he used the wing span as the independent variable. Three of the most useful of these (given in Table 13.3) can be used to predict the wing area, wing moment of inertia, and second moment of area of the humerus (just distal to the pectoralis insertion) of birds and bats. There is some information about the meaning of these quantities in Chapter 5, Box 5.1, in connection with the strength of the wing structure. They were used by Kirkpatrick to estimate "safety factors", meaning the margin of strength over the maximum loads expected to be applied to the wing.

## 13.3.2 ALLOMETRY OF PHYSIOLOGICAL VARIABLES

There are oblique inferences in various chapters of this book about physiological limitations on flight performance, for example it seems that swans and other large birds have trouble in sustaining aerobic flight, even at the minimum power speed, even though their flight muscles are equal to the mechanical challenge. To be quantitative about this, one would need to know what mass of heart and lungs is needed to sustain a given level of mechanical power in the muscles, how much power the heart and lungs themselves require to pump air and blood at the necessary rates, and how all these quantities would be expected to scale with the body mass in geometrically similar birds. There is insufficient theoretical basis to attempt this at present. However, in one of many famous papers, A.V. Hill (1950) outlined a way in which this connection might be made.

Hill's paper was mostly about running and swimming animals, and he began with some generalisations about running and jumping. Geometrically similar animals all run at the same top speed (he claimed) because the stride length increases with the size of the animal, but the stepping frequency decreases with the same slope. Likewise, similar animals all reach the same height in a standing jump, because the work needed to jump to a given height is proportional to the mass, and so is the work done by the jumping muscles in one contraction. Although somewhat counter-intuitive, these "laws" are by no means wide of the mark (Pennycuick 1975b). When it comes to aerobically sustained power, Hill reasoned that the heart has to pump blood at a volume rate that is proportional to the rate at which the muscles require oxygen, and hence to the power. He proposed that stepping frequency varies inversely with the leg length (or with the minus one-third power of the mass) but that the work done in each contraction of the muscles is directly proportional to the mass. The total power thus varies with the two-thirds power of the mass, as does the cross-sectional area of the aorta. In that case, the speed at which the blood flows along the aorta is independent of the mass, and the time for a hormone signal to be carried from one end of the animal to the other is directly proportional to the length.

This line of argument would have to be modified for birds, as it has been seen in this chapter that the wingbeat frequency varies with the minus one-sixth power of the mass (not minus one-third), and the specific work is not constant as Hill assumed, but increases with the one-third power of the mass. The volume rate of flow of blood, and the power output of the heart (pressure increase times rate of flow), would have to increase at least in proportion to the mass, although the heart's own contraction frequency decreases with the mass at an unknown slope. One aspect of Hill's argument deals with the rate at which oxygen is absorbed in the lung, whose surface area is assumed to vary with the two-thirds power of the mass. However, if birds' lungs behave as fractal surfaces, as suggested in Chapter 7, Box 7.7, then the absorptive surface would not exactly have any such property as "area", and determining its scaling properties would be a major challenge in itself. Half a century on, Hill's paper still gives only a tantalising glimpse of the laws that govern physiological limitations on performance.

# 14

# WIND TUNNEL EXPERIMENTS WITH BIRDS AND BATS

The basic principles of wind tunnel design are introduced in this chapter, with examples of different tunnel types that have been used for bird flight experiments. The functions of the main wind tunnel components are explained, with methods for measuring wind speed and turbulence. Types of measurements that can be made on birds and bats are covered, with precautions that need to be observed, including the recent introduction of methods for observing vortex wakes.

It is in the nature of flying birds that they do not stay still to be observed. However, motion is relative, and it is possible for the bird to stay still relative to the observer, while still flying normally, if the air moves. This may happen naturally where the wind blows against a cliff, or it can be made to happen artificially in a wind tunnel. A wind tunnel is a device that produces a stream of air that flows past the observer at a known speed and in a controlled manner. It is an essential tool in aeronautical research, because a model that remains stationary, with the air moving past it, is physically identical to the same model moving through still air, but much more convenient for observation and measurement.

## 14.1 ◦ WIND TUNNEL BASICS

### 14.1.1 REQUIRED ATTRIBUTES FOR A WIND TUNNEL

The air flow in the test section of a wind tunnel, where the bird flies, must satisfy the following basic requirements, if measurements made in the tunnel are to have some relevance to the bird's performance in the open air.

(1) The experimenter must have access to the bird while experiments are in progress.
(2) The tunnel must be capable of maintaining a steady wind speed, which is constant over the whole cross section of the test section, and can be accurately set and measured by the experimenter.
(3) The direction of flow must be constant throughout the test section, and parallel to the centreline.
(4) The level of small-scale turbulence in the test section must be low.
(5) Finally, the facility to tilt the tunnel so as to simulate climb and descent is so useful that it should be considered an essential requirement.

There is a large volume of published literature, especially in flight physiology, about experiments in wind tunnels that do not satisfy any of the above requirements. Before pointing out the advantages and shortcomings of particular wind tunnel layouts, I shall first mention a couple of principles that apply to all wind tunnels, and then outline the functions of the major components (the fan, settling section, contraction and test section) that are found in all but the most primitive wind tunnels. More detailed information can be found in textbooks on low-speed wind tunnel engineering such as Pankhurst and Holder (1965) and Rae and Pope (1984).

### 14.1.2 BASIC FLOW PRINCIPLES

Birds fly at low airspeeds, at which air can be considered to be an incompressible fluid. This means that if air flows along a tube whose cross section varies in area, the airspeed varies in inverse proportion to the cross-sectional area of the tube. Where the tube gets narrower, the flow speeds up, and where the tube widens, it slows down. This in turn leads to changes in pressure. Everyone seems to be aware of Bernoulli's Principle, which says that the pressure decreases when the speed increases, and vice versa. At places where the tunnel gets wider, the air slows down and its pressure increases, and conversely when the tunnel gets narrower, the air speeds up and its pressure drops. Energy is continuously dissipated by the motion of the air along the tunnel, and is replaced by a motor driving a fan, which is a device that abruptly increases the pressure of the air flowing through it.

## 14.2 ⬟ WIND TUNNEL LAYOUTS

### 14.2.1 OPEN-CIRCUIT SUCTION TUNNELS

In its simplest form, a suction tunnel (Figure 14.1) consists of a contraction whose inlet is open to the ambient air, and a downstream fan that draws air through the contraction and the test section, and then discharges it back into the surroundings. This layout is easy and cheap to construct, and is often recommended by engineers as a minimum-cost solution that is capable of excellent performance. So it is, *if the test section is carefully sealed.* As Figure 14.1 shows, the pressure in the test section has to be below ambient in a suction tunnel, and consequently air rushes in through any opened panels or poorly sealed holes in the test section. Because it is impossible in practice to do an experiment on a live bird that is completely isolated in a sealed chamber, experimenters carry on anyway, oblivious to the invisible

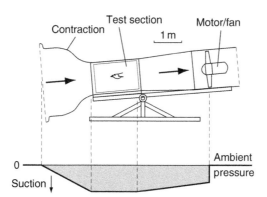

FIGURE 14.1 An open-circuit suction tunnel, as commonly used for physiological experiments. The air is at ambient pressure at the inlet to the contraction, and at the outlet from the fan. The fan's function is to maintain a step increase of pressure as the air flows through it, so the pressure has to be below ambient (grey) everywhere between the inlet and the fan. Air is sucked into the test section through the contraction, but also through any gaps, leaks and holes in the walls of the test section itself. This arrangement works as intended, but only if the test section is carefully sealed. This is practicable with engineering experiments, but usually not with birds, as access to the test section is needed for the experimenter. From Pennycuick et al. (1997).

mayhem caused by open panels and leaks. Suction tunnels can serve as "aerial treadmills" for exercising birds, but quantitative results from such devices should not be taken at face value, however many decimal digits are shown in the results.

## 14.2.2 OPEN-CIRCUIT BLOWER TUNNELS

The shortcomings inherent in the suction arrangement can be avoided by putting the fan at the upstream end of the tunnel, and having the test section in the open air, just outside the outlet of the contraction. This gives unobstructed access to the bird, but requires flow conditioning to straighten the flow, and smooth out the disturbance introduced by the fan. The author's solution for a tilting wind tunnel in a confined space (Figure 14.2) worked quite well, but the turbulence level left something to be desired, as there was too little space to smooth out the disturbance caused by the fan. In principle, the flow in a tunnel of this type can be made as straight and smooth as required, by having a large contraction ratio and many screens in the settling section, but this results in a big machine. The pressure in any open-circuit tunnel is equal to ambient at both the inlet and the outlet, but in contrast to the situation in a suction tunnel, the pressure internally is higher than ambient in a blower tunnel, and returns to ambient at the outlet of the contraction, which is also the beginning of the test section.

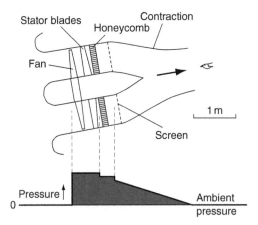

FIGURE 14.2 The pressure in an open-circuit blower tunnel is above ambient (red) everywhere between the fan and the outlet of the contraction. The test section is at ambient pressure, in the open air, outside the end of the contraction. The quality of the flow can be improved with a larger contraction ratio, and a settling section with several screens, but this results in a large machine (after Pennycuick 1968a).

## 14.2.3 CLOSED-CIRCUIT TUNNELS: THE LUND WIND TUNNEL

The Lund wind tunnel was commissioned in 1994, and was the first tunnel to be built specifically for bird flight experiments that also conforms to modern engineering standards of performance (Figure 14.3). It is a conventional low-turbulence closed-circuit wind tunnel, with two special features that are not usually found in engineering wind tunnels. The first is that the whole machine is mounted on a frame that can be tilted by a screw jack at one end. Climbing or descending flight can be simulated by tilting the tunnel to maximum angles of 6° (climb) or 8° (descent). The second is that there is a 50-cm gap at the downstream end of the test section, where the pressure in the tunnel equalises with that of the surrounding air. Everywhere else in the circuit, the internal pressure is higher than ambient (Figure 14.4). This feature allows experimenters free access to the test section without disturbing the flow, an essential characteristic that is not shared by suction tunnels (above). Another tunnel was later built at Andechs, Germany to essentially the same design as the Lund tunnel, by the same contractors (Rollab AB of Solna, Sweden), but the air in the Andechs tunnel circulates horizontally, rather than in the "up-and-over" configuration shown in Figure 14.3. The performance is assumed to be similar, but the Andechs tunnel cannot be tilted, and is therefore limited to horizontal flight.

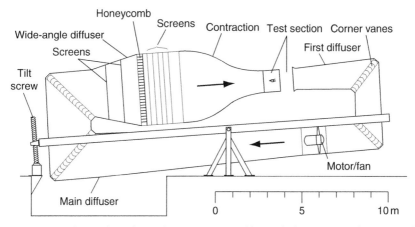

FIGURE 14.3 The Lund wind tunnel is a conventional low-turbulence, recirculating tunnel, modified for use with birds by making the downstream part of the test section open to the surrounding air (see Figure 14.4), and also by mounting the whole machine on a tilting frame. This tunnel has a contraction ratio of 12.25, with a 4-m wide settling section, containing a honeycomb and five screens (after Pennycuick et al. 1997).

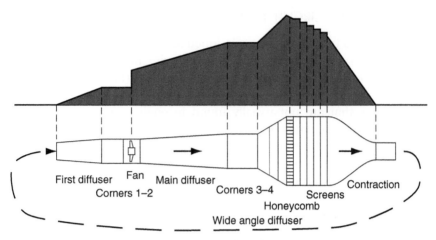

FIGURE 14.4 The Lund wind tunnel (not to scale) is here straightened out, starting and ending at the gap in the test section, where the pressure equilibrates with the surrounding air. The pressure is above ambient (red) everywhere else in the circuit, increasing gradually as the cross section diverges in the diffusers, and abruptly as the air passes through the fan. There is a pressure drop through the honeycomb, and at each screen, followed by a smooth drop back to ambient pressure as the air accelerates in the contraction.

## 14.3 ⬤ WIND TUNNEL COMPONENTS AND THEIR FUNCTIONS

### 14.3.1 DIFFUSERS, CORNERS AND FAN

If we follow the air around a circuit of the Lund wind tunnel, after leaving the test section, it enters the first diffuser, which diverges at a shallow angle, decreasing the wind speed and increasing the pressure. The speed then remains constant round the first two corners, which deflect the air flow through 180°, and deliver it to the fan on the lower level. The fan is a commercial ventilating fan, driven by a 3-phase AC motor, which is itself driven by a sine-wave generator whose frequency can be continuously varied over a wide range. The wind speed does not change as the air passes through the fan, but the pressure increases. After passing through the fan, the air enters the main diffuser on the lower level, which is the structural backbone of the tunnel. Its function is to decrease the wind speed and increase the pressure, converting most of the air's excess kinetic energy into pressure energy. The two "wide corners" turn the air flow back to its original direction on the upper level, using arrays of curved vanes that change the flow direction in the same way as wings. The wide-angle diffuser expands the cross section to 13.7 m² at the entry to the settling section. The air at this point contains turbulence, its speed may be uneven, and there may be swirl left over from the fan, meaning that parts of the flow may be moving at an angle to the tunnel axis. The function of the settling section, which follows, is to remove any such irregularities.

## 14.3.2 THE SETTLING SECTION

As the settling section is the widest part of the tunnel, the speed is lowest there. It is lower than that in the test section by a factor of 12.25, which is the area ratio between the inlet and outlet of the contraction. When the speed in the test section is around 12 m s$^{-1}$, as it often is when birds are flying there, the speed through the settling section is only about 1 m s$^{-1}$, which would be barely perceptible to an observer inside. The settling section contains a honeycomb and five screens, which are placed there because the low airspeed minimises the power that is required to force the air through them. The honeycomb, which comes first, is made of thin sheet metal, and consists of an array of narrow, parallel channels, filling the entire cross section of the tunnel. Its function is to ensure that when the air emerges from the honeycomb, all of it is moving parallel to the axis of the tunnel. Five screens follow, each woven from fine stainless steel wire. Each screen presents some resistance to the flow, so that a pressure drop forms across it. If the speed is higher in one part of the cross section than in another, then the pressure drop across the screen is also higher where the flow is faster, and this produces lateral pressure gradients that tend to even out variations of speed. The screens also break up any vortices that may be present into smaller vortices, which decay more quickly, so that they have time to die out altogether before they reach the test section. The wires of the screens would themselves create small-scale turbulence, but this does not happen if the Reynolds number, based on the diameter of the wires, is less than 40. In the Lund wind tunnel, the Reynolds number of the wires is below this limit when the airspeed in the test section is in the range commonly used for experiments with birds. The screens break up turbulence that was already in the flow, but do not create any new turbulence of their own.

## 14.3.3 THE CONTRACTION AND TEST SECTION

The cross-sectional area decreases in the contraction by a factor of 12.25, and the wind speed increases by the same factor by the time the air is delivered to the inlet of the test section. The contraction itself produces a major reduction of any turbulence that may still be present at the outlet of the settling section, since small fluctuations of speed are carried along unchanged as the air speeds up, and become a smaller percentage of the (increasing) wind speed.

The test section is octagonal, 1.20 m wide and 1.08 m high. This is big enough to accommodate birds with wing spans up to 80 cm, on the basis that the bird's wing span should not exceed two-thirds of the tunnel width. The test section is enclosed by acrylic walls for the first 1.2 m of its length, then there is a 50-cm gap before the air enters the bellmouth of the first, short diffuser (above) leading to the first corner. The function of the gap is to equalise the pressure in the test section with the ambient pressure outside the tunnel, so that experimenters can get at the bird without disturbing the flow. As the wind speed is highest in the test section, the air pressure is lowest there. The pressure is above ambient everywhere else in the circuit, and air rushes out if inspection panels are opened when the tunnel is running (Figure 14.4). Tests with a pitot-static probe (Box 14.1) showed that the wind speed in the test section was within ±1.3% of the mean value over almost the whole of the cross section, while the turbulence was so low that it was difficult to detect it with a hot-wire anemometer (Box 14.2). The root-mean-square value of velocity variations due to turbulence was no higher than 0.04% of the wind speed in the closed part of the test section, and no more than 0.06% in the gap. Both the closed and open parts of the test section can be used for experiments.

**BOX 14.1 Instruments for measuring wind speed.**

The purpose of a wind tunnel is to observe birds flying under known conditions, and the wind speed is the first of those conditions that needs to be measured. Several different methods of wind speed measurement are in common use, based on different physical principles, and each suited to different types of measurements.

**Variable definitions for this box**

| | |
|---|---|
| $C_{d1}$ | Drag coefficient of retreating anemometer cup |
| $C_{d2}$ | Drag coefficient of advancing anemometer cup |
| $h_r$ | Roughness height |
| $k_d$ | Drag constant for anemometer cups |
| $q$ | Dynamic pressure |
| $V_e$ | Equivalent wind speed |
| $V_{h1}$ | Wind speed at height $h_1$ |
| $V_{h2}$ | Wind speed at height $h_2$ |
| $V_r$ | Wind speed due to rotation |
| $V_t$ | True wind speed |
| $\rho$ | Air density |
| $\rho_0$ | Sea level air density in the International Standard Atmosphere |
| $\sigma$ | Density ratio |

**Dynamic pressure and equivalent wind speed**

A pitot-static probe is a standard piece of wind-tunnel equipment which measures the *dynamic pressure*, meaning the pressure difference between an open

BOX 14.1 *Continued.*

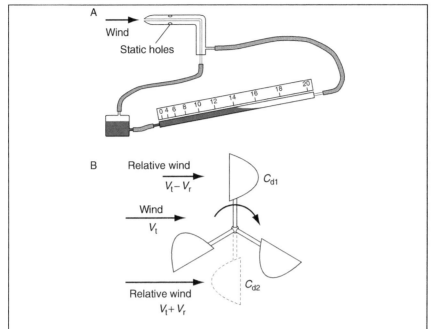

FIGURE 14.5 (A) A pitot-static probe connected to a tilting manometer measures dynamic pressure, but may be calibrated to read airspeed on the basis that the air density is equal to the sea-level density in the International Standard Atmosphere. The instrument then reads Equivalent airspeed, which is less than the True airspeed if the air density is below the sea level value (as it usually is). (B) In a whirling-cup anemometer, the rotation rate depends on the drag coefficient of the cup being higher when the open end faces the relative wind than when the closed end faces the wind. The instrument reads the True airspeed, independently of the air density, so long as the aerodynamic force is large enough to overwhelm the friction in the bearings.

tube pointing into the wind, and a hole or slit in a surface that is parallel to the air flow (Figure 14.5A). Such probes that are intended for measuring the wind speed in low-speed wind tunnels are commonly supplied with a special manometer, in which the tube can be tilted so as to increase its sensitivity. Such manometers, or their electronic equivalents, may be calibrated directly in metres per second for wind speed, but this can be misunderstood. The *true* wind speed ($V_t$) is the speed at which smoke particles are carried along, and it is related to the dynamic pressure ($q$) by:

$$q = \tfrac{1}{2}\rho V_t^2 \tag{1}$$

where $\rho$ is the air density. Conversely, if the dynamic pressure has been measured, and the wind speed is required, it is:

$$V_t = \sqrt{\left(\frac{2q}{\rho}\right)} \tag{2}$$

BOX 14.1 *Continued.*

The air density must be known before the true wind speed can be found from the dynamic pressure. Obviously its value is not known when a manometer is permanently calibrated in terms of wind speed. To do a permanent calibration, a fixed value must be assumed for the air density, and by convention, the value used is $\rho_0$, the air density at sea level in the International Standard Atmosphere, which is 1.226 kg m$^{-3}$. In the wind tunnel, the dynamic pressure (not the wind speed) is actually measured, and the manometer reads the *equivalent* wind speed ($V_e$), defined as:

$$V_e = \sqrt{\left(\frac{2q}{\rho_0}\right)} \tag{3}$$

Only if the ambient air density happens to be equal to $\rho_0$ will the instrument indicate the true wind speed directly. If the true wind speed is required, then the actual air density must be measured, and the "density ratio" ($\sigma$) found as:

$$\sigma = \frac{\rho}{\rho_0} \tag{4}$$

From Equations (2) and (3), the conversion factor from equivalent to true wind speed is "root sigma" ($\sqrt{\sigma}$).

$$V_t = \left(\frac{V_e}{\sqrt{\sigma}}\right) \tag{5}$$

Air density varies from day to day at any particular location, and also decreases with altitude. Only a barometer and a thermometer are required to measure it, (Chapter 2, Box 2.3), and one might suppose that every experimenter would take such an elementary precaution. However, there are numerous examples in the physiological literature of wind tunnel experiments at locations well above sea level, in which "wind speed" was given without specifying the type of instrument used to measure it, or mentioning the air density. This is not a trivial adjustment. For example, at a location 2000 m above sea level, root sigma would be about 0.91 in the Standard Atmosphere, but may vary up and down, depending on the weather conditions.

The instrumentation built into the Lund wind tunnel displays equivalent wind speed, because a bird's minimum power speed ($V_{mp}$), which is the benchmark speed for flight performance (Chapter 3) occurs at a fixed equivalent speed rather than a fixed true speed. Root sigma is also displayed, together with the prevailing air density, so that the true wind speed can be calculated if required. The airspeed used in the displays and output from *Flight* is the true airspeed, and this needs to be borne in mind when using *Flight* as a source of predictions, against which wind tunnel measurements are to be compared. The difference will be substantial at higher locations.

**Whirling cup and turbine anemometers**
At an equivalent wind speed of 8 m s$^{-1}$ or less, the dynamic pressure, at around 4 mm of water or below, is so low as to be difficult to measure.

BOX 14.1 *Continued.*

For such low speeds, other methods of measurement may be more satisfactory. The whirling-cup anemometer, commonly used on weather stations, is a familiar device that will measure wind speed over a wide range, down to a very low minimum. The sensor (Figure 14.5B) consists of several identical cups (at least 3), mounted on arms that project from a central hub. The measured variable is the rotation rate of the shaft, on which the hub is mounted. The device depends on the drag coefficient of the cup being higher when the open end faces the wind than when the closed end is into wind. The rotation rate remains constant when the drag *force* (not the drag coefficient) is the same whether the cup is pointing into wind or downwind. To simplify the matter a little, we can assume that a cup whose open end faces the wind has a drag coefficient $C_{d1}$, and moves downwind at a speed $V_r$, which is directly proportional to the rotation rate, so that its true airspeed is $V_t - V_r$ where $V_t$ is the true wind speed. On the other side of the circle, the drag coefficient ($C_{d2}$) is lower, but the airspeed is higher ($V_t + V_r$). As the drag is the same in both directions,

$$\frac{1}{2}\rho(V_t - V_r)^2 C_{d1} = \frac{1}{2}\rho(V_t + V_r)^2 C_{d2} \tag{6}$$

or:

$$\frac{(V_t - V_r)}{(V_t + V_r)} = \sqrt{\left(\frac{C_{d2}}{C_{d1}}\right)} \tag{7}$$

We can simplify this by introducing a constant $k_d$, where

$$k_d = \sqrt{\left(\frac{C_{d2}}{C_{d1}}\right)} \tag{8}$$

$k_d$ is a "drag constant" (less than 1) that depends only on the shape and arrangement of the cups. Equation (7) can then be rearranged as:

$$V_t = \frac{V_r}{[(1 + k_d)/(1 - k_d)]} \tag{9}$$

Equation (9) shows that for a given set of cups, whose drag characteristics are expressed by the constant $k_d$, the rotation rate is proportional to the true wind speed ($V_t$), and that the air density is not involved. Provided that the friction in the shaft is negligible in comparison with the drag of the cups, a whirling-cup anemometer measures the *true* wind speed, not the equivalent wind speed. This needs to be kept in mind when using an instrument of this type to measure wind speed in the field, and especially when calibrating it in a wind tunnel, where the displayed wind is usually equivalent, not true wind speed. In that case, either the anemometer reading needs to be converted to equivalent wind speed, or else the indicated wind speed needs to be converted to true, before they can be compared [Equation (5) above].

Turbine anemometers work with a small, freely rotating turbine, whose rotation rate is measured by a magnetic or optical sensor. Here too the

BOX 14.1 *Continued.*

rotation rate indicates the true airspeed, provided that the friction in the bearings is negligible compared with the aerodynamic force on the turbine blades. The rotation rate adjusts itself until the lift force on the blades (perpendicular to the air flow) is zero. Being more compact than the whirling-cup type, turbine anemometers are used for airspeed measurement in very slow aircraft, especially hang gliders and paragliders, and they can also be used to measure very low wind speeds in the field or in a wind tunnel. Some small hand-held anemometers are of this type, and measure true wind speed. However, others that work by deflecting a vane against a spring, or lifting a disc in a diverging channel, depend on dynamic pressure, and these instruments measure equivalent wind speed.

**Thermistor anemometers**
Hot wire anemometers are better suited to measuring turbulence than absolute wind speed, and are normally used for that purpose (Box 14.2). However, the same principle is used in some small hand-held anemometers whose sensor element is a thermistor bead at the end of a probe. A thermistor (thermally sensitive resistor) is a device whose resistance varies strongly with temperature. If a thermistor bead is heated, then the rate at which heat is convected away by the air can be used as a measure of airspeed, in much the same way as in a hot-wire anemometer. Thermistor anemometers are popular for measuring very low wind speeds in and around vegetation, but whether they measure true or equivalent airspeed, or something else altogether, is something that would need to be determined by experiment.

**Height correction for a field anemometer**
A portable anemometer mast typically supports the head at some modest height like 3 m, which is usually lower than the flying heights of most of the birds that are tracked. If the wind is blowing over a reasonably flat ground or water surface, the lowest layer of air is retarded by friction with the surface, and the wind speed increases with height, according to a logarithmic relationship. If the wind is measured at a height $h_1$ above the surface, and the bird flies at a measured height $h_2$, then the measured wind speed ($V_{h1}$) at the anemometer head can be corrected to the wind speed at the bird's flying height ($V_{h2}$) by the formula:

$$V_{h2} = V_{h1} \left[ \frac{\ln(h_2/h_r)}{\ln(h_1/h_r)} \right] \tag{10}$$

where $h_r$ is a "roughness length" whose value varies from $10^{-4}$ m for a glassy-smooth surface to $10^{-2}$ m for a very rough surface. $h_r$ is not very critical, and a value of $10^{-3}$ m will serve for medium-rough surfaces.

BOX 14.2 **Measuring turbulence.**

**Variable definitions for this box**

| | |
|---|---|
| $C_D$ | Drag coefficient of test sphere |
| $D$ | Drag of test sphere |
| $k_t$ | Turbulence factor |
| $r$ | Radius of test sphere |
| $Re$ | Reynolds number |
| $Re_{crit}$ | Critical Reynolds number for a sphere in a given wind tunnel |
| $Re_{ref}$ | Critical Reynolds number for a sphere in free air |
| $T$ | Turbulence—rms fluctuations as percentage of wind speed |
| $V$ | True airspeed |
| $\Delta p$ | Pressure drop across a turbulence sphere |
| $\rho$ | Air density |
| $v$ | Air kinematic viscosity |

The term "turbulence", when applied to the air flow in an empty wind tunnel test section, refers to small (millimetre-scale) fluctuations of the wind speed, expressed as deviations from the mean speed. This is conventionally expressed as the root-mean-square (rms) deviation, over a short period of observation, from the mean speed in two directions, parallel to the tunnel axis, and perpendicular to it. A practical criterion for a "low-turbulence" wind tunnel is that the rms value of the fluctuations parallel to the tunnel axis is below 0.1% of the wind speed. The measured value in the Lund wind tunnel (around 0.05%) was so low that the measuring equipment had difficulty in detecting it (Pennycuick et al. 1997). This is a closed-circuit wind tunnel with a contraction ratio of 12.25, and five screens in the settling section (Figure 14.3). Open-circuit suction tunnels of the type shown in Figure 14.1, with contraction ratios of 4 or so, and fewer screens (if any) have turbulence levels at least an order of magnitude higher, and often more. There are two commonly used methods of measuring turbulence, one direct and one indirect.

**Direct measurement of turbulence**
Hot-wire anemometers are commercially available that are capable of the very fast response needed to measure small-scale turbulence directly (Bruun 1995). The sensing device consists of a fine platinum wire which is heated to a temperature far above the air temperature by an electric current (Figure 14.6). The resistance of the wire increases with increasing temperature. The control circuitry continuously monitors the resistance of the wire from the voltage across it and the current through it, and adjusts the current so as to hold the resistance (and hence the temperature) constant at some preset value. Heat is convected away from the wire by the air flow at a rate that depends on the airspeed, and the current needed to maintain the set temperature is a measure of the airspeed. Small fluctuations in the airspeed can be measured at frequencies of several kilohertz. A hot-wire anemometer is not the instrument of choice for absolute measurements of either True or Equivalent airspeed (Box 14.1), but it readily provides the *ratio* of speed fluctuations to the mean speed, as needed for turbulence measurements. Its main limitation is that the sensor elements are delicate, expensive and easily broken.

**BOX 14.2** *Continued.*

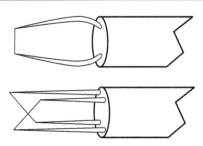

FIGURE 14.6 Two types of hot-wire anemometer sensors made by Dantec Dynamics A/S. The tip of the probe is encased in a 5-mm diameter tube, which is itself mounted in a 20-cm long holder, pointing into wind. Tapered metal prongs are held in a ceramic matrix, and support the anemometer wires at their tips. Dantec Type 55P01 (above) has a single wire of 5 $\mu$m diameter, copper- and gold-plated at each end, while Dantec Type 55P61 (below), is a double sensor with two crossed wires, also of 5 $\mu$m diameter, that is able to measure wind speed fluctuations simultaneously in two directions mutually at right angles.

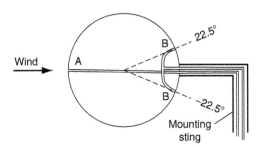

FIGURE 14.7 A turbulence sphere: an external manometer reads the pressure difference between hole (A) at the upwind stagnation point of the sphere, and the ring of holes (B) in the wake, around the support sting. The drag coefficient of the sphere can be determined from this pressure difference.

**The turbulence sphere**
The level of small-scale turbulence can also be measured indirectly by observing the drag coefficient of a sphere (Rae and Pope 1984). An accurately machined, polished metal sphere is supported at the middle of its downwind side by a sting. If the tunnel is equipped with a drag balance, the sting can be mounted on this to give a direct measurement of the drag ($D$), otherwise an equivalent measurement can be obtained from pressure measurements from holes on the surface of the sphere (Figure 14.7 and below). If the drag is measured, the drag coefficient ($C_D$) is:

$$C_D = \frac{2D}{(\rho V^2 \pi r^2)} \tag{1}$$

BOX 14.2 *Continued.*

where $\rho$ is the air density, $r$ is the radius of the sphere, and hence $\pi r^2$ is its cross-sectional area. The value of $C_D$ depends on the position where the boundary layer separates from the surface of the sphere on the downwind side, and this is itself a function of the Reynolds number ($Re$), defined as:

$$Re = \frac{2rV}{v} \tag{2}$$

where $v$ is the kinematic viscosity of the air (Chapter 2, Box 2.2), and the reference length $2r$ is the diameter of the sphere. At low Reynolds numbers, the boundary layer is laminar and readily separates from the surface after passing the widest point, resulting in a wide wake and a high drag coefficient. As the Reynolds number is increased, at some point the boundary layer becomes turbulent, and it then remains attached further round on to the downwind side of the sphere. The wake becomes narrower, and the drag coefficient decreases. The curve of drag coefficient versus Reynolds number dips, and the "critical" value of the Reynolds number ($Re_{crit}$) is defined by convention as the value at which the drag coefficient passes downwards through 0.3 (Figure 14.8). The observed value of $Re_{crit}$ can be used as a measure of the turbulence level in the wind tunnel. The higher the level of turbulence, the lower the value of $Re_{crit}$. The "reference" value of the critical Reynolds number ($Re_{ref}$) is 385,000, because this value is obtained if the sphere is mounted on an aircraft, in free air where there is no turbulence on a scale small enough to affect the boundary layer around the sphere. In a wind tunnel, where there is small-scale turbulence, $Re_{crit}$ is less than $Re_{ref}$, and the more turbulence there is in the air (in terms of percentage of the wind speed), the lower is $Re_{crit}$. Rae and Pope use a "turbulence factor" ($k_t$) which is greater than 1, to define the turbulence level of a particular wind tunnel, where

$$k_t = \frac{Re_{ref}}{Re_{crit}} \tag{3}$$

and they show a graph relating this to the turbulence level ($T$) expressed as a percentage of the wind speed, as above. The lower part of their graph is an approximately straight line:

$$T = 1.25(k_t - 1) \tag{4}$$

If the turbulence factor is 1, meaning that $Re_{crit} = Re_{ref}$, then there is zero turbulence, while if $k_t = 2$, then $T = 1.25\%$, an unacceptably high value. Equation (4) apparently holds up to about $k_t = 2.5$.

**Turbulence factor from pressure measurements**
Since the drag of a sphere is caused by the pressure difference between the upwind and downwind sides, the drag coefficient can be estimated by measuring this difference, using a suitably instrumented "turbulence sphere" as shown in Figure 14.7. The sphere has a single hole (A) in the middle of the upwind side, and a ring of interconnected holes (B) on the downwind side, surrounding the support sting. The tubes are brought out through the support sting, and connected to a manometer outside the tunnel, which registers the

BOX 14.2 *Continued.*

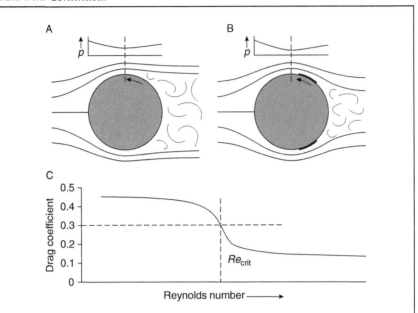

FIGURE 14.8 (A) At low Reynolds numbers the air flow around a sphere (coming from the left), separates from the surface soon after the widest point, where the pressure is minimal. The pressure (*p*, graph at top) drops as the air accelerates around the sphere, until it reaches the widest point (vertical dashed line). Beyond the minimum-pressure point, pressure increases causing a pressure gradient against the wind (curved arrow), which causes the air near the surface of the sphere to reverse direction, and undercut the flow. The boundary layer separates from the surface, leaving a wide, disorganised wake on the downwind side. The drag on the sphere is mostly due to the pressure difference between the upwind and downwind sides. (B) At high Reynolds numbers a thin layer of turbulence forms on the surface, and this helps to keep the boundary layer attached to the surface despite an adverse pressure gradient. The disorganised wake becomes narrower, and the drag coefficient decreases. (C) The formation of the turbulent boundary layer, and resultant narrowing of the wake, results in a dip in the drag coefficient, when plotted against Reynolds number (below). The "critical" Reynolds number ($Re_{crit}$) is defined by convention as the Reynolds number at which the drag coefficient is 0.3. In turbulence-free air, this occurs at $Re = 385,000$, but small-scale turbulence in the air stream triggers the transition at a lower Reynolds number. A measurement of $Re_{crit}$ can be used to estimate the turbulence level (Box 14.2).

pressure difference ($\Delta p$) between holes A and B. This is compared with the dynamic pressure ($q$), defined as:

$$q = \frac{1}{2}\rho V^2 \tag{5}$$

where $\rho$ is the air density and $V$ is the true airspeed. According to Rae and Pope (1984), the drag coefficient is 0.30, and $Re$ [from Equation (2)] is equal to $Re_{crit}$, when the ratio $\Delta p/q$ is equal to 1.220.

BOX 14.2 *Continued.*

**Turbulence measured by PIV**

The recent application of particle imaging velocimetry (PIV) to the study of the wakes of birds flying in the Lund wind tunnel (Spedding et al. 2003a,b) has been outlined in Chapter 4, and this technique can also be used to measure both the airspeed and the turbulence level in the air stream itself. Rather than generating a single number to summarise the turbulence level in the air stream, as a turbulence sphere does, the Lund PIV installation can produce a map of turbulence intensity in its plane of observation. The method was used to assess the effect of a net, which could be installed across the upstream end of the test section during the initial flights of untrained birds, to prevent them from entering the contraction. This was a piece of fishing net of 29 × 29 mm mesh, made from 0.15 mm diameter nylon thread. The PIV system produced an image of the air in a rectangular virtual frame about 180 mm high, whose plane was vertical and aligned with the tunnel axis, and positioned about 1 m downwind of the net, a short distance upwind of the gap between the test section and the first diffuser. The image was a two-dimensional colour-coded map of turbulence intensity in the image plane. It showed almost no turbulence over most of the frame, with sharp horizontal lines of intense turbulence, regularly spaced vertically, corresponding to the individual threads of the net. Such an image could only have been obtained in a tunnel that was initially almost free of turbulence, and it raises a question that could only be investigated in a tunnel of this quality: what is the effect of turbulence in the air stream on the performance of a free-flying bird in gliding or flapping flight? A single vertical wire upstream of the bird could be used to inject a known level of turbulence at a known location on the wing. This might lead to some insights into the behaviour of the boundary layer on a feathered wing.

## 14.3.4 EFFECT OF TUNNEL WALLS ON MEASUREMENTS

There is an extensive literature on the effects of wind tunnel boundaries on the air flow around and forces exerted on a model mounted on a balance. Numerous specific cases are reviewed at length by both Pankhurst and Holder (1965) and Rae and Pope (1984), but the conclusions are not readily adapted to predict the effects on the performance of a free-flying bird in flapping flight. In a closed test section, the wall is a boundary across which the velocity of the flow is zero, whereas in an open test section there is still a boundary, but it is defined by zero pressure gradient across it, rather than zero flow velocity. It seems from Pankhurst and Holder's account that many of the known effects are of equal magnitudes in closed and open test sections, but of opposite sign. Thus the lift coefficient of a model wing is enhanced in a closed test section, as compared to its free-air value, but reduced by a similar amount in an open test section. This raises the possibility that if the same experiment were to be carried out in the closed part of the Lund tunnel's test section, and then repeated in the open part, then the mean of the two sets of performance measurements would be near the free-air value. Where this is not practical, tunnel wall effects can be kept small by restricting experiments to birds whose wing span is no more that two-thirds the width of the test section.

### 14.3.5 WHY DOES TURBULENCE MATTER?

The effect of small-scale turbulence on aircraft models is that it stimulates the formation of turbulence in the boundary layer, that is the layer of air immediately adjacent to the solid surface, where the shear is strong, and the effects of air viscosity are felt. In general, the lifting properties of a wing depend on the boundary layer remaining attached to a curved surface, and these properties break down if the boundary layer separates from the surface. Turbulence in the boundary layer makes it less prone to separation. In the case of a sphere, this effect can be used to measure the turbulence level in the air stream. This method of measuring turbulence is described in Box 14.2, and gives some insight into related effects that occur on wings (Schmitz 1960). The feathered surfaces of the wings and bodies of birds seem to be more resistant to boundary-layer separation than those of inert models, and even those of frozen birds, but it is far from clear how this is achieved. Nothing is known about the effect of turbulence in the air stream on the performance of birds' wings, but these effects could be important. They can only be studied in an air stream that contains a very low level of turbulence.

## 14.4 ◆ BIRDS IN WIND TUNNELS

### 14.4.1 TRAINING AND CONDITIONING

Regardless of the type of measurement, the hypothesis to account for it will include assumptions about what the bird is doing. If the hypothesis comes from a power curve calculated by the *Flight* program, the assumption is that the bird is flying steadily along at a constant airspeed. Although a wide range of birds will fly in a wind tunnel with almost no training, most tend to wander about in the test section, both horizontally and vertically, and to speed up and slow down. Often this behaviour is quite extreme, involving a spurt of activity that takes the bird to the upstream end of the test section, near the roof, after which it coasts back to the downstream end, near the floor, and then repeats the cycle. In general, less energy is consumed by flying steadily at a constant speed, than by maintaining the same average speed, while performing cyclic or irregular manoeuvres. This is not something for which a "correction" can be applied. If the objective of the experiment depends on measuring power, then the bird has to be trained to fly steadily, and maintain its position in the middle of the test section, otherwise the result will be biased upwards by an amount that is unknown, but may be large.

If a small increase in the bird's mass during the experiment can be accepted, then food rewards are a simple and effective method of conditioning (Box 14.3), in which the feeder itself acts as the position reference. As this has to be upwind of the bird's position, it needs to be streamlined, and positioned where its wake will not impinge on the bird's wings. It is easier to train a bird to fly steadily for few seconds than for a few hours, and consequently satisfactory results are more easily obtained from mechanical measurements, based on short video sequences, than from physiological methods that require the bird to fly steadily for hours. Perceiving the necessity of this Rothe and Nachtigall (1987) went to the trouble of selectively breeding a strain of pigeon that naturally flew steadily for hours on end. Other physiologists have

BOX 14.3 **Training birds to fly in a wind tunnel.**

Flying steadily in a wind tunnel is an easy task for a bird, provided that the test section is big enough, the air is smooth, and the wind speed is within the bird's comfortable range, not too fast, and no slower than the bird's minimum power speed. The first step before beginning training is to weigh the bird, and measure its wing span and wing area, following the procedures in Chapter 1, Box 1.3, and then run a power curve calculation in the *Flight* program, the output of which will include an estimate of the bird's minimum power speed ($V_{mp}$). Some experimenters think initially that reducing the wind speed will make it easier for the bird to fly, but this is not necessarily so. It is easier for a bird to fly a little faster than $V_{mp}$ than a little slower. A wind speed below $V_{mp}$ not only makes flight more strenuous (i.e., requires more power from the bird) but also introduces control difficulties (Chapter 9). The wind speed should be set to about 1.1 $V_{mp}$ to start training, and increased rather than reduced if the bird appears to be having difficulty in flying steadily. If the wind tunnel can be tilted to give a downhill gradient of about 1 in 20, this reduces the power required, while still being in a regime (above $V_{mp}$) where speed control is stable (Chapter 9).

The bird needs to be accustomed to the wind tunnel surroundings, and to have a regular routine, in which events happen predictably at the same time each day. If the bird is already tame enough to fly to the experimenter's hand for food when training begins, then a couple of weeks of daily training will usually suffice to train it to fly steadily. The best strategy is to start with several birds, and then concentrate on one or two that respond best to training. If the bird eats something that can be dispensed remotely in small amounts, then food reward is the easiest method of conditioning. Pigeons, for example, eat dry seeds, and are especially fond of rice grains, and small,

BOX 14.3 *Continued.*

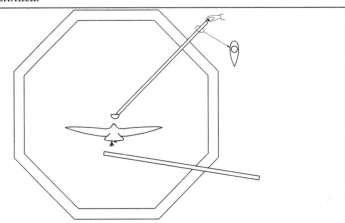

FIGURE 14.9 A feeder used to train pigeons to fly in the blower wind tunnel shown in Figure 14.2. The experimenter rolled dried peas down the tube, and the pigeon retrieved them from the spoon on the end. The tube was encased in a streamlined fairing to minimise its wake. After Pennycuick (1968a).

dried peas. An early conditioning apparatus for pigeons (Pennycuick 1968a) is shown in Figure 14.9. It consisted of a brass tube, passing diagonally downwards from the upper edge to the centre of the outlet of the blower wind tunnel shown in Figure 14.2. A spoon was soldered to the end of the tube, and the experimenter rolled peas down the tube, so that they appeared in the spoon. For the first few days of training, the pigeon stood on a perch, where it could easily reach the spoon, and got its entire daily food ration in this manner. Next, the same procedure was followed with the wind turned on. Then the perch was moved downwards and back, so that the pigeon had to use its wings to reach the spoon. Eventually, the pigeon let go of the perch, and found that it could still reach the spoon while flying, and after that the perch was removed and forgotten. The tube was surrounded by a streamline fairing to keep its wake as narrow as possible, and the diagonal position was designed to keep this wake clear of the pigeon's wings (as well as making peas roll down the tube). A variant of this method, shown in Chapter 6, Figure 6.5, was used in the same wind tunnel to train a fruit-bat whose preferred food was banana. This was supplied through a plastic tube which came up diagonally from below. Training was somewhat more complicated in this case, as the bat's resting position was to hang from its feet below the perch, rather than to stand on it (Pennycuick 1971). The bat had to be trained first to climb over the perch and suspend its weight from its wings, before training could be continued (for 4 months) to the point where it let go of the perch and flew free.

It may be noted that neither pigeons nor fruit-bats normally feed in flight, but both learned to do this without difficulty. Probably just about any flying animal will do this, if a feeder can be designed that supplies food to the head without generating a wake that impinges on the wings. A streamlined

BOX 14.3 *Continued.*

> tube that comes vertically up from below, on the tunnel centre line, is prob-
> ably the best option. It is then up to the experimenter's ingenuity to devise a
> remotely operated food dispenser that can be mounted on the top of the
> tube, and cannot be used by the bird as a perch. There are some birds
> (swifts) that only feed in flight. In this case a vertical tube, ending at the
> centre of the upstream end of the test section, could be connected to a
> chamber containing a swarm of flying insects, which would be carried up
> the tube by a gentle current of air.

usually accepted the erratic behaviour of their birds as "normal", which
it may be. Accountable it is not.

## 14.4.2 WIND TUNNEL MEASUREMENTS ON BIRDS

The types of measurements that can be made on birds in wind tunnels
are mostly different from those that engineers make, because it is not
practical to use balances to measure aerodynamic forces on birds
(below). Measurements on birds flying in wind tunnels fall into three
general categories, physiological measurements on the bird, mechanical measurements on the bird, and measurements on the wake. Whatever the type of measurement, the objective should be to compare the
results with the predictions of a hypothesis. *Flight* can be used as a
source of predictions of power (mechanical or chemical) as a function
of speed in flapping flight, and of glide ratio as a function of speed in
gliding flight. If a discrepancy between an observed and a predicted
quantity proves to be resolvable, by adjusting the values of variables
such as drag coefficients or conversion efficiencies in the *Flight* Setup
screens, this can be used as a means of improving the default values
of quantities that are difficult to measure.

## 14.4.3 MEASUREMENTS OF CHEMICAL POWER

Among early wind tunnel experiments on birds were those of Tucker
(1968b) and Rothe et al. (1987) who measured the rate at which flying bud-
gerigars and pigeons consumed oxygen, using a face mask connected by
tubes to a respirometer outside the test section. A measurement of this
type gives an estimate of the total rate at which the animal consumes fuel
energy, not only for the mechanical requirements of flight, but also for
overheads such as basal metabolism, and the power requirements of the
heart and lungs. It is difficult to separate these components. Also the drag
of the mask and the tube is liable to affect the mechanical power required
by the bird, and this in turn affects the chemical power. The mask changes

the shape of the front end of the body, and may cause flow separation further downstream, which would result in a large drag increase, even if the drag of the mask itself is counterbalanced. The measurements themselves are not easy to interpret, since the conversion from volume of oxygen to fuel energy consumed depends on the type of fuel substrate that is being oxidised. It appears that birds use some carbohydrate as fuel initially, and change progressively to consuming only fat, over the first minutes or hours of flight. Consequently, steady flight has to be maintained for hours to allow the bird to settle into a "physiological steady state", in which the fuel can be identified.

The doubly-labelled water (DLW) method estimates the total amount of carbon dioxide that an animal has produced, in the interval between two blood samples. This can also be used to estimate the chemical energy consumed in the period, if the fuel substrate can be identified. The animal is first injected with a quantity of water, in which some of the hydrogen atoms have been replaced by deuterium ($^2$H), and some of the oxygen atoms by the heavy isotope $^{18}$O. These are stable isotopes, which are not hazardous in any way, and are chemically identical with the common isotopes $^1$H and $^{16}$O. Following the injection of the isotopes, both oxygen and hydrogen are lost as the bird respires, but the labelled oxygen declines faster because oxygen is lost in both water and carbon dioxide, whereas hydrogen is lost only in water. On taking another blood sample after an interval, the amount of carbon dioxide that the animal has produced and lost in the interval can be deduced from the difference in the rate of loss of the two isotopes. Measurement of the isotope abundances requires a mass spectrometer, and in most cases this means that the samples have to be sent to a remote laboratory for analysis. The technique suffers from the same drawback as respirometry as a method of measuring chemical power in the wind tunnel, that the bird has to fly steadily and continuously for several hours.

## 14.4.4 MEASUREMENTS OF FORCE AND WORK

The most common use of wind tunnels in aeronautics is to test small-scale models, that will later be scaled up to full-sized aircraft. This involves mounting the model on a array of balances which, in its most basic form, measures the lift and drag forces on the model, and also its pitching moment, that is the tendency to rotate nose-up or nose-down. This approach has been tried with dead, frozen birds, or parts of birds such as dried wings, but the results are invariably disappointing. As noted in Box 3.2, it is not even possible to get a meaningful drag measurement from a frozen bird body in this way, because the boundary layer will not stay attached to the surface. The same thing happens with isolated wings, which are in any case far too complicated to be set up in a way that simulates a posture used in any identifiable phase of flight. Only the bird knows how to set its wings to achieve a particular result, and consequently wind tunnels intended for experiments on birds have to be designed so that a live bird can be trained to maintain a steady position while flying in the tunnel, without any direct physical restraint.

The basis of mechanical measurements on a free-flying bird begins with steady flight in equilibrium, which only truly occurs in a gliding bird, flying in a tilted air stream. A glide polar (gliding performance curve) can be measured by determining the minimum angle of tilt at which the bird can glide, over a range of different speeds (Chapter 10). This is possible because the bird's weight in gliding equilibrium is balanced by the resultant of lift and drag, allowing the drag, in effect, to be measured from the tilt angle. In horizontal flapping flight, there are cyclic accelerations at the wingbeat frequency and out-of-balance components of force can be measured by observing the accelerations that they cause. This can be combined with observations of wing motions to estimate the work done by the flight muscles in each wingbeat cycle, and hence the mechanical power. A tilting wind tunnel can also be used in flapping-flight experiments to increase or decrease the mechanical power by small amounts, and this can be related to other measurements such as the wingbeat frequency. The observation (from tilting the tunnel) that the wingbeat frequency varies directly with the power allows the wingbeat frequency to be used in turn to identify the minimum power speed in horizontal flight (Pennycuick et al. 1996a).

## 14.4.5 MEASUREMENTS OF MECHANICAL POWER

An early attempt at directly measuring the mechanical power output of the pectoralis muscle (Biewener et al. 1992) was based on measuring the force applied by the muscle at its attachment to the humerus, and the distance through which the muscle shortened at each wingbeat, multiplying these together to get the cycle work, and multiplying by the wingbeat frequency to get the power (Chapter 7). A resistance straingauge was surgically implanted on the upper surface of the deltoid crest of the humerus, to measure the distortion of the crest caused by the downward pull of the muscle, so converting the deltoid crest, in effect, into a spring balance. Shortening of the muscle was estimated from the rotation of the humerus, and examination of superficial fibres of the pectoralis. The work done by the muscle was then estimated by multiplying the supposedly point force by the distance through which some point in the muscle shortened. This is not an ideal way of looking at it, since the pectoralis has a long insertion along the deltoid crest, whose distal end has about twice the moment arm of the proximal end. It would have been better to find the work by the more usual method, as the product of the moment and the rotation angle. The technique was invasive, and suffered from the unavoidable shortcomings that wet bone, being a visco-elastic material, is hardly the material of choice for a spring balance to measure an oscillating force, that implanting the straingauge involved surgery, and that reading its output required a trailing wire to equipment outside the tunnel. The measurements of work and power are at best proportional to the actual values. These experimenters were, however, the first to identify the mechanical power, as opposed to the rate of consumption of fuel energy, as a quantity to be measured in wind tunnel experiments.

The video method of measuring mechanical power does not require any surgery or data wires (Box 14.4). This method has only been tried once, on a swallow flying in the Lund wind tunnel, and the practical implementation left some room for improvement in this case also. The work done by the pectoralis muscle in the interval between successive video frames was calculated by measuring the rotation of the humerus from a video camera that viewed the bird from behind, and

BOX 14.4 **Measuring mechanical power from video.**

The video method of measuring mechanical power requires synchronised sequences from two digital video cameras, while the bird flies steadily for several wingbeats (Pennycuick et al. 2000). One camera views the bird horizontally from the side, and measures the vertical motion of the body at each wingbeat. The other camera is placed downstream of the bird in the tunnel. It views the bird from behind, and measures the angular motion of the humerus.

**Variable definitions for this box**

| | |
|---|---|
| $A$ | Moment arm of lift force about shoulder joint |
| $A_i$ | Moment arm of strip i |
| $a$ | Upward acceleration of body, from video |
| $C_L$ | Lift coefficient of wing |
| $g$ | Acceleration due to gravity |
| $F_u$ | Vertical component of force exerted by humerus on one shoulder joint |
| $L$ | Lift force on one wing |
| $M$ | Moment exerted by pectoralis about shoulder joint |
| $m_b$ | Mass of the body, excluding the wings |
| $P$ | Mechanical power |
| $Q$ | Work done during downstroke |
| $S_i$ | Area of strip i |
| $V_i$ | Local relative air velocity at strip i |

BOX 14.4 *Continued.*

| $\Delta Q$ | Work done in interval between successive video frames |
|---|---|
| $\Delta\varphi$ | Humerus rotation in interval between successive video frames |
| $\varphi$ | Humerus rotation angle above horizontal |
| $\omega$ | Angular velocity of wing |
| $\rho$ | Air density |
| $\tau$ | Wingbeat period |

**Work increments**

Mechanical power is the average rate at which work is done by the pectoralis muscles of both sides, that is, the work done in one wingbeat (the *cycle work*) multiplied by the wingbeat frequency. The work is assumed to be done during the downstroke, defined as the period during which the humerus rotates downwards in each interval between one video frame and the next. During each frame interval, the work done ($\Delta Q$) by the pectoralis muscle of one side is obtained by estimating the moment ($M$) that the muscle exerts about the shoulder joint, and multiplying this by the angle ($\Delta\varphi$) through which the humerus rotates in the interval:

$$\Delta Q = M\Delta\varphi \tag{1}$$

Both $M$ and $\varphi$ are reckoned as negative downwards, making the work done positive. The increment of rotation ($\Delta\varphi$) is measured directly from the rear-view camera, but estimating the moment ($M$) requires information from both cameras.

**Forces and moments**

In Figure 14.10A one wing is shown rotating downwards, and exerting a lift force ($L$), which is perpendicular to the local relative air flow (by definition), and also tilted inwards towards the body. The inward, horizontal component is balanced by the mirror-image inward force from the other wing, but the upward component ($F_u$) is applied to the body at the shoulder joint. The combined force ($2F_u$) on both shoulder joints can be found by measuring the upward acceleration of the body from the side-view camera. The mass of the body without the wings ($m_b$) can be estimated from a published regression by Kirkpatrick (1990), which is given in Chapter 13, Table 13.3. The upward force on one shoulder joint is then:

$$F_u = \frac{[m_b(a + g)]}{2} \tag{2}$$

where $a$ is the measured upward acceleration, and $g$ is the acceleration due to gravity. From Figure 14.10A, the lift force ($L$) is:

$$L = \frac{F_u}{cos\varphi} \tag{3}$$

where $\varphi$ is the rotation angle of the humerus, measured from the horizontal position.

Figure 14.10A shows that it is still necessary to estimate the moment arm ($A$), meaning the distance between the centre of lift and the shoulder joint, before the moment ($M$) itself can be found, since

BOX 14.4 *Continued.*

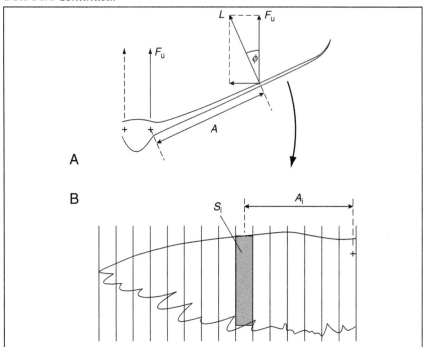

FIGURE 14.10 (A) During the downstroke, the lift on the wing can be considered to be a single force acting in a direction perpendicular to the axis of the wing and to the local relative air flow, at a moment arm $A$ about the shoulder joint. The magnitude of the upward force $F_u$ on the shoulder joint is estimated by measuring the upward acceleration of the body. (B) A wing tracing is ruled into strips, and the moment contributed by strip $i$ is assumed to be proportional to its area $(S_i)$, its moment arm $(A_i)$ and the square of the local relative air velocity (Box 14.4). The sum of the moments for all the strips is divided by the sum of the areas to get the mean moment arm (after Pennycuick et al. 2000).

$$M = LA \qquad (4)$$

The moment arm is found from a tracing of one wing, fully spread, with the position of the centre of rotation of the shoulder joint marked (Figure 14.10B) The tracing is ruled into a number of chordwise strips, for each of which the area and the moment arm is measured. Figure 14.10B shows the area $(S_i)$ and the moment arm $(A_i)$ of strip number $i$. The assumption is that each strip contributes a force proportional to its area, and to the square of the local relative airspeed $(V_i)$, which is the resultant of the forward speed $(V)$ and the upward relative air velocity due to the downward rotation of the wing. For strip $i$:

$$V_i = \surd(V^2 + A_i^2 \omega^2) \qquad (5)$$

where $\omega$ is the angular velocity of the wing. The mean moment arm $(A)$ for the whole wing can be found by summing the moments for all the strips,

BOX 14.4 *Continued.*

and dividing by the sum of the lift contributions for all the strips. Actually, this does not require the actual magnitudes of the forces to be calculated:

$$A = \frac{\sum (S_i A_i V_i^2)}{\sum (S_i V_i^2)} \tag{6}$$

The moment ($M$) can now be found from Equation (4). If it is assumed that the lift coefficient ($C_L$) is the same for all strips, then its value can be found as:

$$C_L = \frac{2L}{[\rho \sum (S_i V_i^2)]} \tag{7}$$

where $\rho$ is the air density.

**Power**
The power is found by summing the increments of work ($\Delta Q$) from Equation (1) for all time intervals during the downstroke. This gives the work ($Q$) done during the downstroke:

$$Q = \Sigma \Delta Q \tag{8}$$

The wing beat period ($\tau$) is the time interval between successive frames when the stroke angle changes from negative to positive, that is, when the humerus passes upwards through the horizontal position. The power ($P$) for the individual wingbeat cycle is the work done during the downstroke, divided by the wingbeat period:

$$P = \frac{Q}{\tau} \tag{9}$$

In steady flight (not sustained for long in these experiments), the power should be the same in each wingbeat. In that case the average power over a series of wingbeats can be estimated by summing the work done in a series of wingbeats, and dividing by the sum of the wingbeat periods.

multiplying this by the moment exerted by the pectoralis muscle about the shoulder joint. The moment was deduced by first measuring the upward force on the shoulder joints, from the acceleration of the body as observed from a side-looking video camera, and then taking account of the geometry of the wing, and its rotation angle. The vertical accelerations were derived from the measured pixel position of a white spot painted on the side the head, which might under-estimate the vertical movements if the bird were to stabilise its head position. Actually wind tunnel birds usually allow the head to oscillate up and down with the body, except when contemplating landing, but it would be better to get the vertical position from a spot in the middle of the bird's back, viewed three dimensionally with a pair of cameras above the bird.

The same method, applied to spots on the upper surface of the wing above the ends of the humerus, would give a better estimate of humerus rotation, which is difficult to measure accurately in the pictures from the rear-view camera.

The measurements (Figure 14.11) showed the measured acceleration oscillating between about $-0.7g$ and $+2g$ meaning that the apparent gravity perceived by the bird oscillated between $+0.3g$ and $+3g$, seven times per second (see Chapter 9). This somewhat rigorous regime would probably be much the same for any bird or ornithopter in steady

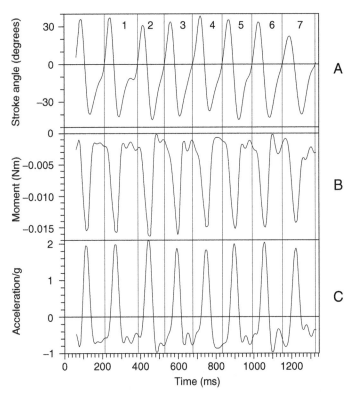

FIGURE 14.11 Results for a series of seven wingbeats in a swallow flying in the Lund wind tunnel. (A) Each wingbeat begins and ends at the point where the humerus passes upwards through the horizontal position. (B) The moment applied by the pectoralis muscle to the humerus is reckoned negative downwards, and reaches a negative peak during each downstroke. It is always below zero, meaning that the wing is never completely unloaded, even during the upstroke. (C) The downward moment during each downstroke produces an upward acceleration of the body. When the acceleration is zero, the bird feels $1g$, as on the ground, while a measured acceleration of $-1g$ corresponds to free fall, when the bird feels weightless. The acceleration varies over a range of more than $2.5g$ at each cycle, at a wingbeat frequency around 7 Hz (after Pennycuick et al. 2000).

flapping flight, except that the frequency would be lower in the larger sizes. It reflects the fact that the wing has to be partially or fully unloaded during the upstroke, allowing the body to accelerate downwards. The apparent gravity therefore has to be more than $+1g$ during the downstroke, so as to maintain the $+1g$ average required for level flight. The observed moment was always negative (directed downwards) but its magnitude dropped almost to zero during the upstroke. The power estimates varied with speed much as expected, but were approximately double the estimates from *Flight*. This was attributed to the fact that the swallow had not been trained to fly steadily in the tunnel, so that measurements could only be obtained when it happened to stay in the camera's field of view for a few wingbeats. Despite its shortcomings, the experiment showed that it is possible to get a measurement of mechanical power from a video sequence lasting only a second or two, without needing surgery, or any direct physical contact with the bird. This is wholly impractical in measurements of chemical power.

## 14.4.6 MEASUREMENTS OF CONVERSION EFFICIENCY

The conversion efficiency of the flight muscles is the ratio of their mechanical power output to the rate at which these muscles consume chemical energy (not to the total chemical power). Despite a copious literature on this subject, only two valid measurements of conversion efficiency in flying birds have ever been published, both by members of the same research group, using the same tilting wind tunnel, and measuring chemical power via the rate of oxygen consumption (Tucker 1972; Bernstein et al. 1973). Although the absolute magnitude of the mechanical power was not measured, a known *increment* of mechanical power could be imposed on the bird, by tilting the tunnel by a small amount. For example, if chemical power is measured initially with the tunnel horizontal, and then again with the tunnel tilted so as to force the bird to climb at, say, $0.1 \text{ m s}^{-1}$, then the mechanical power needed to overcome drag should be the same as before, plus an increment equal to the bird's weight, multiplied by $0.1 \text{ m s}^{-1}$. Dividing this mechanical power increment by the observed increment of chemical power gives the efficiency, without needing to know the absolute values of either the mechanical or the chemical power. Despite some scatter in the efficiency estimates, both experiments gave mean values of 0.23 for the efficiency, which is why this is the default value used in *Flight*. This value is about as expected from the thermodynamics underlying energy conversion, and from classical studies of the energetics of locomotion in mammals (Chapter 7).

Numerous authors have measured chemical power in wind tunnel experiments, but mechanical power has been measured only once (above), and that was in a purely mechanical experiment that did not involve measuring chemical power. To get an estimate of efficiency, both chemical and mechanical power (or increments thereof) have to be measured in the same experiment. It is not valid to estimate the conversion efficiency, as some authors have done, by comparing a measurement of chemical power with a calculated value for mechanical power. In the first place, this overlooks the fact that the chemical power includes other components of chemical energy expenditure, in addition to that due to the mechanical power. Secondly, it neglects anything in the conditions of the experiment that may bias the power, such as shortcomings of the wind tunnel, or erratic behaviour of the bird. If the power is biased upwards, this affects the measured chemical power, but not the calculated mechanical power, and therefore leads to an erroneously low estimate for the efficiency, even if due allowance is made for basal metabolism and other metabolic overheads.

## 14.4.7 OBSERVING VORTEX WAKES

The structure of the "footprints" that a bird leaves in the air, in the form of its vortex wake, is of great interest for the light that it sheds on the aerodynamics of flapping flight, and also as a way to measure the amount of power that the bird is exerting to supports its weight and overcome drag. Being invisible, the wake has to be "visualised" in some way before it can be studied. The first quantitative wake studies were actually not made in a wind tunnel, but on birds flying through still air that had been seeded with tiny soap bubbles filled with helium. If their buoyancy matches that of the surrounding air, the bubbles are not left behind by accelerating air, and can be photographed with multiple flash exposures to measure the local speed and direction of the flow as the air circulates around the vortices in the wake. In this way, Spedding (1986) demonstrated discrete vortex rings in the wake of a jackdaw flying very slowly between two perches a short distance apart, and also measured their momentum and energy. A kestrel flying at a higher speed showed a different type of wake, with continuous wingtip vortices that undulated up and down, and also moved closer together during the upstroke, and further apart during the downstroke (Spedding 1987b). What happens at intermediate speeds remained unknown for another 20 years, despite extensive speculation. A notion arose at this time that birds have discrete "gaits" associated with different wake types, analogous to the walk, trot and canter of a horse, but this idea eventually proved to be erroneous (Chapter 4).

A clearer picture of the structure of bird wakes began to emerge from the application of particle imaging velocimetry (PIV) to the wake of a thrush nightingale flying in the Lund wind tunnel (Spedding et al. 2003b; Rosén et al. 2004). In this technique, the wake is visualised by seeding the air with a thin fog of liquid droplets. As the air flows past the bird, a regularly flashing laser illuminates the fog particles in a thin vertical sheet, which is aligned with the air flow, and photographed with a digital video camera at the side of the tunnel, with its axis perpendicular to the light sheet. Software that compares the pattern of fog particles in two successive frames can determine the average air velocity in the plane of the light sheet, and can also map local variations in the speed and direction of the flow in this plane. A three-dimensional picture of the wake structure was assembled by combining such maps from observations in which the position of the light sheet varied from the bird's centre line to beyond the wing tip.

# 15

# THEORY AS THE BASIS FOR OBSERVATION

This chapter is an attempt to show how the theory outlined in this book provides a backbone that links together a wide diversity of field and laboratory observations, including satellite tracking of migrants, measurements of air speeds, wingbeat frequencies, body drag coefficients and so on. Measurements of one kind affect the interpretations of others, and the *Flight* program has to reconcile them all. The program can be used as an aid in deciding what to measure in a new project.

Many biologists see data gathering as just a matter of collecting a lot of numbers, and doing statistics, but there is more to it. Theory provides a skeleton, to which observations can be attached, making them into a unified model. This chapter is about one thread out of many that can be followed through the apparently heterogeneous collection of topics discussed earlier in this book to show how observations can be used to calibrate theory, how theory can be used to identify what observations are needed, and how measurements should be made. Statistical methods reveal patterns in sets of numbers, but say nothing about what the numbers mean, whereas in the physical approach adopted in this book, the meaning of the numbers has to be clear and explicit before they are measured. For example, one of the commonest types of field observation on bird flight is the measurement of flight speeds of wild birds, but what is "speed" exactly?

# 15.1 ▰ FLIGHT SPEED MEASUREMENTS

## 15.1.1 THE TRIANGLE OF VELOCITIES

Several methods have been used to observe the speeds at which wild birds fly, including ground-based optical measurements, tracking radar, and satellite tracking. Whatever the method, the primary field observation is a *ground speed vector*, that is a measurement of the distance and direction that the bird moves in a measured time, relative to the earth's surface. This vector consists of two numbers, the *ground speed* and the *track*. The ground speed is a *True* as opposed to an *Equivalent* speed (more below on this vital distinction) and the track is the direction in which the bird moves across the map, measured clockwise from true North. To make comparisons with predictions from the *Flight* programme (or actually for just about any purpose), the *airspeed* rather than the ground speed is required. The airspeed is one component of the *airspeed vector*, the other being the *heading*, which is the direction in which the bird is steering (not necessarily the same as the direction in which it is moving). To get an estimated airspeed vector, a measured *wind speed vector* (wind speed and wind direction) has to be vectorially subtracted from the measured ground speed vector. This amounts to solving a triangle of velocities (Figure 15.1A) for each and every speed observation. The Visual Basic functions given in Box 15.1 will solve this triangle, but care is needed because of conflicting conventions for measuring angles (Figure 15.1B). Also, the heading and track are the directions *towards* which the bird is pointing and moving respectively, whereas the wind direction is the direction *from* which the wind is blowing. This is confusing, but there is a lot of history behind it. It was the *north* wind that carried the pharaohs up the Nile, southwards against the current, and so on down the millennia and around the world.

## 15.1.2 METHODS OF MEASURING THE GROUND SPEED VECTOR

### 15.1.2.1 Ornithodolite

An ornithodolite is an optical instrument that the observer aligns by hand on a flying bird. When the observer presses a button, the instrument makes three simultaneous measurements, the distance to the bird, the azimuth angle (direction relative to north), and the angular

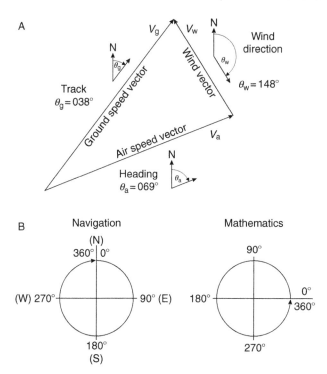

FIGURE 15.1 (A) An example triangle of velocities. Note that the wind direction is by convention the direction *from* which the wind blows, whereas the bird's heading and track are the directions *towards* which the bird points and moves, respectively. The wind direction here is south-east, meaning that the wind vector carries the bird towards the north-west, adding to the north-easterly motion due to its airspeed vector. Addition of the wind and airspeed vectors results in the ground speed vector. Usually, the ground speed vector is measured, and the airspeed vector is found by subtracting the measured wind speed vector from it. (B) Mathematical and navigation conventions measure angles in opposite directions, from different starting points.

altitude (angle above the horizon), and records these, together with the time from the first observation. With a sequence of two or more such three-dimensional positions, the ground distance (both horizontal and vertical) and the track direction can be calculated between each position and the previous one, and divided by the time interval to get the horizontal and vertical components of the bird's ground speed. The original ornithodolite (Figure 15.2) used a 25-cm coincidence rangefinder to measure distance, and home-made 8-bit optical angular encoders to measure the two angles. It worked at distances out to about 100 m, and was first used to collect speed measurements of southern seabirds, both from observing positions on land and also from a moving ship (Pennycuick 1982).

BOX 15.1 **The triangle of velocities.**

The triangle of velocities that relates a bird's ground speed vector to its airspeed vector and the wind speed vector is shown in Figure 15.1A. The usual problem in field studies is to solve the triangle for the airspeed vector corresponding to a measured ground speed vector and wind speed vector. The solution would be straightforward, were it not for difficulties caused by conventions for measuring angles (below, and Figure 15.1B). If the following functions are entered into a Standard Module in Visual Basic, they can be added to any project that involves finding an airspeed vector from ground speed and wind vectors. It is prudent to test these functions thoroughly with input involving track directions around the compass, and winds from around the compass, to make sure that bird and wind directions are the right way round.

### Initialising Pi
Note that Pi is not built in to Visual Basic, and has to be declared. The best place to declare it is at the head of a Standard Module:

```
Option Explicit
Public Pi as Double
```

Its value must be assigned before any procedures that use Pi are called, for example in a Form Load procedure. The following is an easy way to do this:

```
Private Sub Form_Load()
Pi = 4 * Atn(1)
..

..
End Sub
```

The remaining procedures in this box will solve the triangle of velocities according to the conventions that are usual in navigation, with angles in degrees.

### Trigonometric functions in degrees
The following six functions save the hassle of converting angles from degrees to radians and back again, in any project that involves trigonometry.

```
Public Function DegArccos(X As Double) As Double
'Arc Cosine in degrees
DegArccos = 180 * (-Atn(X / Sqr(-X * X + 1)) + (Pi / 2)) / Pi
End Function
```

```
Public Function DegArcsin(X As Double) As Double
'Arc sine in degrees
DegArcsin = 180 * Atn(X / Sqr(-X * X + 1)) / Pi
End Function
```

```
Public Function DegArctan(X As Double) As Double
'Arc tan in degrees
DegArctan = 180 * Atn(X) / Pi
End Function
```

BOX 15.1 *Continued.*

```
Public Function DegCos(X As Double) As Double
'Cosine of an angle in degrees
DegCos = Cos(X * Pi / 180)
End Function

Public Function DegSin(X As Double) As Double
'Sine of an angle in degrees
DegSin = Sin(X * Pi / 180)
End Function

Public Function DegTan(X As Double) As Double
'Tangent of an angle in degrees
DegTan = Tan(X * Pi / 180)
End Function
```

**Vector addition and subtraction**

There is no difficulty in principle in adding two vectors, or subtracting one from another. Each of the input vectors is first converted from polar coordinates (magnitude and angle) into Cartesian coordinates ($X$ and $Y$). The two $X$-components are added (or subtracted as required), and likewise the two $Y$-components. Finally, the resulting $X$ and $Y$ are converted back into polar coordinates, and that is the answer. The problem in navigation is that angles are conventionally defined in a way that differs from the mathematical convention (Figure 15.1B). Navigation angles start at zero (north) and increase clockwise to 360° (also north), whereas zero is east in the mathematical convention, and angles increase *anti*-clockwise from there to 360°. Computer languages normally follow the mathematical convention. More confusion comes from the convention that a bird's heading and track are the directions *towards* which it points and moves, whereas the wind direction is the direction *from* which the wind blows. The wind vector in Figure 15.1A is drawn in the direction towards which the air is moving, so that the motion caused by the wind carrying the bird along is added to the airspeed vector, to get the ground speed vector. However, the wind *direction* is shown 180° from this, because that is the convention. The following two functions take care of converting navigational vectors between polar and Cartesian coordinates, and the next one finds the airspeed if the ground speed and wind are given. Angles are in degrees.

```
Public Sub PolarToCart(R As Double, Theta As Double, _
                  dX As Double, dY As Double)
'Transform a vector from polar coordinates (R, Theta) to Cartesian (dX, dY)
dX = R * DegSin(Theta)
dY = R * DegCos(Theta)
End Sub

Public Sub CartToPolar(dX As Double, dY As Double, _
                  R As Double, Theta As Double)
'Transform a vector from Cartesian coordinates (dX, dY) to polar (R, Theta)
Dim dD As Double
If dY < 0 Then
   dD = 180
   GoTo Skip1
```

BOX 15.1 *Continued.*

```
ElseIf dX < 0 Then
   dD = 360
   GoTo Skip1
Else
   dD = 0
End If
Skip1:
If dY = 0 Then
   Theta = dD + 90
   GoTo Skip2
Else
   Theta = dD + DegArctan(dX / dY)
End If
Skip2:
R = Sqr((dX * dX) + (dY * dY))
End Sub

Public Sub VgToVa(Vg As Double, Trk As Double, _
                  Vw As Double, WindDir As Double, _
                  Va As Double, Hdg As Double)
'Find air speed vector from ground speed and wind vectors
'Vg is the bird's ground speed
'Trk (track) is the direction TOWARDS which the bird is moving
'Vw is the True wind speed
'WindDir is the direction FROM which the wind is blowing
'Va is the bird's True air speed
'Hdg (heading) is the direction TOWARDS which the bird is pointing

Dim Xg As Double 'X component of ground speed vector
Dim Yg As Double 'Y component of ground speed vector
Dim Xw As Double 'X component of wind vector
Dim Yw As Double 'Y component of wind vector
Dim Xa As Double 'X component of air speed vector
Dim Ya As Double 'Y component of air speed vector

Call PolarToCart(Vg, Trk, Xg, Yg)
Call PolarToCart(Vw, WindDir, Xw, Yw)
Xa = Xg + Xw
Ya = Yg + Yw
Call CartToPolar(Xa, Ya, Va, Hdg)
End Sub
```

In a later study (Pennycuick 2001), this same instrument was used to track small birds, but was complemented for larger species by a Leica Vector, which is a pair of 7 × 42 binoculars (excellent for bird watching), with built-in angular encoders for measuring azimuth and altitude, and a laser rangefinder that projects a narrow infrared beam through the

FIGURE 15.2 The author tracking albatrosses and petrels with the original ornithodolite at Bird Island, South Georgia in January 1980. The observer looks through the rangefinder eye-piece (upper right) while steering the instrument with the handle at right to keep it aligned on the bird, and rotating the handle at left to set the rangefinder. Pressing the white button (by right thumb) captured an observation of the bird's azimuth, angular altitude and range, together with the time. The large dial by the observer's right hand is a display of wind speed and direction, coming from a yacht anemometer head, mounted on a mast a few metres away. Data on species and wind were entered with the keyboard and the whole record was recorded on a cassette tape recorder (nearest camera). Photo by J.P. Croxall.

binoculars. This instrument works at ranges out to 600 m or more for medium-sized birds, such as gulls and buzzards (2 km for swans), and only needs a computer interface to make it into an ornithodolite. When used in the same project, the Vector and the ornithodolite were both interfaced to the same computer, which produced data files in a common format from either source. The Vector has the advantage that it will not produce any output unless it is accurately aligned on the bird, and the disadvantages that it is difficult to get rangefinder returns from small or dark-coloured birds, and all too easy to get spurious returns from background objects, including clouds.

## 15.1.2.2 Tracking radar

The principle of tracking radar is the same as that of the ornithodolite, in that the output consists of timed observations of azimuth, angular altitude and range, but the range is measured by timing the echoes from pulses of centimetre-wavelength radio waves, focused into a narrow beam by a parabolic antenna. Once a target is acquired, the antenna is steered automatically to maximise the strength of the echoes. The azimuth and altitude measurements are obtained from the steering mechanism, and are recorded, together with the range, at regular time intervals. Mobile tracking radars are used at weather stations to track balloons, in order to measure winds aloft, and have been acquired by a number of ornithologists, and used to track migrating birds. Individual large birds, or flocks of large or small birds, can be tracked by this method out to distances of 20 km or so. The method works in daylight or at night, and is unaffected by poor visibility, but the consequence of this, and of its long range, is that it is seldom possible to be sure of the target's identity. This means that comparison with a predicted speed is only feasible in circumstances where the target can be visually identified, or is identifiable for some special reason from its radar signature.

## 15.1.3 MEASURING THE WIND VECTOR

To get the bird's airspeed and heading from its ground speed and track, an estimate is required of the wind speed and direction in the immediate vicinity of the bird. In the case of short-range ornithodolite observations, a whirling-cup anemometer is usually mounted on a mast in an exposed position near the ornithodolite, together with a vane to measure the wind direction. This type of anemometer measures true (not equivalent) wind speed, and as the wind speed is slowed down by friction with the ground, the anemometer reading needs to be corrected for the bird's measured flying height. (see Chapter 14, Box 14.1). The anemometer can be interfaced to the computer that is controlling the ornithodolite, so that the wind speed and direction are either recorded directly or entered by hand, immediately after each observation of ground speed. Anemometer readings can be influenced by objects or terrain upwind, and the best anemometer site is on a sea or lake shore, with an onshore breeze. A poorly chosen site results in erratic fluctuations of both wind speed and direction, indicating large-scale turbulence. This also affects the birds, and reduces the reliability of measurements.

Because of the ornithodolite's inherently short range, the wind speed can be observed close to the bird, both in space and time. It is more difficult to observe the wind near to a bird that is observed by tracking radar, as the bird can be kilometres from the observer's position, both horizontally and vertically. The usual solution is to release a helium-filled balloon from time to time, and track it as it rises, so recording a vertical profile of wind speed and direction. This method also measures the true wind speed. Owing to the time required for a balloon ascent, it is not feasible to make this type of wind measurement immediately after each speed observation, and it is normally done at intervals of a few hours. The observer has to assume, in effect, that the wind speed and direction at any particular height changes slowly and linearly with time, which is difficult to verify. Balloon ascents produce reasonably consistent wind measurements so long as the wind is blowing over level terrain, but are useless among mountains, because of the large variations of wind speed and direction, both horizontal and vertical, as the wind finds its way through valleys and around hills.

## 15.1.4 $V_{mp}$ AS A BENCHMARK SPEED

Having measured a sample of airspeeds, what do you do with them? You need a benchmark to measure them against, and the most useful one is the minimum power speed ($V_{mp}$—Chapter 3). Birds like swans, which only have just enough muscle power to fly horizontally, are confined to speeds from just below $V_{mp}$ to just above, but as noted in Section 3.3, speeds below $V_{mp}$ are unstable, and difficult for the bird to maintain steadily, whereas speeds above $V_{mp}$ are stable. It seems to be a valid generalisation (Pennycuick 2001) that birds generally fly about at speeds near or slightly above $V_{mp}$, except when under pressure to maximise their range on long migratory flights, in which case higher speeds are needed (Chapter 8).

The *Flight* programme will generate an estimate of $V_{mp}$ for any bird, but it requires the bird's mass, wing span and wing area (or aspect ratio) as input. These measurements are easy to make on a wind tunnel bird (Chapter 1, Box 1.4), but the field observer cannot measure individual birds, and has to rely on mean measurements from a sample of birds of the same species, which have been caught and measured. *Flight*'s Power Curve calculation will generate a graph showing $V_{mp}$ with uncertainty estimates, if guesses are supplied for the uncertainty of the different variables involved in calculating it (Box 15.2). This is not a statistical calculation, but an idea of the uncertainty of wing measurements can be obtained from the standard deviations of reasonably

BOX 15.2 **Uncertainty estimates for $V_{mp}$.**

---

**Variable definitions for this box**

| | |
|---|---|
| $B$ | Wing span |
| $C_{Db}$ | Body drag coefficient |
| $g$ | Acceleration due to gravity |
| $k$ | Induced power factor |
| $S_b$ | Body frontal area |
| $V_{mp}$ | Minimum power speed |
| $\rho$ | Air density |

The estimates of $V_{mp}$ in the *Flight* programme are found from the stepwise computation of mechanical power (Chapter 3), or analytically by differentiating the expression for the mechanical power of an ideal bird (Equation 1 of Box 3.3) and setting the slope to zero. The numerical result is the same in either case. The analytical version is:

$$V_{mp} = 0.807 k^{1/4} m^{1/2} g^{1/2} \rho^{-1/2} B^{-1/2} S_b^{-1/4} C_{Db}^{-1/4}. \tag{1}$$

As the $V_{mp}$ estimates are not derived from any kind of regression, it is not possible to assign confidence limits to them by statistical methods. Instead, $V_{mp}$ itself is calculated as a function which can be differentiated with respect to each of seven independent variables in turn. The partial differentials are:

$$\partial V_{mp}/\partial k = 0.202 \ k^{-3/4} \ m^{1/2} \ g^{1/2} \ \rho^{-1/2} \ B^{-1/2} \ S_b^{-1/4} \ C_{Db}^{-1/4}. \tag{2}$$

$$\partial V_{mp}/\partial m = 0.404 \ k^{1/4} \ m^{-1/2} \ g^{1/2} \ \rho^{-1/2} \ B^{-1/2} \ S_b^{-1/4} \ C_{Db}^{-1/4}. \tag{3}$$

$$\partial V_{mp}/\partial g = 0.404 \ k^{1/4} \ m^{1/2} \ g^{-1/2} \ \rho^{-1/2} \ B^{-1/2} \ S_b^{-1/4} \ C_{Db}^{-1/4}. \tag{4}$$

$$\partial V_{mp}/\partial \rho = -0.404 \ k^{1/4} \ m^{1/2} \ g^{1/2} \ \rho^{-3/2} \ B^{-1/2} \ S_b^{-1/4} \ C_{Db}^{-1/4}. \tag{5}$$

$$\partial V_{mp}/\partial B = -0.404 \ k^{1/4} \ m^{1/2} \ g^{1/2} \ \rho^{-1/2} \ B^{-3/2} \ S_b^{-1/4} \ C_{Db}^{-1/4}. \tag{6}$$

$$\partial V_{mp}/\partial S_b = -0.202 \ k^{1/4} \ m^{1/2} \ g^{1/2} \ \rho^{-1/2} \ B^{-1/2} \ S_b^{-5/4} \ C_{Db}^{-1/4}. \tag{7}$$

$$\partial V_{mp}/\partial C_{Db} = -0.202 \ k^{1/4} \ m^{1/2} \ g^{1/2} \ \rho^{-1/2} \ B^{-1/2} \ S_b^{-1/4} \ C_{Db}^{-5/4}. \tag{8}$$

If each of the seven independent variables is subject to an error $\Delta k$, $\Delta m$, and so on, each of which is known, and may be positive or negative, then the resulting error in $V_{mp}$ is the sum of the errors contributed by each of the independent variables:

$$\Delta V_{mp} = \Delta k(\partial V_{mp}/\partial k) + \Delta m(\partial V_{mp}/\partial m) + \ldots + \Delta C_{Db}(\partial V_{mp}/\partial C_{Db}). \tag{9}$$

However, if the known errors are replaced by uncertainties, whose magnitude can be estimated, but which are independent of each other, and may be positive or negative, then the uncertainty of $V_{mp}$ is the square root of the sum of the squares of the contributions from each of the variables:

$$\Delta V_{mp}^2 = [\Delta k(\partial V_{mp}/\partial k)]^2 + [\Delta m(\partial V_{mp}/\partial m)]^2 + \ldots + [\Delta C_{Db}(\partial V_{mp}/\partial C_{Db})]^2. \tag{10}$$

BOX 15.2 *Continued.*

---

**Power curve uncertainty in *Flight***
The "Draw graph" option of the *Flight*'s power curve calculation displays graphs of mechanical and chemical power on the screen, with $V_{mp}$ marked. If you click the button for "Plot Uncertainty Bands" it comes up with a form on which you can set the uncertainties of the seven independent variables, on a proportional basis, that is, if you enter 0.05 for the wing span this means that you believe the entered value within ±5%. Actually there is an eighth variable in the list, the wing area, which is included because it has a small effect on the uncertainty of the power, although it has no effect on that of $V_{mp}$ because it does not appear in Equation (1) above. Two preset lists of defaults are provided, one for a wind tunnel bird which has been individually measured, the other for wild birds, where you have to rely on means of samples. These lists are editable. When you have set the values, the programme will add uncertainty bands to the graph, above and below the mechanical power curve, and on either side of $V_{mp}$, and it also shows the magnitude of the uncertainty of $V_{mp}$ directly in metres per second.

   If you set all the uncertainties to zero, then of course the uncertainty of $V_{mp}$ is also zero, but this is not as pointless as it sounds, because you can then set each of the seven variables in turn to a fixed uncertainty of, say, 10%, and see the differing effects of each variable on $V_{mp}$. The result is that a 10% uncertainty in any one of four variables (mass, wing span, gravity and air density) results in a 5% uncertainty in $V_{mp}$, while the same 10% uncertainty in the other three (body frontal area, body drag coefficient, and induced power factor) leads to a 2.5% uncertainty in $V_{mp}$. The effect on the power curve is more interesting, as some variables have much the same effect at any speed, while others have a strong effect at low speeds, but little or none at high speeds, and others again are the other way round. The wing span has a much bigger effect than any other variable at any speed. Moral: always measure the wing span very carefully. More information about the uncertainty calculation, and the thinking behind it, can be found in *Flight*'s online manual, and in Spedding and Pennycuick (2001).

---

homogeneous samples, which are typically 3% of the mean for wing spans, 6% for wing areas, and 12% for masses.

   The estimate of $V_{mp}$ also depends on the values of two environmental variables, the acceleration due to gravity and the air density (Chapter 2). A standard value of 9.81 m s$^{-2}$ for gravity is sufficiently accurate for most purposes, anywhere a bird is likely to go, but the air density is another matter, as it varies strongly with height. In a tracking-radar sample of airspeeds, each bird observed will typically be flying at a different height, and therefore in air of a different density. Even if all birds in a sample of an identifiable species are assumed to have the same morphology, each individual will have a different $V_{mp}$, corresponding to the air density at its flying height. It is not appropriate to use a fixed default value for the air density, as variations of air density are large, and cannot be neglected. The observer should *always* measure the air temperature and the barometric pressure at the observing site, whether in the field or in the laboratory. The air density can be found from this information, and corrected as necessary for any measured difference between the bird's flying height and the observer's position (Chapter 2, Box 2.3).

## 15.1.5 TRUE VERSUS EQUIVALENT AIRSPEEDS

Having got an estimate of the air density for each observation of airspeed, there are two options for pooling the data. One can express the observed True airspeed ($V_{obs}$) for each observation as the ratio $V_{obs}/V_{mp}$, where the minimum power speed ($V_{mp}$) is also a True airspeed, calculated separately for each observation, using the estimated air density for that particular observation. Alternatively, the more usual procedure is to calculate a single value of $V_{mp}$ as an *Equivalent* rather than a True airspeed, by using the sea-level value of the air density from the International Standard Atmosphere. In this case, the observed speeds have to be converted individually from True to Equivalent airspeeds. In other words, speed observations have to be *reduced to sea level* (Box 15.3) before they can be pooled or compared with other data. An estimate of the local ambient air density is required to reduce each individual observation of airspeed. The best idea is to estimate the air density at the bird's flying height in the programme that records the data from the ornithodolite or tracking radar, and record it as part of the field data, rather than doing this retrospectively as an afterthought. *It is not valid to pool measurements of True airspeed from birds that were flying at a variety of different heights.* This is a fundamental point, which has escaped some radar observers.

BOX 15.3 **Reducing observations to sea level.**

For some reason many physiologists seem to have difficulty in grasping the fact that more speed and power are needed to fly at high altitudes than lower down, because the air is less dense up there. This cannot be neglected when comparing field observations of birds that are flying at different heights, or laboratory observations made at wind tunnel sites at different elevations above sea level, or even in the same wind tunnel on days with different weather. This is an everyday problem in aeronautical wind tunnels and flight testing, and the standard solution is to "reduce" all observations to sea level in the International Standard Atmosphere, before comparisons are attempted.

BOX 15.3 *Continued.*

## Variable definitions for this box

$f_e$   Equivalent wingbeat frequency
$f_t$   True wingbeat frequency
$P_e$   Equivalent power
$P_t$   True power
$q$     Dynamic pressure
$V_t$   True airspeed
$V_e$   Equivalent airspeed
$\rho$  Air density
$\rho_0$ Sea level air density in International Standard Atmosphere
$\sigma$ Density ratio

## Dynamic pressure

The first building block for calculating aerodynamic force, work and power is the *dynamic pressure* ($q$), which is the excess pressure (above the ambient atmospheric pressure), measured in an open-ended tube pointing into the incident airflow (Chapter 14, Figure 14.5 A). It is

$$q = \tfrac{1}{2}\rho V_t^2, \tag{1}$$

where $\rho$ (Greek rho) is the air density, and $V_t$ is the "True" airspeed (as distinct from the "Equivalent" airspeed—below). Those aerodynamic forces that are due to the air's inertia (as opposed to those that are due to its viscosity) are directly related to the dynamic pressure. The lift force on a wing is due to the difference in pressure between the lower and upper surfaces, and that pressure difference is proportional to the dynamic pressure. The lift is therefore proportional to the air density, and to the square of the speed, other things (like the angle of attack) being equal. Alternatively, if the lift has to equal the bird's weight, then the speed required to do that, other things being equal, is inversely proportional to the square root of the air density. At higher altitudes, where the air density is less, the bird has to fly faster to maintain constant aerodynamic forces. Since power is drag times speed, it also has to expend more power, in direct proportion to the increased speed.

## Equivalent airspeed

A value must be assigned to the air density before a power curve like that of Figure 3.5 in Chapter 3 can be calculated. The power curve always has a minimum, which occurs at the minimum power speed ($V_{mp}$). If the bird's weight is held constant (i.e. both its mass *and* the acceleration due to gravity are constant), but the air density is varied, $V_{mp}$ will change in inverse proportion to the square root of the air density. The reader can easily verify this by selecting one of the "Preset Birds" in *Flight*, and running the power curve calculation several times, changing the altitude between runs. When you change the altitude on the Setup screen, and press TAB, the programme recalculates the air density and displays it. $V_{mp}$ increases as you climb.

This is not what appears to happen if you have an aeroplane, and check some characteristic speed (like the stalling speed) at different heights. The plane always stalls at the same *indicated* speed, as shown by the airspeed indicator on the instrument panel, regardless of height. That is because

**BOX 15.3** *Continued.*

the airspeed indicator is actually a pressure sensor, which measures the dynamic pressure from a forward-pointing pitot tube. In order to calibrate the instrument in terms of speed (Equation, 1), a value has to be assumed for the air density, and by convention this value is chosen as $1.226$ kg m$^{-3}$, which is the air density at sea level in the International Standard Atmosphere (Chapter 2, Box 2.2). The instrument measures the dynamic pressure, and displays the *Equivalent airspeed* ($V_e$), which is defined as the airspeed that would correspond to the measured dynamic pressure, at sea level in the International Standard Atmosphere. For a bird, so long as its weight does not change, $V_{mp}$ occurs at a constant value of the Equivalent airspeed at any height. In terms of True airspeed, as measured in tracking observations, $V_{mp}$ increases at higher altitudes.

### Reducing measurements of speed and power to sea level

Bird "flight speeds" measured in the field from aircraft, by radar or by optical tracking from the ground are True airspeeds. Two birds of the same species that are flying at the same True airspeed are not necessarily flying at the same point on the power curve, because if the heights are different, then the air density will also be different. The same problem applies to aircraft flight testing, and the conventional solution is to "reduce" all the observations to sea level in the International Standard Atmosphere, where the air density ($\rho_0$) is:

$$\rho_0 = 1.226 \text{ kg m}^{-3}. \tag{2}$$

In the case of airspeeds, this means converting the original measurements from True to Equivalent airspeed. This can be done if the local air density ($\rho$) can be estimated for each individual observation, to give the "density ratio" ($\sigma$) which is:

$$\sigma = \frac{\rho}{\rho_0}, \tag{3}$$

Multiplication by $\sqrt{\sigma}$ converts the observed True airspeed ($V_t$) into the Equivalent airspeed ($V_e$):

$$V_e = V_t\sqrt{\sigma}. \tag{4}$$

Equivalent airspeeds observed at different heights can all be plotted together on the same graph, but True airspeeds cannot. To get *Flight* to calculate a power curve using Equivalent airspeed on the X-axis, just set the altitude to zero (sea level).

Likewise, measurements of True power ($P_t$), whether mechanical or chemical, can be reduced to sea level by calculating an Equivalent power ($P_e$), in the same way as in Equation (4) above, because the dependence of power on air density is the same as that for speed:

$$P_e = P_t\sqrt{\sigma}, \tag{5}$$

where $\sigma$ is the density ratio as before. Other variables do not necessarily depend in the same way on air density, and appropriate formulae have to be used to reduce each variable to sea level, according to its expected

BOX 15.3 *Continued.*

dependence on air density. For example, the wingbeat frequency in cruising flight varies inversely with the 3/8 power of the air density (Equation 2 of Box 7.3). Thus, if the true wingbeat frequency ($f_t$) of a migrant were to be recorded by an accelerometer logger that also records the altitude, the raw observations would need to be reduced to sea level to give the equivalent wingbeat frequency ($f_e$) as:

$$f_e = f_t \sigma^{3/8}. \tag{6}$$

The equivalent wingbeat frequency could be used as a remote fuel gauge to estimate the fat mass remaining (Section 15.3.3).

## 15.1.6 VALIDITY OF $V_{mp}$ ESTIMATES

If a bird's minimum power speed is to be used as a benchmark, against which observed flight speeds are to be measured, then some evidence is needed that the estimates of $V_{mp}$ are accurate. The traditional way in which biologists approach such a question would be to measure a bird's chemical power, by some physiological method such as oxygen consumption, while it is flying horizontally at different steady speeds in a wind tunnel, and then find the minimum in the curve of power *versus* speed. Numerous authors have indeed done experiments of that kind, but their approach invariably was to collect measurements of speed and oxygen consumption and do statistical analyses, without any underlying theory. Against such a background, no purpose is seen to be served by carefully measuring the bird's wing span, still less the air density, and these basic measurements are seldom if ever to be found in physiological papers.

Box 15.4 (Figures 15.3–15.6) outlines an experiment which is in many ways the antithesis of the statistical approach that is usual in biology. $V_{mp}$ was determined for two different birds as they flew in a wind tunnel, without making any measurements of power. Instead, it was established by tilting the tunnel that each bird's wingbeat frequency changed by a small amount in the same *direction* as the mechanical power, up when the power increased, and down when it decreased. In level flight, this was sufficient to determine the speed at which the wingbeat frequency, and hence also the power, passed through a minimum, even though the variation in wingbeat frequency was very small. The estimate of $V_{mp}$ so determined turned out to be considerably faster than that predicted by *Flight* (from Equation 2 of Box 3.3) in both birds, in fact neither bird would fly as slowly as the predicted $V_{mp}$.

BOX 15.4 **Resolution of the body drag anomaly.**

A long-standing anomaly in direct drag-balance measurements of the drag of frozen bird bodies was resolved by measuring the wingbeat frequencies of two birds flying in the Lund wind tunnel (Pennycuick et al. 1996a), without making any direct measurements of drag. Instead, wingbeat frequency measurements provided estimates of each bird's minimum power speed ($V_{mp}$), which did not agree with $V_{mp}$ estimates from *Flight*. The discrepancy was resolved by adjusting the value of the body drag coefficient ($C_{Db}$) used by *Flight* to calculate $V_{mp}$. This has repercussions for any calculations that involve body drag, such as the speed and range of long-distance migrants.

**Variable definitions for this box**

$B$     Wingspan
$C_{Db}$    Body drag coefficient
$g$     Acceleration due to gravity
$k$     Induced power factor
$m$     All-up body mass
$V_{mp}$   Minimum power speed
$S_b$    Body frontal area
$\rho$     Air density

**The observations**

Figure 15.3 shows measurements of the wingbeat frequency of a Teal (*Anas crecca*), flying steadily in the Lund wind tunnel, while the experimenter varied the tilt angle of the tunnel from 1° climb to 6° descent. If the tunnel is tilted to simulate a climb, the effect is to increase the mechanical power required from the muscles, by an amount equal to the bird's weight, multiplied by the rate of climb (vertical component of velocity). The mechanical power needed to overcome drag in level flight was not measured, but is unlikely to change much if the tunnel is tilted by a small amount. It is certainly not likely to decrease in a climb. Thus Figure 15.3 can be understood as showing that if the total power required from the muscles increases, then so does the wingbeat frequency, and *vice versa*. The changes in frequency are small but consistent.

Figure 15.4 shows the wingbeat frequency of the same Teal in level flight, over a range of airspeeds from 10 to 16 m s$^{-1}$. The changes in frequency were again very small (about 3%), but consistent. The graph shows a well-defined "minimum frequency speed" at 12.5 m s$^{-1}$. This is also an estimate of the Teal's minimum power speed ($V_{mp}$), since Figure 15.3 shows that wingbeat frequency varies in the same direction as power (above). Figure 15.5 shows the same curve as that of Figure 15.4 plotted on a scale which goes down to zero. The fitted line looks almost horizontal, and the standard error bars (although plotted) are smaller than the point symbols. On the same graph is a similar curve for a smaller bird, a Thrush Nightingale, which was tested in the same project, and showed a minimum wingbeat frequency at 8.55 m s$^{-1}$.

BOX 15.4 *Continued.*

## The discrepancy

When *Flight* calculated power curves for the two birds, the estimates for $V_{mp}$ were much lower than the experimental values, 8.18 m s$^{-1}$ for the Teal and 5.84 m s$^{-1}$ for the Thrush Nightingale. While the Thrush Nightingale would fly at its supposed minimum power speed, albeit with difficulty and in a pronounced nose-up attitude, the Teal would not fly at all below 10 m s$^{-1}$. *Flight* had underestimated $V_{mp}$. According to the theory in Chapter 3, Box 3.4, which underlies these estimates, $V_{mp}$ is determined by seven variables, the induced power factor ($k$), the bird's all-up mass ($m$), the acceleration due to gravity ($g$), the air density ($\rho$), the wing span ($B$), the body frontal area ($S_b$) and the body drag coefficient ($C_{Db}$):

$$\frac{V_{mp} = (0.807 k^{1/4} m^{1/2} g^{1/2})}{(\rho^{1/2} B^{1/2} S_b^{1/4} C_{Db}^{1/4}).} \tag{1}$$

Three of these ($m$, $B$ and $\rho$) were measured, $g$ was set to 9.81 m s$^{-2}$ (see Chapter 2, Box 2.1), and $S_b$ was calculated from the mass (Chapter 3, Box 3.2). That leaves $k$ and $C_{Db}$ as possible sources for the discrepancy. The default value of $k$ was 1.2, meaning that the induced power is assumed to be 20% higher for a bird's flapping wings than it would be for an ideal actuator disc ($k = 1$). A higher value of $k$ would increase the estimate of $V_{mp}$, but not enough. Even wholly unimaginable values (up to 3) are not sufficient to increase the $V_{mp}$ values up to the experimental values. The default value of $k$ is, in any case, more likely to be too high than too low. Figure 15.6 shows the effect on the estimate of $V_{mp}$ of varying $k$ from 1.0 to 2.2, and $C_{Db}$ from 0.04 to 0.40.

Although there are, at first sight, seven candidate variables, it is clear from Figure 15.6 that actually the choice comes down to $C_{Db}$, the body drag coefficient. The discrepancy means that the default values of $C_{Db}$, used in the calculation, were too high. These values came from a number of apparently careful experiments (Tucker 1973; Prior 1984; Pennycuick et al. 1988), based on mounting frozen bird bodies, whose wings had been removed, on a drag balance in a wind tunnel. This results in drag coefficients that are usually between 0.25 and 0.4, with the lower values in larger birds. The original default formula in *Flight* generated a drag coefficient in this range from the estimated Reynolds number of the body in cruising flight. Such high drag coefficients are characteristic of "bluff bodies", that is shapes that are blunt at the downstream end, rather than tapering to a point, so that the boundary layer separates from the surface somewhere past the widest point, leaving a wide, turbulent wake. Frozen bird bodies visibly generate such a wake, indicated by the feathers at the rear end of the body, which lift away from the body and flutter. This is not seen in living birds flying in the wind tunnel, or in birds like geese or swans, filmed in free flight at close range from an ultralight aircraft.

## Resolution of the discrepancy

To resolve the discrepancy shown in Figure 15.5, $C_{Db}$ was reduced for each bird until the predicted $V_{mp}$ coincided with the experimental value. This required a value of $C_{Db} \approx 0.08$ in both the Teal and Thrush Nightingale.

BOX 15.4 *Continued.*

It was expected that the value of $C_{Db}$ would be higher in the smaller bird, because of the lower Reynolds number at which it flies, but in fact no difference was observed. Following this experiment, the default value of $C_{Db}$ for any bird was reduced in *Flight* to a round value of 0.10.

### The conclusion

The high drag coefficients measured on frozen bird bodies were always seen as anomalous, because bird bodies are faired by their feathered covering in such a way that they appear to taper smoothly to a point at the rear end. This would be biologically meaningless, unless the flow remained attached to the tapering shape. The downward revision of the default body drag coefficient is in effect a hypothesis that the feathered body of a living bird behaves like a streamlined body, with a narrow wake, not like a bluff body. The conclusion is that drag coefficients cannot be measured on dead bird bodies. The boundary layer separates from a frozen body, but does not do this on a living one, for reasons not yet understood. The resolution of this anomaly also resolved other anomalies in which the body drag coefficient is involved, especially the range of certain ultra long-distance migrants, discussed in Chapter 8, Box 8.2. These birds would be able to fly the distances that field observers say they do, if their body drag coefficients are around 0.01, but not if they are in the range 0.25–0.40, as was earlier (wrongly) assumed on the basis of drag measurements on frozen bodies.

This was a major discrepancy, although the theory behind the equation for $V_{mp}$ is simple, and based on principles that have been exhaustively tested in aircraft. It remained possible that a wrong assumed value for one of the seven variables in the equation was the source of the trouble, rather than an error in the theory itself. Five of these variables had been measured (the bird's mass, wing span and wing area, the air density and the strength of gravity), and one of the remaining two (induced power factor) was not capable of resolving the discrepancy even if increased to impossibly high values. The error was traced to an erroneously high value that had been assumed for the remaining variable, the drag coefficient of the bird's body. This is an instructive story in itself, as the erroneous default values came from drag measurements made directly on frozen bird bodies in different wind tunnels, by several different and independent authors. It has always been recognised that these measurements seemed to be improbably high, and the reason appears to be that the air flow separates from a frozen bird body much more readily that it does from the same body when alive—an empirical observation, demonstrated by a statistics-free experiment! The solution was to change the default value of this drag coefficient to 0.10, which is far below any value that has been measured on a frozen bird body, but still on the high side for an artificial streamlined body. This number

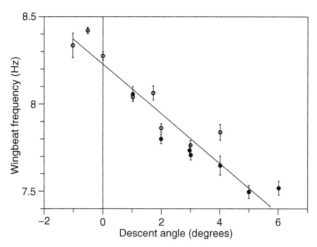

FIGURE 15.3 Wingbeat frequency of a Teal (*Anas crecca*) flying steadily in the Lund wind tunnel at an equivalent airspeed of 13.7 m s$^{-1}$ (open circles) or 15.8 m s$^{-1}$ (solid circles), at different tunnel tilt angles from $-1°$ (climb) to $+6°$ (descent). Zero tilt is level flight. Each point represents the mean of 5 stroboscope observations, and the vertical bars are $\pm 1$ standard error. The correlation coefficient is $-0.946$ for 15 points. After Pennycuick et al. (1996a).

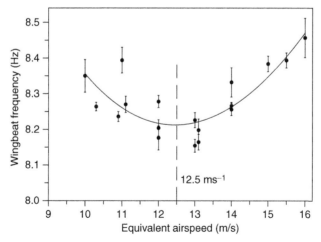

FIGURE 15.4 Wingbeat frequency of the Teal in the Lund wind tunnel as a function of equivalent airspeed. Each point is the mean of 5 stroboscope observations, with bars for $\pm 1$ standard error. Note the suppressed zero on this graph. The variation of wingbeat frequency with speed is small in relation to the mean frequency, but so also is the standard error of the individual points. The fitted line is a generic bird power curve, $f = 5.95 + 21.1/V - 0.000293V^3$, where $f$ is the frequency and $V$ is the speed. After Pennycuick et al. (1996a).

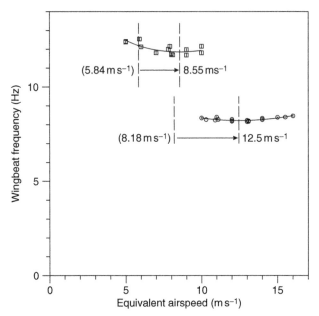

FIGURE 15.5 The same data for the Teal as in Figure 15.4 (circles), compared with similar data for a Thrush Nightingale *Luscinia luscinia* (squares). When the Y-scale is extended to include zero, both curves appear almost straight and horizontal, but the error bars are now so small that they are hidden by the point symbols. In fact both lines are curved, with well-defined minima, as in Figure 15.4. *F*-tests on both curves showed that the variance from the fitted curve is significantly less than from a horizontal line at the mean frequency ($P \ll 0.01$), that is, that the curvature is real. The vertical dashed lines to the left of each curve mark the minimum power speed ($V_{mp}$) as calculated by *Flight* using the original high default values of the body drag coefficient derived from wind tunnel measurements of the drag of frozen bodies. Lowering the assumed drag coefficients to values near 0.08 increased the estimated values of $V_{mp}$, so that both coincided with the minima of the curves as marked. After Pennycuick et al. (1996a).

means that the drag of a bird's body is now assumed to be the same as that of a flat plate that stops the flow completely, over an area that is 10% of the frontal area of the bird's body. It brought the $V_{mp}$ estimates for both of the wind tunnel birds close to the measured values.

## 15.2 ▬ WIND TUNNEL RESULTS RELATED TO FIELD STUDIES

### 15.2.1 BODY DRAG COEFFICIENT AND MIGRATION PERFORMANCE

The body drag coefficient in wind tunnel studies is the same one that is used in *Flight*'s migration simulation, to estimate the work needed to

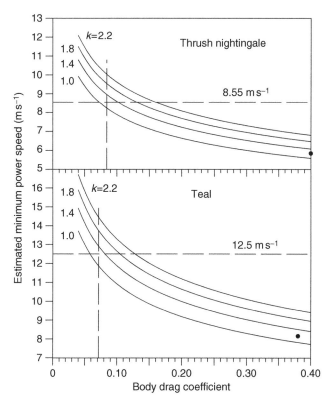

FIGURE 15.6 Calculated curves of the minimum power speed as a function of the body drag coefficient ($C_{Db}$) for the Teal and the Thrush Nightingale, for values of the induced power factor ($k$) from 1.0 to 2.2. The solid circles at the lower right corner of each graph mark the original default values of $C_{Db}$ at the default value for $k$, which is 1.2. The vertical dashed lines are the observed minimum frequency speeds, which correspond to $V_{mp}$, as implied by Figure 15.1. Revision of $k$ does not resolve the discrepancy shown in Figure 15.3, but downward revision of $C_{Db}$ to around 0.08 resolves it in both cases. After Pennycuick et al. (1996a).

propel the body through each kilometre of air, and hence the amount of fuel needed to supply the work. This is not the whole of the work done in migration, but it is a large fraction of it, and this component of the work is directly proportional to the body drag coefficient. The distances flown by ultra-long distance migrants, and the amounts of fuel that they require are constrained by the body drag coefficient. Some field data on two such migrants, a Great Knot migrating from Australia to China, and an Alaskan Bar-tailed Godwit flying non-stop from Alaska to New Zealand, are compared with *Flight*'s simulations in Chapter 8, Box 8.2. One of many points that emerge from these simulations is that neither bird would have been able to fly the distances that field observers say they do, if their body drag coefficients were 0.25 or above, as was formerly believed from measurements on frozen birds' bodies, but that both could reach their known destinations with adequate fuel reserves, if the new default value of 0.1 is assumed (Pennycuick and Battley 2003).

## 15.2.2 GROUND TRACKS FROM SATELLITE TRACKING

Satellite tracking of birds and other animals has become a standard technique in recent years, thanks to the commercial availability of transmitters that can be attached to a bird and tracked by the popular Argos system, and the decreasing size and weight of the transmitters continues to extend this technique to ever smaller birds. Most observers simply track where the animal goes, and where it spends time, which is sufficient to provide a goldmine of data for conservation purposes. However, much more can be learned without a great deal of additional effort, including some details that relate directly to the predictions of the *Flight* programme. Figure 15.7 shows one of four Whooper Swans, a male known from his leg ring as AJU, that were caught and tagged at the Wildfowl and Wetlands trust reserve at Caerlaverock, Scotland, and tracked on their spring migration to Iceland (Pennycuick et al. 1996b). The transmitter is held in place by an elastic belt made of neoprene tape (as used for repairing wetsuits), which retains a constant tension while the swan's body diameter expands and contracts, as it builds up fuel and then consumes it when it migrates. The transmitter is positioned behind the wings, where the belt does not interfere with the patagial membranes (Chapter 5), and where the wings completely cover the box when the swan is on the surface. The curved wire apparently sticking out of the swan's back is the antenna, and the straight probe ahead of it is an air temperature sensor, which was discontinued from later transmitters. Shortly after this picture was taken, the swan was seen dipping his beak in the water and preening the antenna, but not attempting to pull the transmitter off. During the study period, the swan's position was located within a kilometre or two by the Argos system, whenever a satellite passed close enough to pick up the transmissions.

Average ground speed vectors can be obtained directly from the Argos data from the distance, direction and time between successive fixes, but getting wind speed vectors to go with them is more difficult. In this particular project, a meteorologist (Tom Bradbury) was able to interpolate wind vectors for the time and place of each individual Argos fix, because the swans were migrating, mostly very low, across

FIGURE 15.7 Whooper Swan AJU at Caerlaverock in March 1995, shortly after being fitted with a PTT-100 transmitter made by Microwave Telemetry Inc. The transmitter is held in place by an elastic belt made of neoprene tape, which passes twice round the body, with both loops behind the wings. This keeps the belt clear of the patagial membranes, and positions the transmitter where it is completely covered by the wings when the swan is on the surface. The telemetered temperature inside the box then indicates whether the swan is flying or not. The forward probe is an air temperature sensor, and the curved wire is the antenna. Photo by C.J. Pennycuick.

a part of the North Atlantic for which detailed broadcast synoptic charts were available. Airspeeds were concentrated between 20 and 22 m s$^{-1}$, and as wing measurements were obtained for individual swans when they were caught in the winter, these could be compared with $V_{mp}$ estimates, which were 18–20 m s$^{-1}$, depending on the size of the swan. These are retrospective estimates, based on a body drag coefficient which was revised down to 0.1 subsequently to this project (Box 15.4). Less than 5% of airspeed estimates exceeded 22 m s$^{-1}$, and the highest was 27 m s$^{-1}$. The maximum range speed at around 32 m s$^{-1}$ was clearly far beyond these swans' reach, even after their fat reserves were depleted.

## 15.2.3 CONTRARY WINDS AND NAVIGATION

AJU's track when he migrated to Iceland is shown in Figure 15.8. After departing from Caerlaverock, he flew half way across from the Outer Hebrides to Iceland in fine weather, but was then forced down at about 61°N 11°W by poor visibility, low cloud and rain, associated with a warm front. He stayed on the water for about 30 hours. When the front eventually cleared, a west-south-westerly gale was blowing at about 19 m s$^{-1}$, which was near AJU's minimum power speed (about 20 m s$^{-1}$). The swan

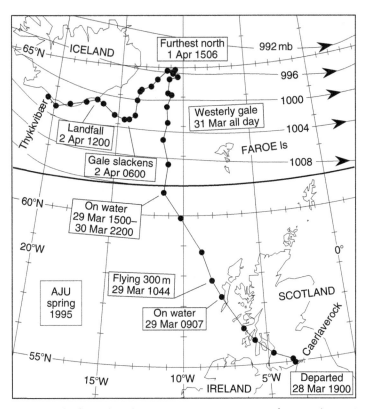

FIGURE 15.8 Track of a male Whooper Swan (AJU) migrating from south-west Scotland to Iceland in Spring, observed by the Argos satellite system. The heavy line at 61°N marks the position where the swan was forced down by poor visibility and rain due to a slow-moving front, and stayed on the surface for 30 hours. After the front cleared, the swan flew all day despite a westerly gale which forced him far to the north of the direct track. When it slackened he returned to his original track, before turning north-west to the Icelandic coast near Höfn. After Pennycuick et al. (1996b).

kept going, but could not make progress against the wind on his original north-westerly track. He was forced far to the north, between the east end of Iceland and the Faroe Islands. When the gale eventually slackened, AJU did not go directly in to the Icelandic coast (which was not far away by then), but cut back south-west to his original track, before turning north-west and making his landfall in the usual place for incoming whoopers, in the middle of the south-eastern coast. Another northbound male (JAP) was forced out to the west by a northerly gale, and appeared in danger of missing Iceland altogether. Instead, he reached the southern tip of Iceland by increasing his airspeed to 27 m s$^{-1}$ for 3 hours, the highest airspeed seen during the project. Both tracks (Figure 15.9) give a strong impression (without actually proving it) that the swans were

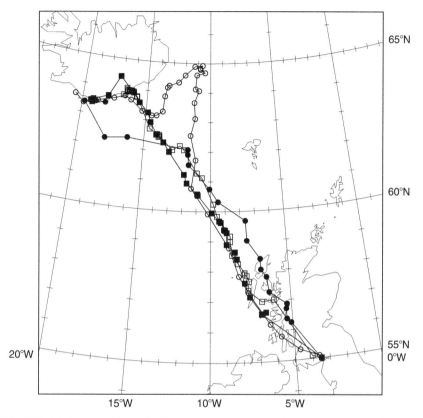

FIGURE 15.9 Four northbound Whooper Swan tracks, including AJU's track shown in Figure 15.8 (open circles). Another swan (JAP—solid circles) was caught by a northerly gale during the sea crossing, and forced far to the west of the direct track. He reached the south coast of Iceland by increasing his airspeed to 27 m s$^{-1}$ and maintaining this for three hours—the highest airspeed observed during the project. After Pennycuick et al. (1999).

aware of their position, relative to the required track, throughout the crossing, and were able to take appropriate corrective action when forced off course. The tracks of two other northbound swans that did not encounter contrary gales are also shown in Figure 15.9, and these were able to maintain a straight track for Iceland over the whole of the 800 km sea crossing.

## 15.2.4 FLYING HEIGHT

The transmitters fitted to the whoopers also included barometric pressure sensors, which can be used as altimeters if the surface pressure is known, as it was from the meteorological analysis. Figure 15.10 shows

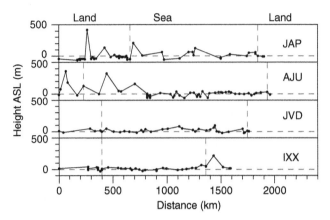

FIGURE 15.10 Heights ASL from telemetered barometric pressure for the four tracks shown in Figure 15.9. The vertical dashed lines are the transitions from land to sea and back to land. After Pennycuick et al. (1999).

the flying heights recorded for the four northbound tracks in Figure 15.9. The swans sometimes climbed a few hundred metres above sea level (ASL) in fine weather when forced to do so by terrain, but mostly flew just above the surface, stopping from time to time on the water. Southbound tracks from the same project showed some flying heights up to nearly 2000 m, by swans that gained height in order to cross high ground in Iceland, and then stayed up there over the sea. The information on one of our whoopers (JAP) in Chapter 7 (Box 7.2 and Table 7.2) makes it clear that his capacity for climbing was marginal even at sea level. This applies to all whoopers, because of their large size, and to a lesser extent to the similar but smaller Bewick's Swan.

Where does this leave the well-known "fact", found in every book except this one, that whoopers climb into the lower stratosphere when migrating? This story actually refers to a single observation by an air traffic controller in 1967, published at second hand 11 years later by Stewart (1978). The controller asked a pilot to investigate an anomalous radar echo, and was informed that it was flock of swans at a height of 8200 m. The air density at that height would be approximately half that at sea level. According to *Flight*, an average whooper's minimum power speed would be 20.9 m s$^{-1}$ at sea level and 32.4 m s$^{-1}$ (1.5 times as much) at 8200 m, and the mechanical power needed to fly at that speed would increase by the same factor from 187 W to 290 W. This would present some problems to a bird that is marginally able to maintain height at sea level (above) even if it had a way to get up to 8200 m.

Elkins (1979) added some meteorological information, including an air temperature estimate of −48 °C, and commented "It seems incredible that sustained flight can occur under such physiologically rigorous environmental conditions". Indeed.

## 15.2.5 MINIMUM CRUISING SPEED

The principle of the minimum power speed, introduced in Chapter 3, implies that if a bird cannot fly much faster than $V_{mp}$, then it cannot fly much slower either. This came to light in a tracking project on Bewick's Swans, which was actually designed for broadcasting purposes, and did not have such detailed weather information as the earlier whooper project. The swans were fitted with transmitters in their breeding area near the mouth of the Pechora River, on the north coast of Russia, and tracked as they flew south-west in autumn to their first prolonged stop at Lake Peipus on the Estonian border. At first sight it appeared that the swans made several stops along the first part of the route, as far as Archangel, and then flew direct to Lake Peipus without stopping. However, the ground speeds were unexpectedly low, and the synoptic weather gave no indication that this was due to headwinds. The calculated $V_{mp}$ for the swans was around 18 m s$^{-1}$, but most airspeed estimates between Argos fixes were well below this, slower than a Bewick's Swan can fly. The terrain is more or less flat, with masses of streams and lakes, and the conclusion is that the swans made frequent short stops, as whoopers do during their sea crossing. The most likely explanation is that these large birds are on the edge of oxygen debt during level cruising flight at $V_{mp}$, and cannot fly for very long periods without stopping to recover. Difficulty in providing the mechanical power required to fly is expected from general scaling considerations (Chapter 7, Box 7.4), but the capacity to meet the resulting oxygen requirements depends on the capacity of the heart and lungs, not on that of the muscles, and is not something that can be predicted from mechanical considerations (see also Sections 7.4.4. and 12.3.2).

## 15.3 ⬤ WINGBEAT FREQUENCY

The frequency with which a bird beats its wings in cruising flight determines the amount of power that it can get out of its flight muscles, and hence is of central importance to calculating performance (Chapter 7). Measurements of a migrant's wingbeat frequency during a long flight could be used (in principle) to monitor its fat reserves.

## 15.3.1 FIELD OBSERVATIONS OF WINGBEAT FREQUENCY

A way of predicting a bird's wingbeat frequency in cruising flight, based on physics (not on a regression), is given in Chapter 7, Box 7.3, and Figure 15.11 is a field test of this prediction, in 16 species of birds observed during the autumn migration season at Falsterbo, Sweden. It shows a relative wingbeat frequency, which is the ratio of the observed wingbeat frequency ($f_{obs}$) to the reference wingbeat frequency ($f_{ref}$), as calculated by *Flight* from Equation 3 in Box 7.3. The relative wingbeat frequency is between 0.81 and 1.05 in all species except the chaffinch, in which it is 1.69. The chaffinches were all bounding, spending an average of 35% of their time flapping, that is, their power fraction ($q$) averaged 0.35. The mechanical effect of this is to increase gravity from its value ($g$) in continuous flapping to $g/q$. According to the formula, the wingbeat frequency should vary inversely with the square root of the power factor, and $\sqrt{(1/0.35)}$ is 1.69, as observed (Chapter 9, Figure 9.3). Figure 15.12 shows the mean airspeeds of the same 16 species in relation to their calculated minimum power speeds. The two passerine species appear to have been flying

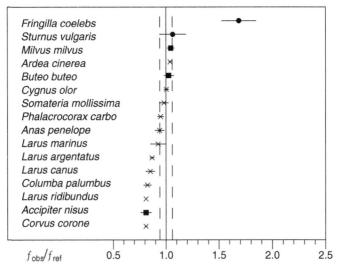

FIGURE 15.11 Wingbeat frequencies of 16 species of birds observed at Falsterbo, Sweden during the autumn migration season, measured from video, and expressed as the ratio of the observed frequency to the calculated reference frequency. Frequencies in intermittent flight styles were measured within a period of flapping, not averaged over flapping and non-flapping phases. Circles indicate species that flew by bounding, at least some of the time, crosses those that flew by steady flapping flight, and squares those that flew by flap-gliding. After Pennycuick (2001).

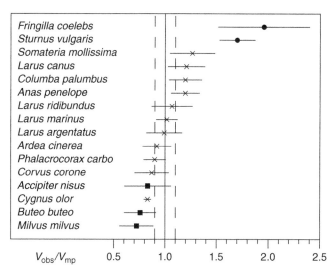

FIGURE 15.12 Airspeeds of the same species as in Figure 15.11, expressed as the ratio of mean observed airspeed to calculated minimum power speed (not adjusted for bounding). Horizontal bars are ±1 standard deviation. Symbols as in Figure 15.11. After Pennycuick (2001).

anomalously fast, but this discrepancy also disappears when the effect of bounding on $V_{mp}$ is taken into account (Chapter 9, Figure 9.2). The effect of the passerine flight style on gravity is not at all obvious, and since observers cannot manipulate the earth's gravity field, there is no way that these effects could have been detected by simply applying statistics to observed wingbeat frequencies, without a physical theory.

## 15.3.2 METHODS OF MEASURING WINGBEAT FREQUENCIES

Although wingbeat frequencies in large birds can be measured with a stopwatch, most people cannot count wingbeats above 4 Hz, which unfortunately excludes most birds. The advent of affordable video camcorders has partially resolved this difficulty, and this is the most commonly used method, but analysis of the video recordings is somewhat laborious and time consuming. A stroboscope gives instant results, and can be made from a liquid crystal shutter, which is a panel than can be made opaque or transparent in response to an electrical signal. Such panels can be obtained in the form of a pair of spectacles with two shutters that can be driven independently. The device is intended for 3D computer displays, which show the left-eye and right-eye views alternately, while the computer opens each shutter during the period when the appropriate picture is on the screen. If both shutters are driven together, the device can be used as a stroboscope. A design for such a device can be found [http://books.

elsevier.com/companions/9780123742995]. This is a quick and accurate method for a bird flying steadily in a wind tunnel under good illumination, but difficult to use in the field.

Accelerometer logging is a method with great potential for studies of long-distance migration. It was noted in Chapter 14 that the acceleration felt by a bird in horizontal flight varies between about zero and $+2g$, in each wingbeat cycle (Figure 14.11). This is independent of the size or mass of the bird, being simply the result of the wings being loaded and unloaded once per cycle. Miniature accelerometer modules are readily available nowadays, which could be mounted in a data logger attached to a bird's back, and used to measure wingbeat frequency in the field. The detector circuit would need to count the number of wingbeats in a measured time interval, counting one wingbeat each time the measured acceleration increases through $+1.5g$ and then decreases through $0.5g$.

### 15.3.3 WINGBEAT FREQUENCY AS A REMOTE FUEL GAUGE

One of the theoretical predictions that was confirmed in the wind tunnel experiments mentioned above (Pennycuick et al. 1996a) was that as the mass of an individual bird increases or decreases, due to feeding or the consumption of fuel, the wingbeat frequency varies with the square root of the mass (Figure 15.13). Thus, if the bird's wingbeat frequency were to be monitored at regular intervals during a long migratory flight, using a recoverable logger or data transmission via satellite, this could be used to monitor its declining mass, as a function of both time and distance. Mass loss does not equate directly to fuel consumed, but it can be used to estimate fuel consumption, by using the migration simulation in the *Flight* programme, as explained in Chapter 8, Box 8.4. This would amount to a fuel gauge, which could be read remotely by the satellite observer. Figure 15.14 is a graph taken from a run of a *Flight*'s migration simulation with the same input data as the one that generated Table 8.3, except that the altitude was set to zero instead of 2000 m, meaning that the air density was set to the sea-level value in the International Standard Atmosphere. The graph is a plot of the flamingo's remaining fat mass (starting at Cell I30 in Table 8.3) against the *equivalent* wingbeat frequency (starting at Cell E57). If the flamingo were carrying a transmitter that recorded GPS positions including height, and also wingbeat counts from an accelerometer, then the observer would first reduce the observed true wingbeat frequency to the equivalent wingbeat frequency at sea level (Box 15.3), and then read the flamingo's remaining fat mass from the sea-level graph of Figure 15.14.

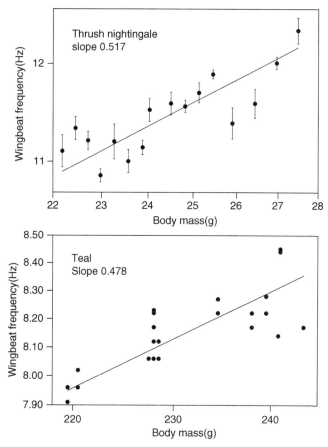

FIGURE 15.13 Variation of wingbeat frequency with body mass in individual birds in level flight. The points for the Thrush Nightingale are means of 5 stroboscope observations with standard error bars, taken during prolonged flights for physiological investigations, while those for the Teal are single stroboscope observations. The expected slope is 0.5. After Pennycuick et al. (1996a).

## 15.4 ● THE THEORETICAL BACKBONE

This chapter may seem something of a miscellany, with speeds measured in the field, wingbeat frequencies in the wind tunnel, satellite tracks, and the body composition of long-distance migrants. The point of the chapter is that all these are linked. Numbers like the body drag coefficient run through them all. A revised value from a wind tunnel experiment calls for a reappraisal of field data on migrants. Speeds measured in the wind tunnel are linked to those observed in the field, but the connection is not apparent until the effect of air density is

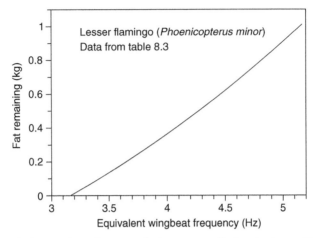

FIGURE 15.14 The wingbeat frequency fuel gauge applied to the Lesser Flamingo featured in the *Flight* migration simulation of Chapter 8. Table 8.3 includes estimates of both the wingbeat frequency and the amount of fat remaining, here plotted so that the fat can be read from the graph if the wingbeat frequency is measured. If a migrating flamingo's wingbeat frequency could be monitored in flight by a GPS logger, the amount of fat remaining could be read from the graph at points in the flight. To make a direct comparison with the observed (true) wingbeat frequency, the programme would have to be run separately for each observation, with the air density corresponding to the observed flying height. The *X*-axis of this graph is the *equivalent* wingbeat frequency, found by setting the air density to the sea-level value. To use it, observed wingbeat frequencies would have to be reduced to sea level before comparing them with the graph, as explained in Box 15.3.

understood and taken into account. Even the effect of varying gravity on wingbeat frequency can be observed from video of small birds in bounding flight, although no experimenter is likely to be able to manipulate gravity in the foreseeable future. The theory outlined in the early chapters of this book is the backbone that ties together heights, speeds, rates of fuel consumption, wingbeat frequencies, and much more that can be observed and measured and predicted by the *Flight* programme. If a number measured by one method cannot be reconciled with another measured in some other way, that is a discrepancy that has to be resolved. The resolution takes the form of a revised default value for some variable, or possibly a change in the structure of the theory itself.

The reason that all of this looks rather strange to many ornithologists is that biological observations are traditionally made in isolation, without a theoretical backbone. Measurements of the rate of oxygen consumption of a particular bird in a particular wind tunnel refer only to that bird under those conditions, however elaborate the statistical analysis. The purpose of having a theory, like the one embodied in the *Flight* program, is that observations on one species can be added to a model that applies to any species, under a range of conditions. The hope (perhaps forlorn) behind this book is that physiologists and field ecologists will acquire the habit of recording essential information such as wing spans and air densities, and will then use the *Flight* programme to incorporate their findings into the communal pool of information.

# 16

# EVOLUTION OF FLIGHT

Flight requires structures (wings) that develop much more lift than drag, and are able to resist bending and torsional loads. The ancestors of birds, bats and pterosaurs are assumed to have taken parallel routes from a parachuting precursor (creating drag only) to a patagial gliding stage corresponding to the level of modern flying squirrels, and then to have found different ways to get past the "squirrel barrier". The evolution of flight feathers was not a simple step, and their presence in *Archaeopteryx* puts the origin of flying birds back into the Triassic.

Soon after Darwin (1859) published his theory of evolution by natural selection, critics started pointing out that flying animals, and especially birds, are awkward to explain. The wings of birds have properties that are not found in the limbs of any animal that does not fly, and have no function except in the context of flight. Natural selection can improve the performance of existing wings, however rudimentary, but how did the very first flying animal originate from an ancestor that had no wings at all, or any other adaptations for flight? The principles of flight were not understood in Darwin's time, but now they give us a basis for understanding the selection pressures needed to bring about the origin of flight, and to drive its further evolution in birds, bats and pterosaurs. This is why these origin problems are addressed in the last chapter of this book, rather than at the beginning. I follow Heilmann (1926) in suggesting that birds (and other flying vertebrates) originated from arboreal gliders, but differ about the stages that led to powered flight, and the nature of the selection pressures that forced progressive change at each stage.

# 16.1 ● EVOLUTION IN ENGINEERING AND IN NATURE

## 16.1.1 THREE FUNDAMENTAL OBSTACLES

As the early history of aeronautics shows, the problem of level, powered flight was a tough nut for engineers to crack, although the engineer is (in principle anyway) free from the constraints of organic evolution. Theory is the DNA of engineering, and the theory of flying machines evolved slowly through the nineteenth century, by its own form of natural selection. Ideas survived if they led to machines that worked, while those that did not were doomed to extinction. The many false starts and blind alleys by which this body of theory developed have been documented in fascinating detail by a modern aeronautical engineer (Anderson 1997). The success of Orville and Wilbur Wright in building and flying the first powered aircraft in 1903 is a historical marker, at the point where aeronautical theory reached a threshold level of coherence. The Wright Flyer was the first working flying machine that natural selection could modify, and from that point on, the further development of practical aircraft was rapid and diverse. It was not always clear to the pioneers themselves what needed to be done in order to make progress, but with hindsight the difficulties that they faced, and eventually overcame, can be resolved into three distinct obstacles, which apply equally to the evolution of flying animals. These were:

(1) The ability to control spatial orientation while being free to move and rotate in three dimensions, without contact with the ground.
(2) The development of a shape that gives a sufficiently high lift:drag ratio, embodied in a structure that is able to withstand the loads imposed by flight.
(3) The development of a source of power that can be used to overcome aerodynamic drag.

A line of animals that do not fly must overcome these same three hurdles, *in that order*, to reach the animal equivalent of the Wright Flyer, and it must do so despite constraints that do not apply to engineers. An engineer can design a wing directly from theory, without necessarily having an existing structure to modify, but animal evolution is constrained by two very restrictive rules. In the first place, any structure with new properties must be developed by modifying some structure which the animal already has, for example a wing with a substantial lift:drag ratio was developed, in at least three cases, by modifying legs that were not exposed to a relative airflow, and had no such property as a lift:drag ratio, zero or not. Secondly, any such change must proceed by infinitely small steps, each of which makes the structure more effective at whatever it currently does. The challenge to the evolutionary biologist is to show how wings could have evolved from the limbs of a non-flying precursor, without violating these rules. The rest of this chapter is about hypothetical paths by which the three known groups of flying vertebrates, birds, pterosaurs and bats, could have evolved from their presumed non-flying ancestors.

## 16.1.2 SPATIAL ORIENTATION AND ATTITUDE CONTROL

Control in flight begins with attitude relative to the horizon (Chapter 9, Box 9.2). The two basic flight controls produce angular accelerations in roll, which means tilting the wings to one side or the other relative to the horizon, and pitch, which means tilting the nose up or down, also relative to the horizon. The third axis is yaw, meaning swinging the nose to one side or the other, but this is of secondary importance in flight, and some aircraft, notably hang gliders, do not even have a yaw control (rudder). The Wright brothers thought carefully about controlling roll and pitch, and tested and improved their system in a series of gliders. The control forces only existed when a relative wind was blowing past the glider, as it would in flight, but as the Wrights' glider did not fly yet, they had to get the relative wind by doing their initial experiments on windy coastal dunes in North Carolina.

Likewise, a system of roll and pitch control can only arise from scratch in an animal that is routinely exposed to a relative wind, while not being restrained from rotating in roll and pitch by contact with the ground. Arboreal animals are natural candidates, because falling out of trees is a hazard inherent in their lifestyle, and this immediately results in a relative wind. When an animal falls, it accelerates downwards, so acquiring an upward relative air flow that in turn produces an aerodynamic force and (possibly) also a moment that produces an angular acceleration around some axis. The squirrel or monkey does not, of course, instantly mutate into a flying animal. Its problem is to minimise the risk of injury when it strikes the ground, and there are two aspects to this that can be addressed by continuous small modifica-tions, as required by natural selection. The first is to set the limbs and tail in positions that rotate the body automatically until its ventral side faces the relative wind. Books on skydiving explain how this is done with a human body, and the same principle also works with other vertebrates. The second is the principle of the aviator's emergency parachute, which is a shape with enough area and a high enough drag coefficient to hold the terminal velocity to a level that is survivable on impact (Box 16.1).

**BOX 16.1 Wing loading and parachute loading.**

Wing loading is usually defined as the ratio of the all-up weight to the wing area, and it is directly related to the range of speeds over which a glider or gliding animal can fly, and to the radius when gliding in circles (Chapter 10, Box 10.2). This is a fixed-wing concept that does not have any simple or direct relationship to performance in flapping flight, but it does apply to both parachutes and fixed wings, including gliding wings.

**Variable definitions for this box**

| | |
|---|---|
| $C_D$ | Drag coefficient |
| $C_{Lmax}$ | Maximum lift coefficient of a glider |
| $D$ | Drag |
| $g$ | Acceleration due to gravity |
| $m$ | Animal's all-up mass |
| $V_s$ | Glider's stalling speed |
| $V_t$ | True airspeed |
| $V_{term}$ | Terminal velocity |
| $S$ | Area of wing or parachute |
| $W$ | Area loading of wing or parachute |
| $\rho$ | Air density |

**Terminal velocity of a parachute**

A parachute is a device that generates a sufficient amount of drag to limit its terminal velocity to below some maximum acceptable value. For example aviators' emergency parachutes hold the descent speed to a value that gives the user a good chance of surviving the impact with the ground. Likewise, an ordinary squirrel falling out of a tree spreads its legs and tail to make a rudimentary parachute (Figure 16.1A). As it falls, its true airspeed ($V_t$) increases and the drag ($D$) builds up with the square of the speed:

$$D = \tfrac{1}{2}\rho V_t^2 S C_D, \tag{1}$$

where $\rho$ is the air density, $S$ is the area of the parachute, and $C_D$ is its drag coefficient. A value of $C_D = 1$ means that the parachute reduces the relative air flow to zero (i.e. stops the air) over the whole of the cross section that it presents to the incident airflow, and practical parachutes actually do have

BOX 16.1 *Continued.*

drag coefficients close to 1. When the falling squirrel's drag is equal and opposite to its weight, the speed stabilises at the terminal velocity ($V_{term}$), which is

$$V_{term} = \sqrt{\left[\frac{(2mg)}{(\rho S C_D)}\right]},\qquad(2)$$

where $m$ is the mass and $g$ is the acceleration due to gravity. If we define the *area loading* ($W$) as the ratio of the weight to the parachute area,

$$W = \frac{mg}{S},\qquad(3)$$

then Equation (2) can be expressed as

$$V_{term} = \sqrt{\left[\frac{(2W)}{(\rho C_D)}\right]},\qquad(4)$$

in other words, the terminal velocity is proportional to the square root of the area loading. Likewise the expression for the minimum (stalling) speed ($V_s$) in gliding flight, given in Chapter 10, Box 10.1 can be expressed as

$$V_s = \sqrt{\left[\frac{(2W)}{(\rho C_{Lmax})}\right]},\qquad(5)$$

where $W$ is the ratio of the weight to the wing area, i.e. the *wing loading*. The difference is that in this case the weight is supported mostly by a lift force coming from a wing, rather than by a drag force coming from a parachute. If the lift coefficient is set to its maximum value ($C_{Lmax}$), the stalling speed is proportional to the square root of the wing loading.

### Significance of area loading

The area loading determines the minimum speed of a fixed wing or parachute, regardless of whether the weight is supported by almost pure lift, as in an efficient glider, or by almost pure drag, as in an emergency parachute, or by the resultant of lift and drag. The measure of a wing's efficiency is its lift:drag ratio, and the wing loading says nothing about this. Those who believe that a low wing loading is the key to gliding performance should visit a gliding competition, and ask the pilots why they ballast their gliders with vast quantities of water before they take off (or read Chapter 10, Section 10.2.1.5).

It should be obvious that one cannot work out a helicopter's minimum speed from its wing loading. When the helicopter is hovering its airspeed is zero, but the relative airspeed over the blades is not zero, because the blades are rotating relative to the fuselage. A helicopter blade has a maximum lift coefficient like any other wing, but the minimum speed at which it will support the weight is not the speed of the helicopter as a whole. It is the resultant of the helicopter's speed and the local relative airspeed due to the blades' rotation, which is different at each point along the blade. The same is true of the flapping wings of a bird. Generalisations that involve wing loading are fixed-wing concepts that apply to gliding birds, but not to helicopters, or to flapping flight. This has not deterred some biological

BOX 16.1 *Continued.*

authors from transferring such notions to flapping birds or bats, especially in relation to turning radius, which is, of course, not limited by wing loading in flapping flight.

### Scaling of wing loading

In geometrically similar animals or aircraft, the weight varies with the cube of the length, while the wing area varies with the square. Scaling an aircraft or animal up isometrically to a larger size results in an increase in wing loading in proportion to the length, or to the one-third power of the mass. In the case of a patagial glider, the bigger the animal, the higher its wing loading, and the faster it has to fly. The hypothetical bird precursor shown in Figure 16.2D is shown with a smaller wing area but higher aspect ratio than the pterosaur precursor of Figure 16.2C, because the membrane is attached posteriorly to the side of the body instead of the legs. If both animals were of similar size, this would result in the bird precursor having a better glide ratio than the pterosaur precursor, but gliding faster—perhaps too fast. We know that modern flying squirrels, whose general shape is similar to that proposed for the pterosaur precursor, work up to a mass over 1 kg, but the proposed bird precursor might have to be smaller, to keep the gliding speed down to a manageable value.

## 16.1.3 FROM PARACHUTE TO WING

Bailing out of a stricken aircraft, or falling out of a tree, is not exactly flying in itself. Emergency parachutes and falling squirrels do nothing more than maintain a stable attitude relative to the horizon, and a low enough terminal velocity to minimise the risk of injury. The parachutes used by skydivers and paratroopers differ in that they have a limited amount of roll and pitch control, sufficient to execute a controlled spiral, and are able to check their descent speed momentarily before touchdown. This greatly improves their ability to land into wind, on a chosen spot, and to soften the impact. The essential difference is that the aerodynamic force developed by a simple emergency parachute is pure drag, in line with the relative air flow, whereas that developed by a steerable parachute has a component (lift) that is perpendicular to the relative air flow. Beyond the steerable parachute comes the paraglider, which has a flexible canopy like a parachute but a higher aspect ratio, that is, the canopy extends out to the sides. The lift can be as much as five times the drag in a paraglider, which is enough to soar on slopes and in thermals, and even to fly level with a minimal back-pack engine.

The limbs of monkeys, and also those of squirrels, are rounded in cross-sectional shape, and generate drag when held out in a jump. The limbs themselves do not provide a starting point from which natural selection could modify them into a shape that would develop some lift. However, there is such a starting point in animals that develop a patagial membrane along each side of the body, stretched between the front and back legs. Initially, a narrow patagium serves to increase the animal's cross-sectional area as it falls, increasing its parachute area, and reducing its terminal velocity. As the membrane becomes wider, it bulges upwards as the squirrel falls, and this fortuitously changes its shape into a thin, curved sheet, which develops a small amount of lift, in addition to a massive amount of drag. A flying squirrel's patagial membrane not only slows it down, but also deflects the flight path so that it is no longer vertical, as in a steerable parachute. This is the crucial feature that can be further improved by natural selection (Figure 16.1).

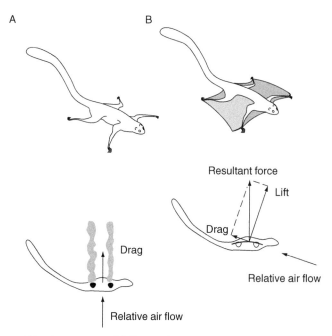

FIGURE 16.1 (A) When a squirrel falls out of a tree, the legs, being bluff bodies in cross section, generate turbulent wakes (grey), and the body and tail do likewise. These wakes result in a high drag coefficient, which limits the terminal velocity, and the energy that has to be dissipated when the animal strikes the ground. (B) A patagium spread between the front and back legs increases the cross-sectional area of the falling animal, and further reduces its vertical velocity. Fortuitously, the patagium bulges upwards in response to air pressure, and is deformed into a rudimentary aerofoil shape, which develops some lift in addition to the drag. Instead of falling vertically, the animal now glides on a gradient that is equal to the ratio of lift to drag.

By providing a non-zero glide ratio, the flexible patagium starts the process by which the glide ratio can be further increased through natural selection. No amount of running on the ground, seizing prey with the arms, or jumping about in trees will start a monkey's or dinosaur's arms on the path towards becoming wings, because they do not have a glide ratio to start with, that natural selection can improve. A thin, flexible patagium must be added before that process can begin. It is my contention in this book that all flying vertebrates, including birds, started from arboreal ancestors that were patagial gliders. The initial stage of this hypothetical evolutionary path can be seen in modern flying squirrels, colugos and possums. Beyond this level, different lines of arboreal gliding animals followed different paths that led to birds, bats and pterosaurs, each of which groups evolved its own characteristic type of wing.

## 16.1.4 RAISING THE LIFT:DRAG RATIO TO THE SQUIRREL BARRIER

Patagial gliding is an adaptation that seems to arise easily in small arboreal animals. Nowak (1991) lists 13 genera and 34 living species of flying squirrels, plus four species of marsupial gliding possums, and two of colugos (whose affinities are uncertain). These animals are very similar anatomically to their non-gliding relatives, but usually differ in being nocturnal, whereas non-flying squirrels and possums are mostly diurnal. The patagium is useful to animals that need to travel distances of tens of metres when they jump from tree to tree, whereas those that live in thicker vegetation, or at lesser heights, no doubt find a patagium more of an encumbrance than an advantage. Flying squirrels have enough roll control to steer the glide path towards a chosen tree, and enough pitch control to flare and land head-up on a vertical tree trunk. They move about in the forest by alternately running vertically up a tree trunk, and then gliding to another tree, descending typically with a glide ratio no greater than 3, meaning that the squirrel glides up to 3 metres horizontally for each metre of height lost. Most of the time and energy required for this type of travel is expended in the vertical climbs. There is selection pressure to flatten the glides, since this allows the animal to travel further for a given expenditure of both time and energy. Flattening the glides is the same as increasing the ratio of lift to drag, and this in turn translates into pressure to lengthen the wings, increasing the wing span and the aspect ratio (not the wing area). Flattening the glides has nothing to do with wing loading, which determines the speed at which the animal glides, not the angle of descent (Box 16.1).

The wing membrane in a flying squirrel stretches to the wrist, so the wingspan can be increased initially by lengthening the arms. There is usually a cartilaginous spur that increases the span a little by holding the tip of the patagium out from the wrist or elbow. There is limited scope for lengthening the arm bones, because this hinders the typical motion by which squirrels climb tree trunks. They do this by alternately pulling up with the arms, at the same time hunching the back and pulling the hind legs up between the body and the tree, and then pushing up with the hind legs, to get a new grip, higher up, with the front claws. For this to work, the limbs cannot be too long, otherwise the front and back limbs interfere with each other. Flying squirrels, and gliding arboreal mammals of other kinds, do have noticeably elongated front legs, compared with animals without patagial membranes, but the selection pressure to flatten the glides fails at the point when the limbs cannot be any longer, without impairing the squirrel's ability to climb up trees. This is the *squirrel barrier*. It limits a flying squirrel's aspect ratio to a value well below 2, which is enough for a simple gliding wing that speeds up travel through the forest, but not enough to make level flight a practical possibility.

## 16.2 ☞ PAST THE SQUIRREL BARRIER

Three vertebrate groups, the birds, bats and pterosaurs, most probably started as arboreal gliding animals, which may have paused in their evolution at the squirrel barrier, but found different ways around it. Once past the barrier, each of these groups evolved wings with aspect ratios of 6 and beyond, thus opening the way to level flight. Lengthening the wings beyond the nearly-square shape typical of flying squirrels requires one or more spars to carry the bending and torsional loads, as the centre of lift of each wing is moved outwards from the body (Chapter 5). The selection pressure that drives the early stages of this process is derived strictly from the flattening of the glides that results from increasing the aspect ratio of the wings. There is no question of adding power to an animal shaped like a flying squirrel, because the muscles that depress the front limbs are adapted for climbing trees, not for flapping flight. A squirrel's arm muscles are in the wrong place, and not big enough to flap its inefficient wings. The hypothetical precursor for each group of flying animals has to be a patagial glider that glides from tree to tree, resembling a flying squirrel in its general body form, but differing in some minor and fortuitous anatomical feature that gives natural selection a way to increase its aspect ratio in response to selection pressure for flatter glides.

## 16.2.1 PTEROSAUR ORIGIN

The pterosaur route for lengthening the wing seems to be the simplest. We may imagine the pterosaur progenitor as a small, arboreal, Triassic archosaur, looking and behaving much like a flying squirrel, with a patagium attached to the outer edge of its hand only, not enclosing the fingers (Figure 16.2C). On reaching the squirrel barrier, this creature continued to increase its aspect ratio, but not by lengthening the humerus and radio-ulna. Instead, it lengthened the last finger, which was actually Digit 4 according to palaeontologists. The four phalanges of this "wing finger" became the spar, carrying all of the bending load developed by the outer part of the wing membrane, and transmitting it to the outer end of the metacarpus. Being able to flex back at the meta-carpal joint, the wing finger could be laid parallel to the metacarpus, so keeping it out of the way during quadrupedal locomotion, either on the ground or on the vertical surface of a tree trunk. Further increase of the wing span was effected simply by lengthening the phalanges of the wing finger, and thickening them to carry the increased bending load, which was developed by a flexible patagial membrane that stretched between the wing finger and the hind leg. Once the wing was long enough to flatten the glide further by absorbing a small amount of muscle power, the animal would have had a second way to respond to selection pressure for flatter glides, by modifying the axial musculature so as to depress the humerus, supplying some work in the process. This does not prevent the original lengthening process from continuing without interruption, all the way to an animal that was capable of horizontal, powered flight, as in the fully developed rhamphorhynch wing shown in Figure 6.10.

## 16.2.2 BAT ORIGIN

Only one difference, trivial in itself, is needed to make a flying squirrel into a bat precursor. Instead of ending at the wrist, the membrane needs to enclose all the fingers, as shown in Figure 16.2B. An arrangement very close to this occurs in the living colugos (*Cynocephalus* spp.). Here again, the initial selection pressure is to increase the aspect ratio so as to lengthen the glides, and this can be done without impairing climbing ability by lengthening the fingers. Once the wing is long enough for incipient flapping to be possible, there is no obstacle to continued lengthening of the wing, together with modification of the axial musculature to provide an (initially small) amount of power by flapping. This leads directly to the bat wing, which first appeared in Eocene times, apparently fully evolved, and more or less as illustrated in Figures 6.3–6.5.

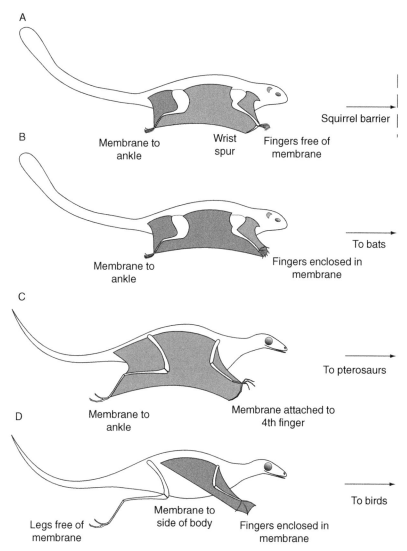

FIGURE 16.2 (A) Flying squirrel, in which the main patagium is stretched between the front and back legs, but does not enclose the fingers or toes. A cartilaginous spur increases the span (only a little) by holding the wing tip further out. (B) Hypothetical bat precursor, which differs from a flying squirrel in having all five fingers enclosed in the membrane. Further evolution is based on lengthening all the Digits 2–5. (C) Hypothetical pterosaur precursor, with three free fingers. The patagium extends beyond the wrist by being attached to the posterior side of the last finger (Digit 4). Further evolution is based on lengthening this finger only. (D) Hypothetical bird precursor resembles bat precursor in having the fingers enclosed in the membrane, but the patagium is attached posteriorly to the side of the body, and not to the hind legs. This results in a wing of smaller area, but higher aspect ratio.

## 16.3 ⬭ EVOLUTION OF THE BIRD WING

It is easy to see how either a bat or a pterosaur could evolve from a patagial glider similar to a flying squirrel, but the bird wing is different, in that it does not depend on tensioning a flexible membrane. The bending and torsional moments due to the aerodynamic force on the wing are carried by the radiating, stiff shafts of the flight feathers, and transmitted through their bases to the wing bones (Chapter 5). Mechanically, this is entirely different from a patagial wing, and it depends on the evolution, starting from scratch after the patagial gliding stage, of an entirely new set of structures, the flight feathers. This must have taken a lot longer than the changes of shape needed to change a patagial precursor into a bat or a pterosaur, because flight feathers are not simple structures. Even if the precursor of *Archaeopteryx* had a body covering of proto-feathers, the subsequent development would not be much quicker, because flight feathers (only) have the complex structure needed to collect up moments and deliver them to the base of the shaft (Figure 5.6). Contour feathers and down feathers may have originated by simplifying flight feathers, but the reverse process would be no quicker than starting from simple reptilian scales.

### 16.3.1 HYPOTHETICAL PATAGIAL PRECURSOR TO *ARCHAEOPTERYX*

I propose that in the early stages of their evolution, the ancestors of birds were arboreal gliders that resembled the ancestors of bats and pterosaurs, but differed from them in that the patagium was attached posteriorly to the sides of the body, but not to the legs. My hypothesis (Pennycuick 1986) is that the bird wing developed in two stages from a gliding precursor like the one shown in Figure 16.2D. In the first (hypothetical) stage, the purely patagial wing developed into a form like that of Figure 16.3, which has some features in common with bat wings, and others that are seen either in modern birds or in the Upper Jurassic *Archaeopteryx* fossils. The second stage was the development of flight feathers, which were already present in *Archaeopteryx*, although the modern system that transfers forces and moments from the bases of the feather shafts to the wing skeleton was not.

The main features of the hypothetical patagial glider of Figure 16.3 can still be seen, albeit much modified, in the modern bird wing, which has a patagium that joins the side of the body to the elbow joint, and continues as a narrow strip along the posterior side of the ulna, and of the reduced hand skeleton. The mechanical arrangement in modern

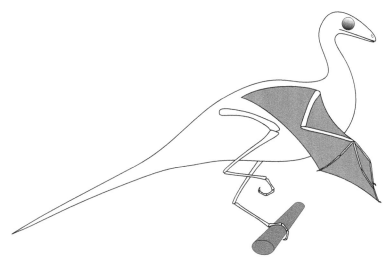

FIGURE 16.3 Hypothetical patagial precursor to *Archaeopteryx*, following lengthening of the fingers, but before the development of flight feathers. The three long fingers (still seen in *Archaeopteryx*) support a patagial wing, with digits 1 and 2 forming a Norberg panel as in bats (Chapter 6). The skin of the patagium is assumed to be covered with smooth, lizard-like scales, the precursors of the flight feathers of *Archaeopteryx*.

birds is explained in Chapter 5, and illustrated in Figure 5.10. The follicles of the scapular feathers, which provide the innermost part of the lifting surface between the body and the elbow, are embedded in the inner patagium, with no close mechanical connection to the humerus. Distal to the elbow, the follicles of the secondary flight feathers, which are embedded in the post-patagial strip, are hinged to the posterior side of the ulna, and connected together by the post-patagial tendon that controls their spreading and depression when the wing is extended. The follicles of the primary feathers are more robustly bound to the reduced metacarpals and phalanges, in a way that transmits the larger moments developed by the primaries to the bones.

The *Archaeopteryx* fossils show a set of 9 primary and 14 secondary flight feathers that look very like those of modern birds (Swinton 1960) but the hand skeleton is not modified to accept concentrated bending loads from the primaries. It consists of three long and slender fingers, with curved claws at their tips. These are usually reconstructed as climbing fingers unconnected with the lifting surface, but they look too long and thin for that. They look much more like the fingers of bats, and we may note that Digit 2 of the wings of fruit bats also carries a claw, although it is not used for climbing. I propose that the fingers of *Archaeopteryx* supported a patagium somewhat like that of a bat,

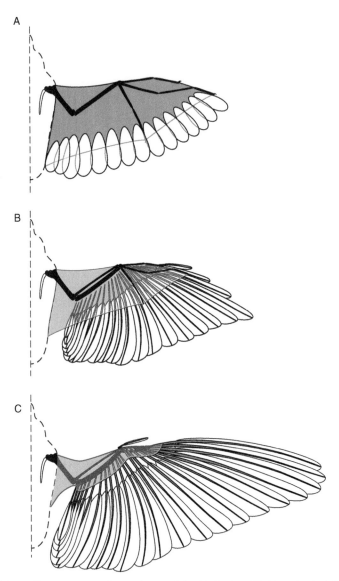

FIGURE 16.4 Three stages in the evolution of the bird wing. (A) Hypothetical *Archaeopteryx* precursor. The patagial wing of Figure 16.3 has been enlarged and lengthened by a single row of large scales, which overlap the posterior and outer edge of the membrane. The anterior edge of each scale is held in a follicle in the patagium. (B) *Archaeopteryx*. The skeleton is the same as in A, but the border scales have become feathers by developing a stiffening rhachis, with vanes on either side of it. The base of the rhachis is held in a follicle, which is bound into the patagium, as in the scapular feathers of modern birds. (C) Modern bird. The post-patagium is reduced to an inner portion supporting the scapular feathers (not shown), and a narrow border along the posterior side of the ulna and manus, holding the follicles of the flight feathers, and the tendon that

with the primary and secondary feather shafts embedded in it, in the same way that the scapular feather shafts are still embedded in the wing-root patagium next to the body in modern birds. Figure 16.3 represents a hypothetical stage of evolution after the patagial glider of Figure 16.2D had extended its wing span by lengthening the three fingers that were enclosed in the membrane, but before the development of feathers. The animal is shown with a small, bat-like wing that is complete with a Norberg panel (allowing the fingers to be more slender than a pterosaur's wing finger) and with a membrane that extends to the sides of the body, but not to the legs.

Such a wing would have less area but a higher aspect ratio than a bat or pterosaur precursor of similar size, at a similar stage of evolution. For a given mass, the animal would glide faster, but at a flatter angle (higher glide ratio). Since the gliding speed increases with the wing loading, which itself increases with the size of the animal, such a wing would most likely evolve in a small animal, keeping gliding speeds down. There is a functional resemblance to the tiny *Draco* gliding lizards of south-east Asia, whose wings are supported by extended ribs that can be folded back along the sides of the body. The *Draco* wing is an oddity with no potential for further evolution, but it works well enough to show that a big patagium is not the only possible way to make a gliding wing.

## 16.3.2 EVOLUTION OF FEATHERS

In the modern bird wing, as described in Chapter 5, the bending and torsional moments developed by the outer and posterior parts of the wing surface are collected by a *distributed spar* consisting of the radiating shafts of the flight feathers. Everyone seems to agree that feathers originated from reptilian scales, but how exactly did the intricate and complex structure of the flight feathers (Figure 5.6) develop from simple scales? One may imagine the patagium of the creature in Figure 16.3 as resembling the skin of modern lizards, covered with thin scales made of keratin, anchored in follicles at their forward edges. My hypothesis for the next stage of bird evolution, from the patagial glider of Figure 16.3 to the feathered wing of *Archaeopteryx*, is illustrated in Figure 16.4A and B, leading on to the modern bird arrangement in

---

connects them together (see also Figure 5.10). The diameter and curvature of the ulna have increased. Digit 1, which formerly supported the leading edge of the patagium and the outermost primary feather, becomes the alula, while Digits 2 and 3 are reduced and thickened, their fused metacarpals forming a robust unit to which the primary follicles are bound. This reinforced connection between the primary feather shafts and the skeleton allows the wing to be longer, and its aspect ratio to be increased.

Figure 16.4C. In Figure 16.4 A, the patagial glider has extended its wing span without much change of aspect ratio, by enlarging the row of border scales along the trailing and outer edge of the patagium. In the *Archaeopteryx* stage of Figure 16.4B the bat-like wing skeleton is still there, supporting the enlarged border scales, but these have now evolved into flight feathers, whose follicles are embedded in the membrane, like those of the scapular feathers of modern birds, but not yet connected to the arm skeleton.

One of the most startling features of *Archaeopteryx* is that the flight feather impressions look very similar to those of modern birds, even though the skeleton is reptilian and lacks the most characteristic bird features. Flight feathers are not variants of thermal insulating material, but structures that have evolved over a long period, enabling the span and aspect ratio of the wings to be progressively increased as they developed. The structure of a flight feather (Figure 5.6) is dedicated wholly to withstanding the bending and torsional moments caused by aerodynamic forces, and delivering these moments through the follicle to the wing skeleton. Figure 16.5 shows some hypothetical stages of evolution that must have preceded *Archaeopteryx*. Simple scales projecting past the edge of the patagium, as shown in Figure 16.5A would have to be thick and unduly heavy to resist the tendency of the air pressure to bend them upwards. A stiffening ridge along the axis of the feather (Figure 16.5B) would help to resist this, and would form the precursor of the tubular feather shaft, which resists

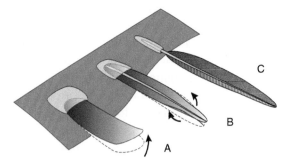

FIGURE 16.5 Three stages in the evolution of flight feathers from simple scales. (A) A simple scale attached at its anterior end to the patagium (pink) tends to bend upwards in response to air pressure. (B) A longitudinal stiffening ridge resists upward bending, and eventually becomes the hollow rhachis of Figure 5.6. The vanes still tend to curl upwards at the edges. (C) Stiffening ridges in the vanes resist the curling tendency. The ridges eventually separate to become the barbs, and are held together in modern birds (and possibly in *Archaeopteryx*) by a system of interlocking hooks, the barbules and barbicels. The base of the feather is bound to the patagium through its follicle.

both bending and torsion, but does not prevent the vanes from curling up at the edges. An array of parallel ridges or corrugations as in Figure 16.5C would stiffen the vanes, forming the precursors of the barbs of modern flight feathers, which eventually (possibly before *Archaeopteryx*) separated to form a porous vane, held together by the matching hooks of the barbules and barbicels.

### 16.3.3 MODIFICATIONS TO THE WING SKELETON

The forearm skeleton of *Archaeopteryx* differs from the straight radio-ulna of bats in that the radius and ulna are separate, and the ulna shows the beginnings of the forward curvature that is so characteristic of modern birds, and is not seen in bats or pterosaurs. This implies (Chapter 5) that the secondary follicles were already joined together by a post-patagial tendon that controlled the fanwise spreading and depression of the feather shafts as the elbow and wrist joints were extended. The outermost three primaries were short, and the hand skeleton was not yet modified to support the bases of long wing-tip primaries. My reconstruction (Figure 16.4B) shows the most anterior primary attached to Digit 1 (the thumb), which itself combines with Digit 2 to form the Norberg panel. In modern birds (Figure 16.4C), the stiffening action of the panel is taken over by the fused metacarpals of Digits 2 and 3, while Digit 1, still with its feather attached, becomes the alula. The more robust support provided by the metacarpal unit allows the wing-tip primaries to be longer, and the aspect ratio to be higher.

## 16.4 ◄ ADDING AN ENGINE

In all flying animals that are capable of horizontal flight, insects included, the work needed to overcome drag comes from muscles that flap the wings. Flapping is not the only way to add power to a glider, and indeed this solution has not found favour with aeronautical engineers. Ornithopters (aircraft that flap their wings) have achieved very little success, despite extensive experimentation. The realisation that lift and propulsion are two different functions, and that it is easier from an engineering point of view to separate them, was one of several major insights that contributed to the Wright brothers' achievement of powered flight. Their "Flyer" and most of its successors used a wing that supported the weight but did no work, and a separate, engine-driven propeller to provide a thrust force that was much less than the weight, and did work against drag. Flying animals use their wings for both functions, and suffer from severe size restrictions (Chapter 7 Box 7.4), due to the fact that the power output of the flight muscles depends on their contraction frequency, which has to be the same as the wingbeat frequency. This restriction can be evaded to some degree in man-powered aircraft by using a geared drive, so allowing the pilot's muscles to contract at a frequency that is independent of the propeller revolution rate (or the flapping frequency in an ornithopter), but no such system is known in animals.

The addition of power to a gliding wing is the last stage of the evolution of level flight, since the animal must have wings that work, before it can flap them. In birds and bats, the main engine for powered flight is the pectoralis muscle, which rotates the humerus ventrally about the shoulder joint, and the same was true of pterosaurs, to judge by their skeletons. Birds differ from the other two groups in that the pectoralis muscles of the two sides originate on opposite sides of a prominent bony keel projecting from the sternum. Obviously, this is not a necessary adaptation for flight, as neither bats nor pterosaurs exhibit any such structure. In bats, the left and right pectoralis muscles pull against one another in the ventral mid line, and no doubt the same was true in pterosaurs. The keel is a structure that resists compression of the pectoralis muscles against the sternum, and performs a function associated with evaporative cooling, that is necessary in birds, but not in bats, and presumably not in pterosaurs (Chapter 5, Section 5.4.2). The pectoralis muscle becomes an engine (however inefficient) when some fraction (however small) of the work done in rotating the humerus downwards is transferred to the air, in a way that replaces energy that has been lost in overcoming drag. Once that happens, accidentally or otherwise, a gliding animal can flatten its glide angle (just a little) by muscular exertion. That in turn creates selection pressure to get a flatter glide for less power, a process that can continue in infinitesimal steps until the flight path is horizontal, and beyond.

## 16.5 ▪ SIZE RESTRICTIONS

Just as the Wright brothers achieved level flight by adding an engine to a glider that had already been extensively tested, so the transition to powered flight in an animal involves adding muscle power (in infinitely small steps) to a precursor that can already glide. The flatter the angle at which the gliding precursor can descend, the less power is required to raise the flight path to the horizontal, and because of the scaling relationship discussed in Chapter 7 Box 7.4, this threshold is more easily reached in a small animal than in a large one. The first animal in any particular line to attain a positive "power margin", meaning that the muscle power available from the muscles exceeded that required for level flight, would most probably have had a body mass in the range 10–100 g (Pennycuick 1986). In the earlier gliding stage, the need to keep gliding speeds down also favours small animals, as gliding speed depends on wing loading, which in turn depends on size (above).

## 16.6 ➧ TIME SCALE OF EVOLUTION

A flying animal does not necessarily take long to evolve, in terms of the geological time scale, especially in an ecological vacuum. Bats are not among the mammals that are known from Mesozoic times, yet fossil bats such as *Palaeochiropteryx* appeared in the Eocene, having apparently completed the transition from non-flying animal to almost-modern bat during the Palaeocene period, when life appeared to be suspended after the demise of the dinosaurs and many other groups of animals. During the great Eocene radiation that followed, identifiable members of most modern orders of mammals appeared, including bats. Both pterosaurs and birds started much earlier, some time after the earlier and bigger extinction event that ended the Permian period. The first known pterosaur is *Eudimorphodon* from the upper Triassic, and it was already typical of the long-tailed pterosaurs that continued to flourish until the upper Jurassic. Its gliding precursors are unknown, but they must have started on the road to evolving flight in the middle or early Triassic.

Birds are different, because their evolution involved more than increasing the span of a patagial gliding wing. The evolution of flight feathers, as opposed to contour feathers or down feathers, and their mechanical integration with the arm skeleton, was an entirely new, and very complex development. The Solnhofen *Archaeopteryx* specimens are the earliest fossils with flight feathers, and this is only known because they were preserved in a fine-grained lithographic limestone in which detailed impressions of the feathers were recorded. Were it not for that fortunate chance, there would be little if anything to distinguish their skeletons from those of small, saurischian dinosaurs. Whether or not *Archaeopteryx* actually was a saurischian dinosaur, as opposed to a close relative, is taxonomic hair-splitting that need not concern us. *Archaeopteryx* lacked most of the skeletal features that distinguish birds from dinosaurs, but it had flight feathers, which are not only the key structural feature that defines the bird wing, but also complex structures that must have taken tens of millions of years to evolve from ordinary reptilian scales. They could not have evolved in any animal but one that was already capable of gliding flight. That means that the patagial precursor of *Archaeopteryx* (upper Jurassic) must have been at least an early Jurassic, or more likely a Triassic animal, indistinguishable in its skeleton from other small, bipedal dinosaurs.

There has been much interest in recent years in the Jehol fauna from the Liaoning province of China, which includes an array of small dinosaurs that are clearly related to *Archaeopteryx* (Milner 2002). Some of these have structures that can be interpreted as feathers, but not flight feathers like those of *Archaeopteryx* or modern birds. The Liaoning formation is of lower Cretaceous age, around twenty million years younger than the upper Jurassic Solnhofen limestone. Interesting though the Liaoning fossils are, they do not shed a great deal of light on the development of flight in birds, as the key innovation (flight feathers) was already present in the earlier *Archaeopteryx* fossils.

# REFERENCES

Abbott, I.H., and von Doenhoff, A.E. (1959). *Theory of Wing Sections.* New York: Dover.

Alexander, R.McN. (1985). The maximum forces exerted by animals. *Journal of Experimental Biology* 115:231–238.

Alexander, R.McN., and Bennet-Clark, H.C. (1977). Storage of elastic strain energy in muscle and other tissues. *Nature* 265:114–117.

Anderson, J.D. (1991). *Fundamentals of Aerodynamics.* New York: McGraw-Hill.

Anderson, J.D. (1997). *A History of Aerodynamics.* Cambridge: Cambridge University Press.

Bäckmann, J., and Alerstam, T. (2002). Harmonic oscillatory orientation relative to the wind in nocturnal roosting flights of the swift *Apus apus. Journal of Experimental Biology* 205:905–910.

Bailey, N.T.J. (1995). *Statistical Methods in Biology.* Cambridge: Cambridge University Press.

Beatty, J.K., O'Leary, B., and Chaikin, A. (1981). *The New Solar System.* Cambridge: Cambridge University Press.

Bernstein, M.H., Thomas, S.P., and Schmidt-Nielsen, K. (1973). Power input during flight of the Fish Crow *Corvus ossifragus. Journal of Experimental Biology* 58:401–410.

Biewener, A.A., Dial, K.P., and Goslow, G.E. (1992). Pectoralis muscle force and power output during flight in the starling. *Journal of Experimental Biology* 164:1–18.

Boettiger, E.G. (1957). Triggering of the contractile process in insect fibrillar muscle. In *Physiological Triggers.* American Physiological Society, T.H. Bullock Ed. pp. 103–116.

Bradbury, T. (1989). *Meteorology and Flight.* London: Black.

Breguet, L. (1922). Aerodynamic efficiency and the reduction of air transport costs. *Aeronautical Journal* 26:307–313.

Britting, K.R. (1971). *Inertial Navigation Systems Analysis.* New York: Wiley.

Bruun, H.H. (1995). *Hot-Wire Anemometry: Principles and Signal Analysis.* Oxford: Oxford University Press.

Budyko, M.I., Ronov, A.B., and Yanshin, A.L. (1985). *History of the Earth's Atmosphere.* Berlin: Springer.

Campbell, K.E., and Tonni, E.P. (1983). Size and locomotion in teratorns (Aves: Teratornithidae). *Auk* 100:390–403.

Carmi, N., and Pinshow, B. (1995). Water as a physiological limitation to flight duration in migrating birds: The importance of exhaled air temperature and oxygen extraction. *Israel Journal of Zoology* 41:369–374.

Chatterjee, S., and Templin, R.J. (2004). Posture, locomotion and paleoecology of pterosaurs. *Geological Society of America Special Paper* 376:1–64.

Chatterjee, S., Templin, R.J., and Campbell, K.E. (2007). The aerodynamics of *Argentavis*, the world's largest flying bird from the Miocene of Argentina. *Proceedings of the National Academy of Sciences* 104:12398–12403.

Carey, S.W. (1976). *The Expanding Earth.* Amsterdam: Elsevier.

Darwin, C. (1859). *On the Origin of Species by Means of Natural Selection.* London: Murray.

de Margerie, E., Sanchez, S., Cubo, J., and Castanet, J. (2005). Torsional resistance as a principal component of the structural design of long bones: Comparative multivariate evidence in birds. *Anatomical Record Series A* 282:49–66.

Dudley, R. (1998). Atmospheric oxygen, giant paleozoic insects and the evolution of aerial locomotor performance. *Journal of Experimental Biology* 201:1043–1050.

Duncker, H.R. (1985). Respiratory system. B. Campbell and E. Lack Eds. In *A Dictionary of Birds*, Calton, pp. 503–509.

Dunning, J.B. (1993). *CRC Handbook of Avian Body Masses*. Boca Raton: CRC Press.

Elkins, N. (1979). High altitude flight by swans. *British Birds* 72:238–239.

Gnaiger, E. (1989). Physiological calorimetry: Heat flux, metabolic flux, entropy and power. *Thermochimica Acta* 151:23–34.

Gordon, A.M., Huxley, A.F., and Julian, F.J. (1966). The variation in isometric tension with sarcomere length in vertebrate muscle fibres. *Journal of Physiology* 184:170–192.

Greenwalt, C.H. (1962). Dimensional relationships for flying animals. *Smithsonian Miscellaneous Collections* 144:1–46.

Grodzinski, U., Spiegel, O., Korine, C., and Holderied, M. (2008). Context dependent flight speed: Bats minimize energy expenditure per time while foraging and per distance while commuting. Submitted to *Journal of Experimental Biology*.

Haubenhofer, M. (1964). Die Mechanik des Kurvenfluges. *Schweizer Aero-Revue* 39:561–565.

Hedenström, A. (2006). Vortex wakes of bird flight: Old theory, new data and future prospects. In *Flow Phenomena in Nature*, R. Liebe Ed. WIT Press, Vol. 2, pp. 706–734.

Hedenström, A., and Alerstam, T. (1992). Climbing performance of migrating birds as a basis for estimating limits for fuel-carrying capacity and muscle work pp. *Journal of Experimental Biology* 164:19–38.

Hedenström, A., Rosén, M., and Spedding, G.R. (2005). Vortex wakes generated by robins *Erithacus rubecula* during free flight in a wind tunnel. *Journal of the Royal Society Interface* 3:263–276.

Hedenström, A., Johansson, L.C., Wolf, M., von Busse, R., Winter, Y., and Spedding, G.R. (2007). Bat flight generates complex aerodynamic tracks. *Science* 316:894–897.

Heilmann, G. (1926). *The Origin of Birds*. New York: Appleton.

Hervieu, R. (1960). Détermination expéditive du point observé par le droite de vitesse ascensionelle d'un astre. Navigation, *Paris* 7:57–68.

Hill, A.V. (1938). The heat of shortening and the dynamic constants of muscle. *Proceedings of the Royal Society Series B* 126:136–195.

Hill, A.V. (1950). The dimensions of animals and their muscular dynamics. *Science Progress* 38:209–230.

Huxley, A.F. (1957). Muscle structure and theories of contraction. *Progress in Biophysics and Biophysical Chemistry* 7:255–318.

Huxley, H.E. (1985). The crossbridge mechanism of muscular contraction and its implications. *Journal of Experimental Biology* 115:17–30.

Jenni, L., and Jenni-Eiermann, S. (1998). Fuel supply and metabolic constraints in migrating birds. *Journal of Avian Biology* 29:521–528.

Kirkpatrick, S.J. (1990). The moment of inertia of bird wings. *Journal of Experimental Biology* 151:489–494.

Kirkpatrick, S.J. (1994). Scale effects on the stresses and safety factors in the wing bones of birds and bats. *Journal of Experimental Biology* 190:195–215.

Kramer, G. (1952). Experiments on bird orientation. *Ibis* 94:265–285.

Kyte, F.T., and Wasson, J.T. (1986). Accretion rate of extraterrestrial matter: Iridium deposited 33 to 67 million years ago. *Science* 232:1225–1229.

Lasiewski, R.C., and Dawson, W.R. (1967). A re-examination of the relation between standard metabolic rate and body weight in birds. *Condor* 69:13–23.

Lilienthal, O. (1889). *Der Vogelflug als Grundlage der Fliegekunst.* Berlin: Gärtner.

Lindström, Å. (1991). Maximum fat deposition rates in migrating birds. *Ornis Scandinavica* 22:12–19.

Lindström, Å., and Kvist, A. (1995). Maximum energy intake rate is proportional to basal metabolic rate in passerine birds. *Proceedings of the Royal Society Series B* 261:337–343.

Lindström, Å., and Piersma, T. (1993). Mass changes in migrating birds: The evidence for fat and protein storage re-examined. *Ibis* 135:70–78.

Mandelbrot, B.B. (1982). *The Fractal Geometry of Nature.* New York: Freeman.

Mangold, O. (1946). Die nase der segelnden Vögel ein Organ des Strömungssinnes? *Naturwissenschaften* 33:19–23.

Margaria, R. (1976). *Biomechanics and Energetics of Muscular Exercise.* Oxford: Oxford University Press.

McMahon, T.A. (1984). *Muscles Reflexes and Locomotion.* Princeton: Princeton University Press.

Milner, A. (2002). *Dino-Birds: From Dinosaurs to Birds.* London: Natural History Museum.

Minasian, S.M., Balcomb, K.C., and Foster, L. (1984). *The World's Whales.* Washington: Smithsonian Books.

Norberg, R.Å. (1968). Physical factors in directional hearing in *Aegolius funereus* (Linné) (Strigiformes), with special reference to the significance of the asymmetry of the external ears. *Arkiv för Zoologi Series 2* 20:181–204.

Norberg, U.M. (1969). An arrangement giving a stiff leading edge to the hand wing in bats. *Journal of Mammalogy* 50:766–770.

Norberg, U.M. (1970). Functional osteology and myology of the wing of *Plecotus auritus* Linnaeus (Chiroptera). *Arkiv för Zoologi Series 2* 22:483–543.

Norberg, U.M. (1972a). Functional osteology and myology of the wing of the Dog-faced Bat *Rousettus aegyptiacus* (É. Geoffroy) (Mammalia, Chiroptera). *Z. Morph. Tiere* 73:1–44.

Norberg, U.M. (1972b). Bat wing structures important for aerodynamics and rigidity (Mammalia, Chiroptera). *Z. Morph. Tiere* 73:45–61.

Norberg, U.M. (1990). *Vertebrate Flight.* Berlin: Springer.

Novick, A., and Leen, N. (1969). *The World of Bats.* Lausanne: Edita.

Nowak, R.M. (1991). *Mammals of the World.* 5th Edition. Baltimore: Johns Hopkins University Press.

Nudds, R.L., Taylor, G.K., and Thomas, A.L.R. (2004). Tuning of Strouhal number for high propulsive efficiency predicts how wingbeat frequency and stroke amplitude relate and scale with size and flight speed in birds. *Proceedings of the Royal Society Series B* 271:2071–2076.

Officer, C.B., and Drake, C.L. (1985). Terminal Cretaceous environmental events. *Science* 227:1161–1167.

Pankhurst, R.C. (1944). *A method for the rapid evaluation of Glauert's expressions for the angle of zero lift and the moment at zero lift.* Reports and Memoranda No. 1914, British Aeronautical Research Council.

Pankhurst, R.C., and Holder, D.W. (1965). *Wind Tunnel Technique.* 2nd Edition. London: Pitman.

Pennycuick, C.J. (1960). The physical basis of astro-navigation in birds: Theoretical considerations. *Journal of Experimental Biology* 37:573–593.

Pennycuick, C.J. (1967). The strength of the pigeons wing bones in relation to their function. *Journal of Experimental Biology* 46:219–233.

Pennycuick, C.J. (1968a). A wind tunnel study of gliding flight in the pigeon *Columba livia. Journal of Experimental Biology* 49:509–526.

Pennycuick, C.J. (1968b). Power requirements for horizontal flight in the pigeon *Columba livia. Journal of Experimental Biology* 49:527–555.

Pennycuick, C.J. (1969). The mechanics of bird migration. *Ibis* 111:525–556.

Pennycuick, C.J. (1971). Gliding flight of the dog-faced bat *Rousettus aegyptiacus,* observed in a wind tunnel. *Journal of Experimental Biology* 55:833–845.

Pennycuick, C.J. (1972). Soaring behaviour and performance of some East African birds, observed from a motor-glider. *Ibis* 114:178–218.

Pennycuick, C.J. (1973). Wing profile shape in a fruit bat gliding in a wind tunnel, determined by photogrammetry. *Periodicum Biologorum* 75:77–82.

Pennycuick, C.J. (1975a). Mechanics of flight. In *Avian Biology,* D.S. Farner and J.R. King Eds. New York: Academic Press, Vol. 5, pp. 1–75.

Pennycuick, C.J. (1975b). On the running of the gnu (*Connochaetes taurinus*) and other animals. *Journal of Experimental Biology* 63:775–799.

Pennycuick, C.J. (1982). The flight of petrels and albatrosses (Procellariiformes), observed in South Georgia and its vicinity. *Philosophical Transactions of the Royal Society Series B* 300:75–106.

Pennycuick, C.J. (1983). Thermal soaring compared in three dissimilar tropical bird species, *Fregata magnificens, Pelecanus occidentalis* and *Coragyps atratus. Journal of Experimental Biology* 102:307–325.

Pennycuick, C.J. (1986). Mechanical constraints on the evolution of flight. *Memoirs of the California Academy of Sciences* 8:83–98.

Pennycuick, C.J. (1987a). Flight of seabirds. In *Seabirds: Feeding Ecology and Role in Marine Ecosystems,* J.P. Croxall Ed. Cambridge University Press, pp. 43–62.

Pennycuick, C.J. (1987b). Flight of auks (Alcidae) and other northern seabirds compared with southern procellariiformes: Ornithodolite observations. *Journal of Experimental Biology* 128:335–347.

Pennycuick, C.J. (1987c). Cost of transport and performance number, on Earth and other planets. In *Comparative Physiology: Life in Water and on Land,* P. Dejours, L. Bolis, C.R. Taylor, and E.R. Weibel Eds. Liviana/Springer, pp. 371–386.

Pennycuick, C.J. (1988a). *Conversion Factors: SI Units and Many Others.* Chicago: University of Chicago Press.

Pennycuick, C.J. (1988b). On the reconstruction of pterosaurs and their manner of flight, with notes on vortex wakes. *Biological Reviews* 63:299–331.

Pennycuick, C.J. (1989). Span-ratio analysis used to estimate effective lift:drag ratio in the double-crested cormorant *Phalacrocorax auritus,* from field observations. *Journal of Experimental Biology* 142:1–15.

Pennycuick, C.J. (1990). Predicting wingbeat frequency and wavelength of birds. *Journal of Experimental Biology* 150:171–185.

Pennycuick, C.J. (1991). Adapting skeletal muscle to be efficient. In *Efficiency and Economy in Animal Physiology*, R.W. Blake Ed. Cambridge University Press, pp. 33–42.

Pennycuick, C.J. (1992). *Newton Rules Biology.* Oxford, Oxford University Press.

Pennycuick, C.J. (1996). Wingbeat frequency of birds in steady cruising flight: New data and improved predictions. *Journal of Experimental Biology* 199:1613–1618.

Pennycuick, C.J. (1997). Actual and "optimum" flight speeds: field data reassessed. *Journal of Experimental Biology* 200:2355–2361.

Pennycuick, C.J. (1998a). Computer simulation of fat and muscle burn in long-distance bird migration. *Journal of Theoretical Biology* 191:47–61.

Pennycuick, C.J. (1998b). Field observations of thermals and thermal streets, and the theory of cross-country soaring flight. *Journal of Avian Biology* 29:33–43.

Pennycuick, C.J. (2001). Speeds and wingbeat frequencies of migrating birds compared with calculated benchmarks. *Journal of Experimental Biology* 204:3283–3294.

Pennycuick, C.J. (2002). Gust soaring as a basis for the flight of petrels and albatrosses (Procellariiformes). *Avian Science* 2:1–12.

Pennycuick, C.J. (2003). The concept of energy height in animal locomotion: Separating mechanics from physiology. *Journal of Theoretical Biology* 224:189–203.

Pennycuick, C.J., and Battley, P.F. (2003). Burning the engine: A time-marching computation of fat and protein consumption in a 5420-km flight by Great Knots (*Calidris tenuirostris*). *Oikos* 103:323–332.

Pennycuick, C.J., and Rezende, M.A. (1984). The specific power output of aerobic muscle, related to the power density of mitochondria. *Journal of Experimental Biology* 108:377–392.

Pennycuick, C.J., and Webbe, D. (1959). Observations on the fulmar in Spitsbergen. *British Birds* 52:321–332.

Pennycuick, C.J., Alerstam, T., and Larsson, B. (1980). Soaring migration of the common crane *Grus grus* observed by radar and from an aircraft. *Ornis Scandinavica* 10:241–251.

Pennycuick, C.J., Obrecht, H.H., and Fuller, M.R. (1988). Empirical estimates of body drag of large waterfowl and raptors. *Journal of Experimental Biology* 135:253–264.

Pennycuick, C.J., Heine, C.E., Kirkpatrick, S.J., and Fuller, M.R. (1992). The profile drag of a hawk's wing, measured by wake sampling in a wind tunnel. *Journal of Experimental Biology* 165:1–19.

Pennycuick, C.J., Fuller, M.R., Oar, J.J., and Kirkpatrick, S.J. (1994). Falcon versus grouse: Flight adaptations of a predator and its prey. *Journal of Avian Biology* 25:39–49.

Pennycuick, C.J., Klaassen, M., Kvist, A., and Lindström, Å. (1996a). Wingbeat frequency and the body drag anomaly: Wind tunnel observations on a Thrush Nightingale (*Luscinia luscinia*) and a Teal (*Anas crecca*). *Journal of Experimental Biology* 199:2757–2765.

Pennycuick, C.J., Einarsson, O., Bradbury, T.A.M., and Owen, M. (1996b). Migrating whooper swans (*Cygnus cygnus*): Satellite tracks and flight performance calculations. *Journal of Avian Biology* 27:118–134.

Pennycuick, C.J., Alerstam, T., and Hedenström, A. (1997). A new wind tunnel for bird flight experiments at Lund University, Sweden. *Journal of Experimental Biology* 200:1441–1449.

Pennycuick, C.J., Bradbury, T.A.M., Einarsson, Ó., and Owen, M. (1999). Response to weather and light conditions of migrating Whooper Swans *Cygnus cygnus* and flying height profiles, observed with the Argos satellite system. *Ibis* 141:434–443.

Pennycuick, C.J., Hedenström, A., and Rosén, M. (2000). Horizontal flight of a swallow (*Hirundo rustica*) observed in a wind tunnel, with a new method for directly measuring mechanical power. *Journal of Experimental Biology* 203:1755–1765.

Piersma, T. (1998). Phenotypic flexibility during migration: Optimization of organ size contingent on the risks and rewards of fueling and flight. *Journal of Avian Biology* 29:511–520.

Piersma, T., and Gill, R.E. (1998). Guts dont fly: Small digestive organs in obese bar-tailed godwits. *Auk* 115:196–203.

Piersma, T., Hedenström, A., and Bruggemann, J.H. (1997). Climb and flight speeds of shorebirds embarking on an intercontinental flight; do they achieve the predicted optimal behaviour? *Ibis* 139:299–304.

Prince, P.A., Wood, A.G., Barton, T., and Croxall, J.P. (1992). Satellite tracking of wandering albatrosses (*Diomedea exulans*) in the South Atlantic. *Antarctic Science* 4:31–36.

Pringle, J.W.S. (1948). The gyroscopic mechanism of the halteres of Diptera. *Philosophical Transactions of the Royal Society Series B* 233:347–384.

Prior, N.C. (1984). *Flight energetics and migration performance of swans.* Ph.D. University of Bristol.

Pumphrey, R.J. (1948). The sense organs of birds. *Ibis* 90:171–199.

Pumphrey, R.J. (1961). Sensory organs: Hearing. In *Biology and Comparative Physiology of Birds*, A.J. Marshall Ed. Academic Press, Vol. 2, pp. 69–86.

Rae, W.H., and Pope, A. (1984). *Low-Speed Wind Tunnel Testing.* 2nd Edition. New York: Wiley.

Rayleigh, Lord. (1883). The soaring of birds. *Nature* 27:534–535.

Rayner, J.M.V. (1979a). A vortex theory of animal flight. I. The vortex wake of a hovering animal. *Journal of Fluid Mechanics* 91:607–730.

Rayner, J.M.V. (1979b). A vortex theory of animal flight. II. The forward flight of birds. *Journal of Fluid Mechanics* 91:731–763.

Rayner, J.M.V. (1979c). A new approach to animal flight mechanics. *Journal of Experimental Biology* 80:17–54.

Rayner, J.M.V. (1985). Linear relations in biomechanics: The statistics of scaling functions. *Journal of Zoology* 206:415–439.

Rosén, M., and Hedenström, A. (2001). Gliding flight in a jackdaw: A wind tunnel study. *Journal of Experimental Biology* 204:1153–1166.

Rosén, M., Spedding, G.R., and Hedenström, A. (2004). The relationship between wingbeat kinematics and vortex wake of a thrush nightingale. *Journal of Experimental Biology* 207:4255–4268.

Rothe, H.J., and Nachtigall, W. (1987). Pigeon flight in a wind tunnel. I. Aspects of wind tunnel design, training methods and flight behaviour of different pigeon races. *Journal of Comparative Physiology Series B* 157:91–98.

Rothe, H.J., Biesel, W., and Nachtigall, W. (1987). Pigeon flight in a wind tunnel. II. Gas exchange and power requirements. *Journal of Comparative Physiology Series B* 157:99–109.

Scheid, P. (1982). Respiration and control of breathing. In *Avian Biology*, D.S. Farner, J.R. King, and K.C. Parkes Eds. Academic Press, Vol. 6, pp. 405–453.

Schmitz, F.W. (1960). *Aerodynamik des Flugmodells*. 4th Edition. Duisberg: Carl Lange.

Simons, M. (1987). *Model Aircraft Aerodynamics*. 2nd Edititon. Hemel Hempstead: Argus.

Slonaker, J.R. (1918). A physiological study of the anatomy of the eye and its accessory parts of the English sparrow (*Passer domesticus*). *Journal of Morphology* 31:351–459.

Spedding, G.R. (1986). The wake of a jackdaw (*Corvus monedula*) in slow flight. *Journal of Experimental Biology* 125:287–307.

Spedding, G.R. (1987a). The wake of a kestrel (*Falco tinnunculus*) in gliding flight. *Journal of Experimental Biology* 127:45–57.

Spedding, G.R. (1987b). The wake of a kestrel (*Falco tinnunculus*) in flapping flight. *Journal of Experimental Biology* 127:59–78.

Spedding, G.R. (1992). The aerodynamics of flight. *Advances in Comparative Physiology* 11:51–111.

Spedding, G.R., and Pennycuick, C.J. (2001). Uncertainty calculations for theoretical power curves. *Journal of Theoretical Biology* 208:127–139.

Spedding, G.R., Rayner, J.M.V., and Pennycuick, C.J. (1984). Momentum and energy in the wake of a pigeon (*Columba livia*) in slow flight. *Journal of Experimental Biology* 111:81–102.

Spedding, G.R., Rosén, M., and Hedenström, A. (2003a). Quantitative studies of the wakes of freely-flying birds in a low-turbulence wind tunnel. *Experiments in Fluids* 34:291–303.

Spedding, G.R., Rosén, M., and Hedenström, A. (2003b). A family of vortex wakes generated by a thrush nightingale in free flight in a wind tunnel over its entire natural range of flight speeds. *Journal of Experimental Biology* 206:2313–2344.

Stewart, A.G. (1978). Swans flying at 8000 metres. *British Birds* 71:459–460.

Swinton, W.E. (1960). The origin of birds. In *Biology and Comparative Physiology of Birds*, A.J. Marshall, Ed. New York: Academic Press, Vol. 1, pp. 1–14.

Thomas, A.L.R. (1993). On the aerodynamics of birds' tails. *Philosophical Transactions of the Royal Society Series B* 340:361–380.

Tickell, W.L.N. (2000). *Albatrosses*. Robertsbridge: Pica Press.

Tucker, V.A. (1968a). Respiratory physiology of house sparrows in relation to high-altitude flight. *Journal of Experimental Biology* 48:55–66.

Tucker, V.A. (1968b). Respiratory exchange and evaporative water loss in the flying budgerigar. *Journal of Experimental Biology* 48:67–87.

Tucker, V.A. (1972). Metabolism during flight in the Laughing Gull, *Larus atricilla*. *American Journal of Physiology* 222:237–245.

Tucker, V.A. (1973). Bird metabolism during flight: Evaluation of a theory. *Journal of Experimental Biology* 58:689–709.

Tucker, V.A. (1998). Gliding flight: Speed and acceleration of ideal falcons during diving and pull out. *Journal of Experimental Biology* 201:403–414.

Tuite, C.H. (1979). Population size, distribution and biomass density of the Lesser Flamingo in the Eastern Rift Valley, 1974–76. *Journal of Applied Ecology* 16:765–775.

Videler, J.J. (2005). *Avian Flight*. Oxford: Oxford University Press.

von Mises, R. (1959). *Theory of Flight*. New York: Dover.

Wainwright, S.A., Biggs, W.D., Currey, J.D., and Gosline, J.M. (1976). *Mechanical design in organisms*. London: Arnold.

Walls, G.L. (1942). *The Vertebrate Eye and its Adaptive Radiation*. Bloomfield Hills, MI: Cranbrook Press.

Welch, A., Welch, L., and Irving, F. (1977). *New Soaring Pilot*. 3rd Edition. London: Murray.

Wellnhofer, P. (1975). Die Rhamphorhynchoidea (Pterosauria) der Oberjura-Plattenkalke Süddeutschlands. *Palaeontographica Part A* 148:1–30.

Wellnhofer, P. (1991). *The Illustrated Encyclopedia of Pterosaurs*. London: Salamander.

White, D.C.S., and Thorson, J. (1975). *The Kinetics of Muscle Contraction*. Oxford: Pergamon.

Wilkie, D.R. (1968). *Muscle*. London: Arnold.

# INDEX

Printed and bound by CPI Group (UK) Ltd, Croydon, CR0 4YY

03/10/2024

01040415-0015